Applications of AI and IoT in Geotechnical Engineering for Sustainable Infrastructure

Artificial intelligence (AI) and the Internet of Things (IoT) are transforming the way geotechnical engineers monitor, analyze, and manage soil-structure systems. *Applications of AI and IoT in Geotechnical Engineering for Sustainable Infrastructure* explores how these emerging digital technologies can enhance decision-making, improve infrastructure resilience, and support sustainable development in civil engineering projects. It examines the integration of AI and IoT in geotechnical engineering, including slope stability assessments, data-driven techniques for soil classification and characterization, foundation and liquefaction analysis, optimizing earthwork practices and operations, tunnels, and earth-retaining structures using real-time sensor networks and data-driven models. This book also discusses machine learning techniques for predicting soil behavior, real-time monitoring of pile foundations, automated site investigation methods, and digital tools for risk assessment and infrastructure health monitoring/assessment.

This book is intended for researchers, graduate students, and practicing engineers in geotechnical and civil engineering. It will also benefit professionals working in smart infrastructure, disaster risk reduction, and infrastructure asset management who are interested in integrating digital technologies into sustainable engineering practice.

Applications of AI and IoT in Geotechnical Engineering for Sustainable Infrastructure

Edited by
Vedprakash C. Maralapalle,
Jayatheja Muktinutalapati, Bogireddy Chandra,
Gangadhara Reddy Narala, and Abdullah Ansari

CRC Press
Taylor & Francis Group
Boca Raton London New York

CRC Press is an imprint of the
Taylor & Francis Group, an **informa** business

Designed Cover Image: Shutterstock

First edition published 2027
by CRC Press
4 Park Square, Milton Park, Abingdon, Oxon, OX14 4RN

and by CRC Press
2385 NW Executive Center Drive, Suite 320, Boca Raton FL 33431

CRC Press is an imprint of informa UK Limited

British Library Cataloguing-in-Publication Data
A catalogue record for this book is available from the British Library

ISBN: 9781041199175 (hbk)
ISBN: 9781041199182 (pbk)
ISBN: 9781003714125 (ebk)

DOI: 10.1201/9781003714125

Typeset in Times
by codeMantra

Contents

Preface

Geotechnical engineering plays an important role in the development of safe and reliable infrastructure. Engineers often deal with complex soil behavior, uncertain ground conditions, and environmental challenges. In recent years, new digital technologies such as artificial intelligence (AI) and the Internet of Things (IoT) have started to change how geotechnical problems are studied, monitored, and managed. These technologies allow engineers to collect real-time data, analyze large datasets, and make better decisions for the design and maintenance of infrastructure.

This edited book, *Applications of AI and IoT in Geotechnical Engineering for Sustainable Infrastructure*, brings together recent research and practical applications of these technologies in geotechnical engineering. The chapters in this volume cover several important areas, including AI-based soil classification, prediction of soil compaction parameters, modeling of soil–water characteristic curves, and machine learning approaches for predicting geotechnical failures. This book also includes studies related to slope stability analysis, landslide susceptibility mapping, and the use of AI for assessing geotechnical hazards. In addition, some chapters focus on the application of AI in improving construction practices such as earthwork optimization and the prediction of tunnel boring machine performance.

Another key focus of this book is the use of IoT technologies for monitoring geotechnical structures. Modern sensors and connected monitoring systems allow engineers to track the performance of foundations and other infrastructure in real time. This book includes examples such as monitoring pile foundations and using data-driven tools for infrastructure health assessment. Some chapters also explore the use of AI in seismic data analysis and earthquake-related hazard assessment.

This edited book brings together contributions from researchers and engineers from different parts of the world who are working in geotechnical engineering and related fields. Their work shows how new digital tools can help improve the understanding of soil behavior, reduce risks in infrastructure projects, and support more sustainable development.

This edited book is intended for graduate students, researchers, and practicing engineers interested in modern approaches in geotechnical engineering. It may also be useful for professionals working in infrastructure planning, hazard assessment, and smart infrastructure systems.

The editors would like to sincerely thank all the authors who contributed their work to this book. Their efforts and collaboration made this volume possible. We also thank the editorial team at CRC Press for their guidance and support during the preparation of this book.

About the Editors

Vedprakash C. Maralapalle is an associate professor at the L. S. Raheja School of Architecture, Mumbai, India. He holds a Ph.D. in Civil Engineering from NMIMS University, Mumbai, with research focused on the physical modeling and numerical analysis of vertical piles. He has over 12 years of teaching experience and industry exposure as a site engineer in construction projects. His research interests include geotechnical engineering, pile foundations, physical and centrifuge modeling, numerical analysis using advanced software tools, artificial intelligence applications in civil engineering, smart infrastructure systems, and sustainable urban development. He has contributed to several research publications, book chapters, and conference presentations in areas related to infrastructure resilience, smart cities, and environmental monitoring using geospatial and machine learning techniques. In addition to his academic work, he is involved in structural and material testing consultancy and is a licensed structural engineer associated with municipal authorities. His academic and professional expertise supports interdisciplinary research integrating civil engineering, geospatial technologies, and data-driven approaches for sustainable infrastructure planning.

Jayatheja Muktinutalapati is a researcher and academician in the domain of civil engineering. He earned his Ph.D. in Civil Engineering from Birla Institute of Technology and Science (BITS) – Pilani and his master's degree in Geotechnical Engineering from Jawaharlal Nehru Technological University Hyderabad (JNTUH). He has three years of teaching experience and also worked as an institute postdoctoral fellow (iPDF) at the Indian Institute of Technology, Dharwad (IIT – Dh) for a year. He is currently working as an assistant professor in the RICS School of Built Environment, Amity University Maharashtra, Mumbai. He is a life member of the Indian Geotechnical Society (IGS). He has published ten research papers in SCI/Scopus-indexed journals and four book chapters, and he has filed a patent in the domain of sustainable geotechnics. Additionally, he serves as a reviewer for four international journals. He has received prestigious fellowships, including the Junior Research Fellowship (2017), nominated by DST – SERB, India. His areas of research include recyclable materials, retaining structures, life cycle assessments, and AI applications in civil engineering.

Bogireddy Chandra is currently a foreign scientist at the ALT University in Almaty, Kazakhstan, and a visiting scholar in the Department of Civil and Intelligent Construction Engineering at Shantou University in China. Previously, he served as an assistant professor in the Department of Civil Engineering at Vardhaman College of Engineering, Hyderabad. In 2019, he earned his Ph.D. in Civil Engineering from the Department of Civil Engineeringat SVNIT, Surat, Gujarat, India, for his thesis titled "Rapid Assessment of Compressibility Behavior of Clayey Soil from Experimental and Finite Element (FE) Simulations." In 2012, he earned his M.Tech. in Geotechnical Engineering from Sri Venkateswara University College of Engineering, S V University, Tirupati, Andhra Pradesh, India, where he achieved the top position in his class. Following his master's degree, he worked as an assistant professor in the Department of Civil Engineering at Sree Vidyanikethan Engineering College (Autonomous), Tirupati, A.P., India, from 2013 to 2014. He earned his B.Tech. in Civil Engineering from JNTU Anantapur, Andhra Pradesh, India, in 2010. During his research, he received two Best Paper Awards at national conferences and delivered 35 guest lectures in national and international seminars. He published 30 research papers, six patents, and two books: *Proceedings of the 1st Indo-China Research Series in Geotechnical and Geoenvironmental Engineering* and *IOT & AI Innovations in Civil Engineering: Transforming Infrastructure*. Furthermore, he is an associate member of ASCE, SM Institution of Civil Engineers, and the Institute of Engineers (India), and a fellow of FSIESRP Life Member. His areas of research interest include basic geomechanics, numerical modeling, image analysis, site characterization, earthquake engineering, biogeotechnology, sustainable and carbon reduction ground engineering (MICP, Biochar), consolidation, pipeline geotechnical engineering, and the study of atmospheric aerosols in geoenvironmental systems.

Gangadhara Reddy Narala is an academician and researcher in civil engineering with a specialization in geotechnical engineering. He earned his Ph.D. in Civil Engineering from IIT Bhubaneswar and a master's degree in Geotechnical Engineering from NIT Bhopal. He is currently serving as an assistant professor at the Fiji National University, Suva, Fiji. Prior to this role, he worked as a researcher at Nanjing University and Shantou University, China. He has made significant contributions to research, with over 32 SCI/Scopus-indexed journal publications and 15 conference papers. He has also co-edited the book *Advances in Sustainable Materials and Resilient Infrastructure* (2022). He currently serves as an associate editor for the journal *Innovative Infrastructure Solutions* and is a member of the editorial boards of *Discover Civil Engineering, Soils and Rocks*, and *Discover Geosciences*. His research interests lie in geotechnical and geoenvironmental engineering, particularly the beneficial utilization of industrial waste materials, pavement geotechnics, climate-resilient infrastructure, sustainable soil stabilization, the application of biopolymers and biochar in landfill engineering, slope stabilization, and climate change adaptation. He has received several prestigious fellowships, including the Chinese Government Scholarship (CSC) in 2019, nominated by the Ministry of Human Resource Development, India, to conduct research in China. More recently, he was awarded the GHD Foundation Visiting Engineering Scholarship (2026), which supports his visit to the Humanitarian Engineering group at the University of Sydney, Australia.

Abdullah Ansari currently serves as a research professor at the Earthquake Monitoring Center (EMC), Sultan Qaboos University, Muscat, Oman. His research focuses on seismic risk assessment, geotechnical earthquake engineering, tunnel and underground structure performance, and infrastructure resilience under multi-hazard conditions. He has extensive experience in developing analytical, numerical, and data-driven models for evaluating the seismic vulnerability of critical infrastructure systems. A key aspect of his research involves integrating artificial intelligence and machine learning techniques with traditional geotechnical engineering approaches to improve the prediction of seismic damage and infrastructure performance. His work has contributed to the development of innovative frameworks for seismic fragility modeling, risk quantification, and resilience assessment of transportation networks and underground infrastructure. He has published more than 30 SCI-indexed journal articles in internationally recognized journals, such as *Scientific Reports*, *Soil Dynamics and Earthquake Engineering*, and *Tunnelling and Underground Space Technology*. In addition to his research contributions, he actively participates in international scientific collaborations and multidisciplinary research initiatives. He has been involved in several research projects supported by international organizations and research foundations, including collaborations with institutions in Oman, South Korea, China, and India. He also serves as a reviewer and editorial contributor for numerous international journals in civil engineering and earthquake engineering. Through his research, he aims to bridge the gap between advanced scientific modeling and real-world engineering applications, contributing to the development of safer, more resilient infrastructure systems in earthquake-prone regions.

Contributors

Asiqur Rahman Abir
Department of Civil Engineering
International University of Business Agriculture and Technology
Dhaka, Bangladesh

Priyanka Ahirwar
Department of Civil Engineering
Maulana Azad National Institute of Technology
Bhopal, Madhya Pradesh, India

C. H. Ajay
Department of Civil Engineering
Aditya University
Surampalem, Andhra Pradesh, India

Shima Al-Balushi
Earthquake Monitoring Center
Sultan Qaboos University
Muscat, Oman

Abdullah Ansari
Earthquake Monitoring Center
Sultan Qaboos University
Muscat, Oman

Anas Ansari
Department of Electrical Engineering
National Taiwan University of Science and Technology
Taipei City, Taiwan

Sudhir Bhatt
Department of Civil Engineering
Hitkarini College of Engineering and Technology
Jabalpur, Madhya Pradesh, India

J. Y. V. Shiva Bhushan
Department of Civil Engineering
VNR Vignana Jyothi Institute of Engineering and Technology
Hyderabad, Telangana, India

Ravindra Budania
Department of Civil Engineering
Gyan Vihar School of Engineering & Technology
Suresh Gyan Vihar University
Jaipur, Rajasthan, India

Chatrabhuj
Department of Civil Engineering
Kamla Nehru Institute of Technology
Sultanpur, Uttar Pradesh, India

Kaustav Chatterjee
School of Civil and Environmental Engineering
Oklahoma State University
Stillwater, Oklahoma

Abubakar Dayyabu
Department of Civil Engineering
Nile University of Nigeria
Abuja, Nigeria

Mohak Desai
Senior Data Analyst
NIQ
Vadodara, India

Arindam Dey
Department of Civil Engineering
Indian Institute of Technology Guwahati
Guwahati, India

Rakesh Kumar Dutta
Department of Civil Engineering
National Institute of Technology
Hamirpur, Himachal Pradesh, India

V. Gayathri
Department of Civil Engineering
Kumaraguru College of Technology
Coimbatore, Tamil Nadu, India

Tammineni Gnananandarao
Department of Civil Engineering
Aditya University
Surampalem, Andhra Pradesh, India

Kumar Harsh
Department of Civil Engineering
National Institute of Technology Jamshedpur
Jamshedpur, Jharkhand, India

Pradeep Kumar Jain
Department of Civil Engineering
Maulana Azad National Institute of Technology
Bhopal, Madhya Pradesh, India

Govindarajan Kannan
Department of Civil Engineering
Mepco Schlenk Engineering College
Sivakasi, Tamil Nadu, India

Amit Kumar
Department of Civil Engineering
National Institute of Technology
Hamirpur, Himachal Pradesh, India

B. A. V. Ram Kumar
Department of Civil Engineering
School of applied Engineering
GMR Institute of Technology (GMRIT) - Deemed to be University, Rajam
Andhra Pradesh, India

Abdulhameed Danjuma Mambo
Department of Civil Engineering
Nile University of Nigeria
Abuja, Nigeria

Vedprakash C. Maralapalle
L. S. Raheja School of Architecture
Bandra, Mumbai, India

Shital Marlapalle
Department of Civil Engineering
Gyan Vihar School of Engineering & Technology
Suresh Gyan Vihar University
Jaipur, Rajasthan, India.
Amity School of Architecture and Planning
Amity University, Mumbai, India

Balendra Mouli Marrapu
Department of Civil Engineering
Anil Neerukonda Institute of Technology Visakhapatnam
Andhra Pradesh, India

Gloria Nkeiruka Matthew
Department of Chemical Engineering
University of Abuja
Abuja, Nigeria

Kundan Meshram
Department of Civil Engineering
Guru Ghasidas Vishwavidyalaya
Bilaspur, Chhattisgarh, India

Kamalanandhini Mohan
Department of Civil Engineering
SRM Institute of Science and Technology
Kattankulathur, Tamil Nadu, India

Surendar Natarajan
Department of Civil Engineering
Sri Sivasubramaniya Nadar (SSN) College of Engineering
Kalavakkam, Tamil Nadu, India

Frank Nnebe
Department of Civil Engineering
Nile University of Nigeria
Abuja, Nigeria

Ifeyinwa Ijeoma Obianyo
Department of Civil Engineering
Nile University of Nigeria
Abuja, Nigeria

Azikiwe Peter Onwualu
Department of Mechanical Engineering
African University of Science and Technology
Abuja, Nigeria

R. S. V. Rashma
Faculty of Construction Management
Global Business Studies (GBS)
Dubai, United Arab Emirates

K. V. N. S. Raviteja
Sustainable Engineering Research Laboratory
Department of Civil Engineering
SRM University-AP
Amaravati, Andhra Pradesh, India

Ippili Saikrishnamacharyulu
Department of Civil Engineering
GIET University
Gunupur, Odisha, India

Rayhan A. Shium
Department of Applied Technology and Mathematics
Immaculata University
Chester County, Pennsylvania

S. K. Shreyas
Department of Civil Engineering
Indian Institute of Technology Guwahati
Guwahati, India

Kumar Shubham
Department of Civil Engineering
Indian Institute of Technology Guwahati
Assam, India

Abdhesh Kumar Sinha
Department of Civil Engineering
National Institute of Technology Jamshedpur
Jamshedpur, Jharkhand, India

Evangelin Ramani Sujatha
Centre for Advanced Research in Environment
School of Civil Engineering
SASTRA Deemed University
Thanjavur, Tamil Nadu, India

Md Fahim Faisal Tanha
Department of Computer Science and Engineering
University of Liberal Arts Bangladesh
Dhaka, Bangladesh

Utkarsh
Department of Civil Engineering
Sharda University
Greater Noida, Uttar Pradesh, India

Murugesh Vedanayagam
Department of Civil Engineering
JCT College of Engineering and Technology
Coimbatore, Tamil Nadu, India

Hemanth Kumar Yerrabolu
Department of Civil Engineering
Dadi Institute of Engineering & Technology - An Autonomous Institute
Anakapalle, Visakhapatnam
Andhra Pradesh, India

A Review on Integration of AI and IoT in Geotechnical Engineering

1

Govindarajan Kannan and
Evangelin Ramani Sujatha

1.1 INTRODUCTION

Geotechnical engineering is the backbone of civil engineering infrastructure. It is increasingly vulnerable to challenges arising from variability and uncertainty. Several factors such as urban growth, climate change, social and economic risks, ageing infrastructure and sustainability policy drivers necessitate the demand for resilient infrastructure. The heterogeneity and uncertainty associated with geomaterials and the technology associated with assessing the geomaterial suitability for the infrastructure projects also remain considerable challenges. The inherent nature of soil deposits and rocks and the influence of factors like climate and groundwater level again contribute to uncertainty and heterogeneity. In addition, limitations of conventional techniques for investigating the suitability of the construction site compound the problem of uncertainty, constraining the full assessment of subsurface behaviour. These complexities create a gap between predicted and actual performance that amplifies associated risks under extreme events or failures. The advances in artificial intelligence (AI) and Internet of

DOI: 10.1201/9781003714125-1

Things (IoT) are now laying the groundwork for digital transformation in geotechnical engineering practices towards adaptability and sustainability, addressing the concerns of variability and uncertainty. The potential of AI and IoT in transforming geotechnical engineering is gradually recognized in research and also in emerging practice. IoT provides the formwork for continuous monitoring of decisive parameters such as pore pressure, deformation and vibration, and AI offers the computational platform to interpret large data streams, identify patterns and forecast failures or hazards. Their integration creates data-driven decision systems that can reduce uncertainty and improve adaptability of the civil and geotechnical infrastructure contributing to long-term resilience of infrastructure promoting sustainable growth.

Conventional monitoring and prediction methods adopted for geotechnical infrastructure rely on discrete measurements, periodic field inspections and empirical correlations to a large extent. These observations fail to capture the continuous and non-linear nature of the geomaterial and geomaterial–structure interaction. Instruments like inclinometers, gauges to measure settlement and piezometers provide valuable data but their temporal coverage and spatial coverage are limited. These sensors or instruments record point-based data and are sparsely distributed, missing local variations in stress, strain or pore pressure between monitoring points. Also, measurements are periodic, failing to reflect the rapid or transient changes. In case the readings are manually recorded, human error also cannot be ruled out. The time delay in data acquisition leads to the absence of real-time data analytics, impeding corrective measures during the development of failures. Also, the conventional monitoring systems and the models developed based on these systems do not provide mechanisms for adaptive decision-making based on evolving ground conditions. Monitoring data stored in stand-alone systems is difficult to integrate across different instruments or projects, restricting the development of a unified data framework for predictive analysis and risk modelling. Conventional models are also largely reactive, identifying damage only after failure with no inherent capacity to forecast failure progression or issue early warnings. Also, the threshold-based alarms are simplistic, lacking learning ability. Existing systems do not adapt to changing boundary conditions caused by climate variability or allow dynamic recalibration in response to field data. Further, conventional analytical and numerical models are constrained by assumptions including soil homogeneity, simple boundary conditions and loading paths, resulting in significant discrepancies between observed and predicted values. Existing systems are cost intense and pose logistic difficulties in maintaining large and dense sensor networks. Also, they are not scalable for large areas or remote sites such as long embankments or underground networks. These limitations restrict the choice of the conventional systems in their ability to detect early signs of failure or distress, particularly in case of dynamic events or extreme climates, mandating the need for intelligent, data-scarce systems that can provide continuous assessment and predictive insight (Suryasentana & Shiel, 2023).

The integration of AI and IoT provides a progressive framework to overcome the inherent limitations of the conventional predictive and monitoring systems. IoT enables continuous data acquisition that is autonomous through dense wireless sensor networks that are capable of measuring geotechnical parameters like pore pressure, strain, displacement and vibration in real time, enabling spatially distributed monitoring and immediate data transmission. This ensures that dynamic changes within the ground are recorded and analysed as they occur. Storage can be accommodated in cloud or edge platforms. AI techniques, including machine learning (ML), deep learning and hybrid data models, help process large and complex datasets, extract hidden correlations, identify early signs or distress, and forecast potential failures (El-Abbasy, 2025). The combination of AI and IoT together creates an intelligent decision-support environment that augments the precision of geotechnical assessment and proactively enables adaptive design strategies and their maintenance (Ma & Guo, 2020). The data-driven approach facilitated by integrated AI–IoT systems marks the shift from reactive monitoring to predictive geotechnical infrastructure management that is resilient and aligns with the emerging vision of smart civil systems that are sustainable (Sheil et al., 2020).

The integration of AI and IoT establishes a continuous workflow that links field sensing, data transmission, intelligent analysis and predictive decision-making into a single adaptive framework. This convergence reduces uncertainty and improves the reliability of infrastructure performance assessment. Despite the growing recognition of these technologies, their application in geotechnical engineering is limited compared to other civil engineering disciplines like structural engineering or transportation engineering. This limited adaptation can be attributed to the complexity of the subsurface system and difficulties in data standardization. This review addresses the gap by exploring the potential of AI–IoT integration in developing intelligent data-driven geotechnical systems capable of learning and adapting from field behaviour to site-specific conditions. The flowchart in Figure 1.1 shows the organization of this chapter.

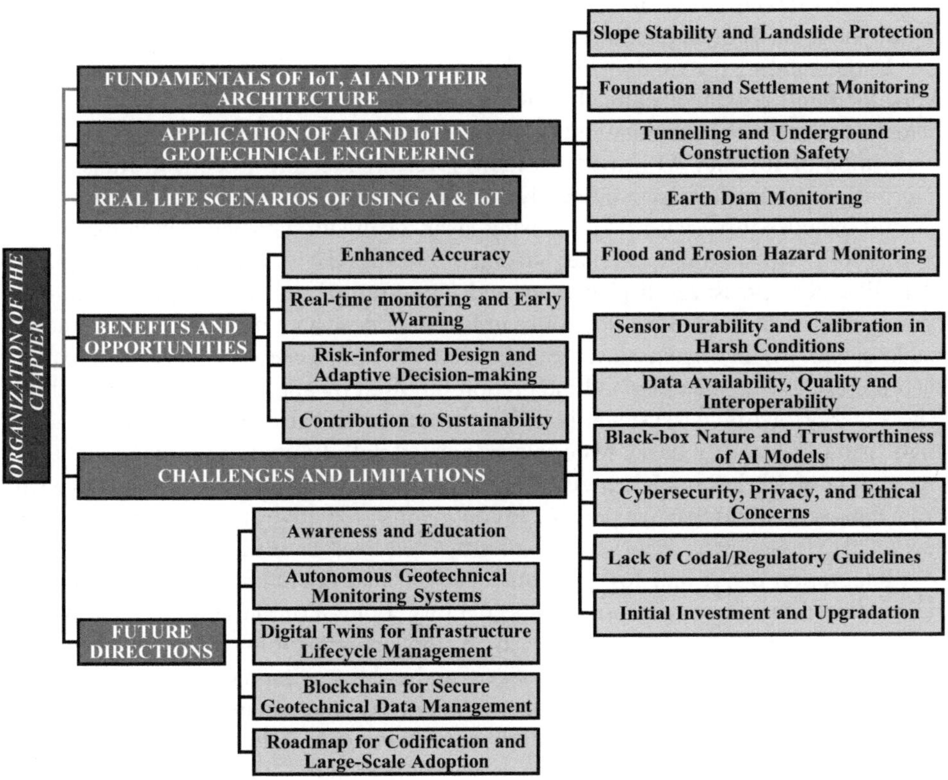

FIGURE 1.1 Organization of the chapter.

1.2 FUNDAMENTALS OF IoT, AI AND THEIR ARCHITECTURE

IoT is a network of interconnected sensors, devices and communication systems that enable continuous data acquisition, while AI is a computational approach that enables systems to learn from data, identify patterns and predict scenarios. Both IoT and AI complement each other within the context of intelligent systems. IoT provides the infrastructure that enables sensing and communication for capturing field data like pore pressure, displacement and soil moisture across the space, and AI serves as an analytical and learning component for interpreting the collected data and detecting instability/failure/distress. Liang et al. (2023) used embedded IoT sensors in earth dams to monitor internal deformations and moisture levels that help in monitoring dam safety. AI includes ML, robotics and computer vision. AI aims to mimic human-like intelligence and enables the machine to perceive, reason, learn and make informed decisions from data through patterns (Russell & Norvig, 2010). AI techniques have the ability to model non-linear and complex processes and perform modelling without initial assumptions on the relation between input and output variables (Baghbani et al., 2022). AI can predict the relationship between factors even if the physical relationship between them is not known (Sarker, 2021) and can also handle incomplete datasets and decipher meaningful patterns (Bello et al., 2015).

Understanding the fundamentals of IoT and AI is essential for developing a framework for intelligent geotechnical systems that enable seamless connectivity between field data acquisition, computational analysis and decision support (Menon et al., 2025). Integration of IoT and AI provides a multi-layer architecture to collect, transmit and interpret geotechnical data intelligently and continuously. The combined architecture in general includes three layers – (i) sensing layer for in-situ data acquisition that mostly consists of distributed sensors like piezometer, strain gauges, etc. that measure parameters like pore pressure, displacement and vibration; (ii) network layer for real-time transmission and communication for seamless data flow with minimal latency and signal loss; and (iii) intelligence layer for data analytics and decision support encompassing data preprocessing, cleaning and integration followed by application of AI algorithms. The synergy of IoT–AI systems is a boon in geotechnical engineering owing to the high spatial variability of geomaterials associated with limited or uncertain data. The integration of AI–IoT in civil engineering applications (El-Abbasy, 2025) marks a shift from data-scarce and reactive practice to continuous predictive and evidence-based geotechnical assessment as reported in studies by Piciullo et al. (2025).

Integration of AI and IoT presents technical and operation challenges that must be addressed to ensure reliability and scalability. Long-term durability of sensors and power management are critical concerns in implementing IoT–AI systems. Difficulty also can be encountered in setting up secure and stable communication in underground sites or remote sites which is a major limitation. The huge volume of data generated by dense sensor networks create additional challenges particularly related to storage, noise filtering and transmission bandwidth, mandating the need for standardized protocols for data formatting and quality assurance. AI models must be designed to handle incomplete, noisy and uncertain dataset without compromising prediction accuracy. The core of addressing these challenges lies in developing adaptive algorithms and resilient network architecture for realizing the full potential of the IoT–AI systems. Numerous design strategies are explored to improve the sustainability and reliability of the IoT–AI systems. Energy-efficient hardware and renewable power sources are adopted to enhance life of sensors in remote deployments. Edge computing can be used to enable preliminary data processing close to sensing nodes to reduce transmission loads and ensure faster local decision-making during critical events. Standardized communication protocols must be established to improve system integration and scalability across different sites. Data uncertainty can be handled by using hybrid or physics-based AI models which also make data interpretation easier. The above strategies can help in developing self-sustainable, resilient and adaptative monitoring frameworks for geotechnical infrastructure.

The IoT–AI architecture is particularly suitable for geotechnical engineering owing to the inherent complexity, non-linearity and time-dependent behaviour of the geomaterials including their spatial variability, path dependency and coupled hydro-mechanical response. Continuous data acquisition is a boon for the geotechnical engineering infrastructure as it can accommodate evolving field conditions, and AI can supplement processing and interpreting these data to recognize patterns and trends to model non-linear interactions and make predictions. The IoT–AI framework bridges the gap between computational understanding and physical behaviour and thereby forms the foundation for intelligent geotechnical systems applicable to foundations, retaining structures, underground excavations and slopes.

1.3 APPLICATIONS IN GEOTECHNICAL ENGINEERING

1.3.1 Slope Stability and Landslide Prediction

Slope failure and associated landslides are one of the most devastating disasters that pose threat to life and would destroy the infrastructure. Although this is a more common phenomenon of natural slopes (Santhosh Kumar & Sennimalai Chandrasekaran, 2022), evidences show that such failures happened in engineered slopes as well (Hyodo et al., 2012; Raj & Sengupta, 2014). Traditional slope stabilization involves a complex process of identifying the most critical slip surface by several trial-and-error methods which has been simplified with the incorporation of software-based finite element techniques. Yet, slope failure is more common, and this disaster remains life-threatening because the soil–soil interaction, state of stress, response to external factors such as the weather and rainfall conditions vary, and hence, a real-time monitoring of the slope became necessary. The incorporation of IoT in slope monitoring along with the digital twin-based real-time simulations and responses made a breakthrough in the early warning systems.

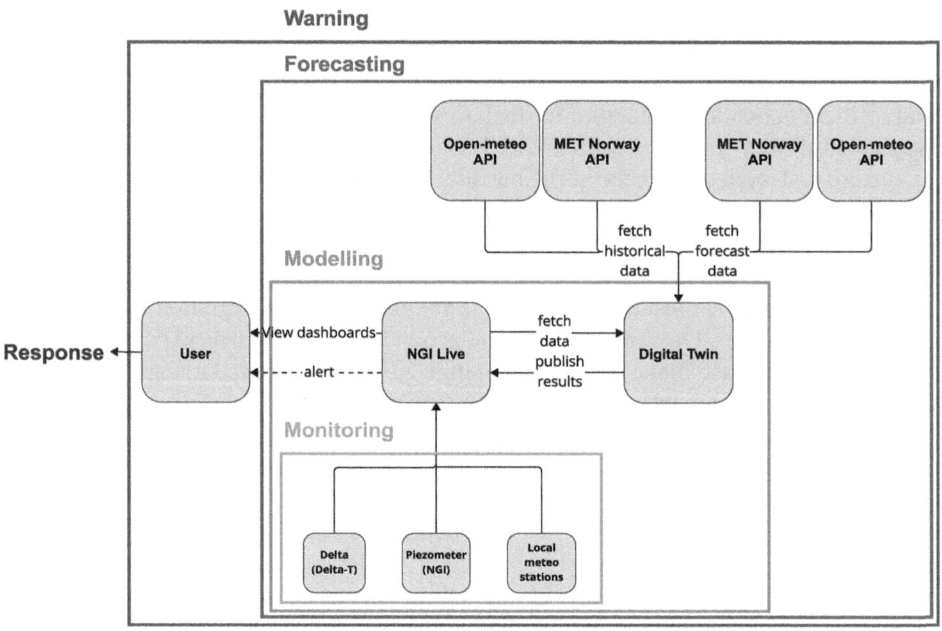

FIGURE 1.2 Operational architecture of the IoT-based early warning system (Piciullo et al., 2025).

Piciullo et al. (2025) attempted to incorporate IoT and digital twin-based responses in real-time slope monitoring at Norway. Based on the hydrological, meteorological and vegetation indicators from past data and future projections, they attempted to determine the factor of safety (FoS) using a commercial slope stabilization software and calculated the FoS on a regular basis to create a fully automatic early warning system that would alert if these parameters do not meet the threshold value. The operational architecture of IoT system used by Piciullo et al. (2025) is shown in Figure 1.2.

Xu and Zhang (2025) proposed an AI-based method to determine the risk factor of engineered highway slopes empowered with digital twin technique. The method used instant segmentation and random forest method to identify slope with high susceptibility of failure. Further, the artificial neural network (ANN) technique was used to determine soil type based on existing drilling data. These data were used to determine the FoS and alert if there is a potential for slope failure with the help of an early warning system.

1.3.2 Foundation and Settlement Monitoring

In general, a foundation design will consider both shear and settlement criteria to ensure the structure is safe against failure by all means. However, in high-rise structures and other important infrastructure like bridges (Yi et al., 2019), settlement will be progressive, and under unforeseen situations like earthquakes and liquefaction, the structure may sink, and hence, a real-time monitoring is now seen as essential for such high-rise buildings, bridges, etc. Razip and Daud (2022) performed laboratory model tests to detect ground surface and building monitoring using ultrasonic and tilt sensors. Wu et al. (2025) attempted to use IoT in monitoring pile installation by assessing factors like length, position and verticality of piles, pile drilling and lifting speeds, volume and water–cement ratio of grout, etc. Using suitable techniques, the indicators were assigned weight coefficients which were used in the quality monitoring of the pile installation process as shown in Figure 1.3.

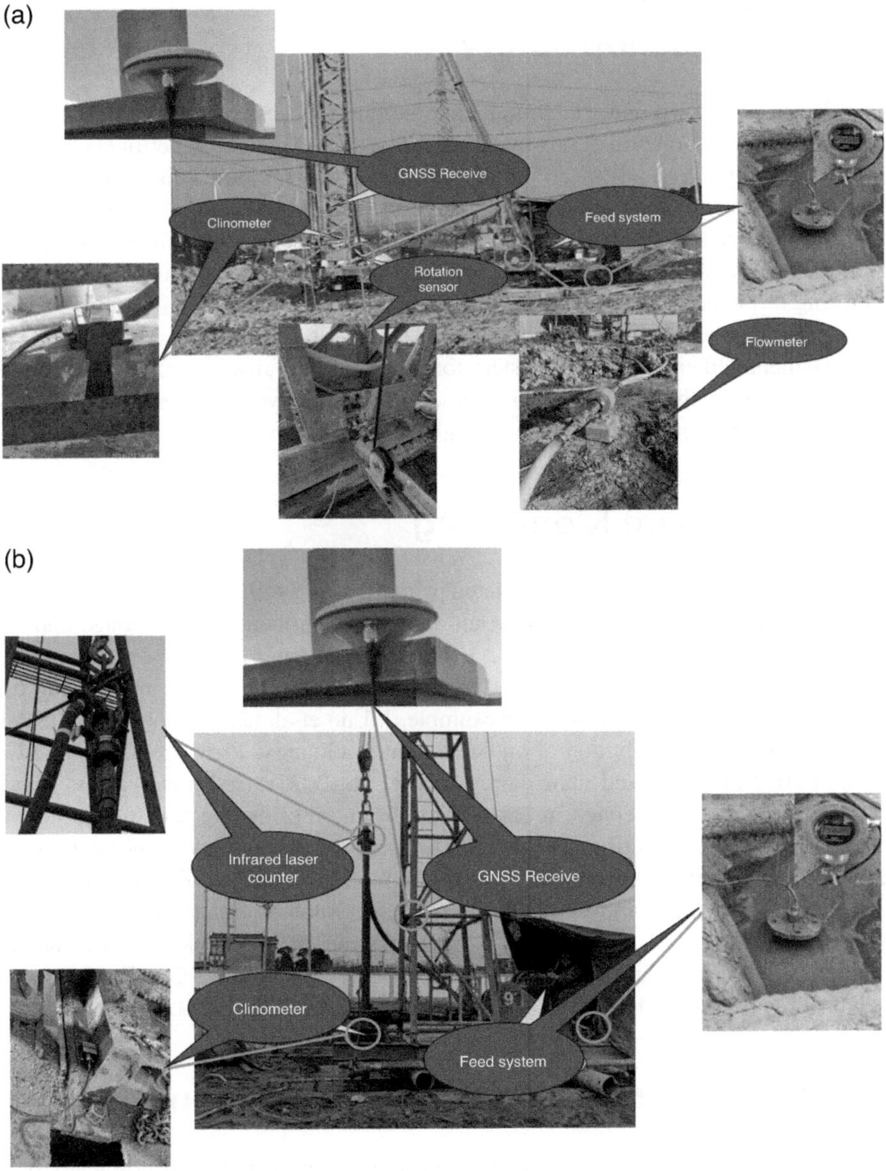

FIGURE 1.3 Installation of IoT system in (a) soil–cement mixing system and (b) bored pile construction system (Wu et al., 2025).

There are similar such literature that incorporates IoT and AI techniques in bridge settlement monitoring as well. For instance, Yi et al. (2019) used IoT-based sensors and data processing system to monitor settlement of a high-speed railway bridge in China. Abdu et al. (2023) used three AI algorithms to predict the pier settlement-based train acceleration response based on the data generated using simulation.

1.3.3 Tunnelling and Underground Construction Safety

Tunnelling is a relatively older technique that facilitates access between different regions for transportation, extraction of raw materials, ores, etc. Although tunnelling became simpler in recent times with advanced tunnel boring machines, monitoring such tunnels and other underground construction is essential, as collapse of tunnel or underground structure would incur a huge loss of life and property. With similar technologies mentioned in previous cases, IoT and AI were useful in various associated applications such as temperature monitoring of underground cable tunnels (Chen et al., 2019) and void detection in tunnel grouting (W. Luo et al., 2024). For instance, Dui et al. (2024) designed a cost-effective risk assessment and early warning system for underground railway tunnels due to ground settlement. Further, they claimed their method to be effective based on the effectiveness achieved while running simulations on Ningbo, China, rail transit data.

1.3.4 Earth Dam Monitoring

Earth dams are barriers that are constructed primarily with soil as their core. Unlike dams made of concrete, these earth dams would be greatly influenced by seepage and earth pressure parameters. Instrument-based earth dam monitoring is not a new finding in the field. In fact, there are several studies that show earth dam behaviour monitoring using piezometers for long time periods. For example, J. Luo et al. (2019) monitored the deformation and pressure distribution of an earth dam in Guangxi Province, China, between 1999 and 2012 and analysed their data. Liang et al. (2023) used underground wireless sensors to monitor the seepage in earth dams by monitoring the soil moisture in real time. B. Liu et al. (2024) utilized a vibration-sensing IoT technology to monitor the compaction process during an earth dam construction. Additionally, IoT plays a crucial role in other dam monitoring and maintenance processes such as water level monitoring for dams (Aditya et al., 2021; Siddula et al., 2018), and gate open and closure of dams based on rainfall conditions (Hamzah et al., 2024; Hussen Hajjaj et al., 2020). Further, several AI algorithms were used to predict the seepage behaviour of dams (Garsole et al., 2023).

1.3.5 Flood and Erosion Hazard Monitoring

One of the most widely researched applications of AI and IoT is their efficiency in predicting floods and the associated erosion. The reason is that AI helps in understanding and studying the complex past flood records and the prediction of floods, and IoT has a crucial role in real-time monitoring of the flood and erosion data. Also, digital twin models will be helpful in simulating the effects of flood based on the present ground conditions. For example, Ahmed et al. (2024) used data from satellite images to detect flood using deep learning algorithms and achieved a segmentation accuracy of around 87%. Also, several researchers have worked on flood risk assessment and protection measures based on AI data analysis (Jones et al., 2023; Z. Liu et al., 2025; Pham et al., 2021). Simultaneously, several IoT-based research proposed methods or instruments to quantify soil erosion. For example, Lin et al. (2019) proposed an IoT-based smart stick that would measure soil erosion; Mendes et al. (2024) proposed a rainfall-triggered sensor with cameras and rain sensors to measure erosion in field.

1.4 REAL-LIFE SCENARIOS USING AI AND IoT

The capability of AI in engineering applications is limitless. Till date, several advanced studies are in progress to expand the boundaries. Although the concept of using AI in civil engineering is not widely brought into practice as an aid to handle the large data developed in IoT-based sensors, AI has been used for several important purposes. Some of the notable applications of AI and IoT in infrastructure development projects are listed in Table 1.1.

TABLE 1.1 Practical implementations of AI and IoT in operation in various nations

SL. NO.	*NAME AND LOCATION OF THE STRUCTURE*	*AI/IOT INTEGRATION*	*REMARKS*	*REFERENCES*
1	Jakarta, Indonesia	Flood and rainfall sensors	Made the data available to public through Jakarta Kini app	Setiawan et al. (2025)
2	Thailand	Flood tracking and mitigation, damage detection	Earth observing mission one, using moderate imaging spectrometer	Chien et al. (2013)
3	150 countries	Google flood forecasting	Using hydrologic and inundation model	*Google Flood Forecasting* (2025); Nevo et al. (2022)
4	Msikaba bridge, South Africa	IoT sensors to monitor strain, displacement and temperature changes	Indirectly monitors settlement	Ajayi et al. (2025)

(Continued)

TABLE 1.1 (*Continued*) Practical implementations of AI and IoT in operation in various nations

SL. NO.	*NAME AND LOCATION OF THE STRUCTURE*	*AI/IOT INTEGRATION*	*REMARKS*	*REFERENCES*
5	Over 100 dams in Bulgaria	IoT sensors to monitor water level	Helpful in alerting in times of flood	Tsiukhai (2024)
6	Gudai-Darri iron ore mine, Western Australia	Digital twin model	Aimed in transferring automated water trucks and monitoring wastage	Rio Tinto (2022); Zvarivadza et al. (2024)
7	Burj Khalifa, Dubai	IoT sensors to monitor strain, displacement and temperature changes	Indirectly monitors settlement	Yusuf et al. (2023)

1.5 BENEFITS AND OPPORTUNITIES

The applications of AI and IoT in geotechnical engineering are not limited to the ones in discussion. Only typical applications and some crucial examples were highlighted in the discussion, and several such studies are in progress around the world. The following are some of the advantages of incorporating AI and IoT in geotechnical engineering applications.

1.5.1 Enhanced Accuracy

The first and foremost reason to adapt IoT in any civil engineering application is its ability to generate data at periodic intervals without any compromise in accuracy. For applications such as slope stability and building health monitoring (including foundation settlement monitoring), it is essential to maintain a record of data. It is highly impossible to document physical measurements at regular cycles at higher degree of accuracy where sensors with real-time synchronization would be more beneficial (Alam et al., 2024). Also, the incorporation of AI will be very much effective in analysing these data which otherwise would be practically impossible with manual tools.

1.5.2 Real-Time Monitoring and Early Warning

The collected data need proper analysis to design an early warning system. Yet, in order to truly understand the behaviour of such slope (or) dam (or) foundation (or) tunnel, rather than mere analysis, it is essential to know the real-time behaviour of such structures to know their influence on the external factors and how they respond to real-world scenario. Digital twins are helpful in recreating the behaviour of the original structure, such that it would be easier to create an early warning system in case if any of the abovementioned structure is on the verge of failure (Mwanza et al., 2024; Shao et al., 2025).

1.5.3 Risk-Informed Design and Adaptive Decision-Making

Integrating IoT-based continuous sensing is made possible with AI-driven analytics which enhances risk-informed design. IoT sensors provide real-time measurements of deformation, pore pressure, moisture, and vibration, etc., allowing the designer/engineer to quantify evolving site conditions rather than rely on static data (i.e.) one-time investigation. AI–IoT-enabled systems therefore help in making decisions, based on real-time data and forecasted data, thereby providing more reliability in the design of risk assessment and early warning system for flood hazard mapping, slope failures, tunnel instability, etc. (Goodarzi et al., 2019; Singh & Manoharan, 2024) where the parameters rely on several uncertainties such as the rainfall, runoff conditions and state of stress.

1.5.4 Contribution to Sustainability

Although contribution to sustainability is not a direct application of such AI and IoT, in civil engineering, they greatly help in preventing many disasters, structural failures, loss of life and reduction in carbon emissions. Consequently, it will help in enhancing the lifespan of structures and also in developing models that would efficiently predict the material and climatic behaviour. AI–IoT interventions in geotechnical engineering can help in optimizing resources, particularly materials, as it helps in predicting soil behaviour accurately and minimizes overdesign and redundant consumption of materials on a large scale, particularly for ground improvement, all of which leads to sustainable development. Similarly, these interventions help in predictive maintenance through early detection that enables the extended lifespan of infrastructure which is a considerable step towards sustainability.

1.6 CHALLENGES AND LIMITATIONS

1.6.1 Sensor Durability and Calibration in Harsh Conditions

An important factor in introducing IoT-based hardware in the field also comes with a challenge of its longevity. Under such harsh environments of continuous pressure, rainfall and vibrations, it will be very difficult to maintain the life and accuracy of such a sensor in the long run. Also, calibration of sensors that are embedded in the ground or structures will be highly impossible if such an issue arises in the future which would damage the structure. Another important issue is the inability to replace the embedded sensors.

1.6.2 Data Availability, Quality and Interoperability

The major challenge in civil engineering field is the availability of data for such AI and ML operations (Harle, 2024). Considering the fact that such technologies are new to the domain, dataset for advanced techniques is not easy to access, or in some cases, they are limited. Especially in geotechnical engineering, the uniqueness of soil properties in each place makes it very complicated; the dataset sourced from one location may not be suitable or reusable for other similar situations (Brinkgreve, 2005). In most cases, the quality of data is an arguable hindrance for a good AI model generation (W. Liang et al., 2022). If disturbances in data are left unnoticed, that would also create an inaccurate model.

1.6.3 Black Box Nature and Trustworthiness of AI Models

The most challenging part of using AI is debugging the results. Although it is possible to track the inputs and outputs in most AI models, the internal decision-making process and the behaviour of the algorithms are neither transparent nor easily understandable, which is known as the 'black box nature' (Abioye et al., 2021; Nyokum & Tamut, 2025). Another important aspect is the question of accuracy with the model prediction; while the models can give effective performance with the trained, validated and tested dataset, it will not guarantee that the performance will be similar for a new set of data.

Another important factor about this AI-based model development, specifically in geotechnical engineering, is the question of adaptability. Several experimental studies have been conducted to determine the soil properties (Ankah et al., 2025; Pham et al., 2019) or soil-structure behaviour (Jong et al., 2021) using AI techniques. Unlike concrete, the characteristics and behaviour of each soil are unique. Hence, the adaptability and versatility of AI models are still questionable.

1.6.4 Cybersecurity, Privacy and Ethical Concerns

The applications of AI and ML in geotechnical engineering are at an early stage where the data are merely used for experimental purposes. However, expecting a promising growth in the future, the data sourcing will be a great issue. Most of the ML models are generated from IoT-based sensors that belong to a third party, from whom sourcing the data will not be easy. Without those data, understanding the ML model is highly impossible. In another perspective, the access of data pertaining to slope or dam at a particular geographical location or even accessing the subsoil profile of such locations without proper ethical permission is a threatening concern, as it would lead to several serious results if the data fall into the wrong hands.

1.6.5 Lack of Codal/Regulatory Guidelines

At present, AI is mostly in the experimental phase in every domain, and there is no well-defined guideline that advocates the incorporation of AI and IoT in industries. The field of geotechnical engineering is no exception to this. Even a well-trained model may tend to deviate from the actual outcome if there is a flaw in the AI model training (Plevris & Hosamo, 2025). For instance, in critical scenarios where a human life (such as medicine and diagnostic field) and where life along with infrastructure are under risk (such as disaster management, building foundation failure, dam/slope failure), decision-making is crucial, and in such scenarios, the AI model cannot be blamed for inaccurate prediction or decision-making. In other words, the boundaries and limitations on the usage of AI should be made clear.

1.6.6 Initial Investment and Upgradation

Studies show that in certain applications like bridge damage detection, the AI incorporation exhibited up to 99% efficiency (Nyokum & Tamut, 2025), but the cost of procuring the essential sensors and implementing the technology will be much higher in the current scenario (Omar et al., 2022). Another important concern is that, although there is flexibility in establishing a smart AI and IoT-enabled modern infrastructure, the situation may not be the same in existing structures. The upgradation of structures that were already built to meet the trends of new monitoring system will be utmost difficult. Also, such modern techniques need additional skilled employees to handle the data and interpret the outcomes which will again result in increased expenses.

1.7 FUTURE DIRECTIONS

1.7.1 Awareness and Education

The first and foremost step in establishing these technologies universally is by creating awareness about the application of such methods and educating the engineers to integrate AI and IoT into their design and execution. The hesitation in the implementation of AI and IoT in any civil engineering domain is the lack of awareness to adopt modern technology. For instance, in their study on assessing the hurdles in incorporating AI and IoT in the Malaysian construction industry, Omar et al. (2022) inferred from their survey that some practising engineers are of the mistaken view that AI will replace them in the market. Also, another important factor is that some practitioners felt that they have insufficient time to learn a new technology, and even if they employed some specialists, an additional amount has to be spent on training and paying the specialists. However, as stated earlier, a proper framework should be established to safeguard life and structures from disasters; hence, creating awareness on a different level of approach is much essential.

1.7.2 Autonomous Geotechnical Monitoring Systems

In its present state, there are sufficient examples to show that automated data collection, processing and decision-making is possible through AI, yet the effort to make a universally compatible model is the need of the hour. Future research should highly focus on generating universal models that would work on any soil/environmental conditions capable of decision-making to trigger early warning systems.

1.7.3 Digital Twins for Infrastructure Lifecycle Management

Although sensors are installed in many newly constructed modern structures, the data remain merely a storage for future references, no simulation-based real-time monitoring is done on a regular basis. In the future, to create a sustainable infrastructure and community, a proper prediction with digital twin technology for such structure should be made mandatory. In another perspective, the development of physical model for prediction is outdated and would incur a huge investment to understand the behaviour at every possible crucial condition, where digital twin technology would be handy. A great example of this approach is virtual Singapore, an initiative by the Singapore Land Authority where a digital twin model of the city has been used for multiple applications including disaster management and emergency services (Shahat et al., 2021). Similarly, a digital twin of Zurich in Switzerland has been developed in view of urban planning (Schrotter & Hürzeler, 2020). Although the digital twin models of these two cities are still in their preliminary phase of usage and have scope for development, the initiative benefits the end-users and policymakers greatly in effective planning.

1.7.4 Blockchain for Secure Geotechnical Data Management

If this technology grows as it is in its present state, a major hurdle may be expected in the future: how to store and manage the huge volume of data generated by such IoT sensors. Hence, a blockchain establishment would be necessary to securely handle the generated data. Although not fully discussed in a geotechnical engineering perspective, Waqar et al. (2024) discussed the challenges in implementing the blockchain technology in construction industry which are mostly applicable to geotechnical engineering as well. Some of the major concerns include the complexity in the interoperability with the existing systems, return of investments upon implementation, concern on the energy used in operating the blockchain technology, the reliability of the technology on a longer run and few other minor concerns regarding the industry standards and stakeholders.

1.7.5 Roadmap for Codification and Large-Scale Adoption

The universal adoption of this technology is not a simple or easy process. Since this technology is in its early phase, a huge investment is the essential leap to explore more, followed by the establishment of benchmark datasets. The funded research should primarily focus on generation of either universal models or adaptive models that can perform both real-time monitoring and automated decision-making to trigger the early warning systems. The next step will be the development of a collaborative ecosystem and parallelly the guidelines to secure, share and use the data. Upon the success of the abovementioned factors, real-time infrastructure projects should be established incorporating the universal AI model and monitoring system to establish its capability to the world. Once this is universally accepted, a regulatory body should be established to oversee the liability of projects using AI and IoT automation. Another simultaneous process is the expansion of sensors and the digital tool market to ensure that this technology can be accessed with ease. The next step will be the development (or revision) of codal provisions to incorporate AI and IoT ethically. Also, a legal framework should be formed to enforce the rules and regulations for the usage of AI and IoT tools and the penalty if it is not properly followed. More importantly, updating every aspect discussed to meet the contemporary need is essential to successfully establish this technology globally.

1.8 CONCLUSIONS

This chapter summarizes various AI and IoT applications in geotechnical engineering, ranging from slope stabilization to earth dam monitoring. Additionally, it discusses the opportunities, limitations and the roadmap for the universal adoption of these technologies. Based on the available literature and the current growth trajectory, it is evident that AI holds significant potential in structural health monitoring, disaster management and related areas. However, unlike other materials, soil behaviour is unique and complex, making it difficult to develop a universal model that accurately represents its characteristics. Furthermore, there are still limited practical implementations of AI-based decision-making systems in this domain. As a result, the practical application of these technologies remains underdeveloped. Future research should focus on both the development of AI-based models incorporating IoT and the demonstration of their practical implementation in real-world scenarios.

REFERENCES

Abdu, D. M., Wei, G., & Yang, W. (2023). Assessment of railway bridge pier settlement based on train acceleration response using machine learning algorithms. *Structures*, *52*, 598–608. https://doi.org/10.1016/j.istruc.2023.03.167

Abioye, S. O., Oyedele, L. O., Akanbi, L., Ajayi, A., Davila Delgado, J. M., Bilal, M., Akinade, O. O., & Ahmed, A. (2021). Artificial intelligence in the construction industry: A review of present status, opportunities and future challenges. *Journal of Building Engineering*, *44*, 103299. https://doi.org/10.1016/j.jobe.2021.103299

Aditya, V. M. V. S., Tanishq, Ch. T. S., Badri Sai, V. C., Dhuli, S., & Atik, F. (2021). IoT based water level monitoring system for dams. *2021 12th International Conference on Computing Communication and Networking Technologies (ICCCNT)*, Indian Institute of Technology: Kharagpur, India on July 6–8, 2021, 1–6. https://doi.org/10.1109/ICCCNT51525.2021.9579945

Ahmed, I., Ahmad, M., Jeon, G., & Chehri, A. (2024). An internet of things and AI-powered framework for long-term flood risk evaluation. *IEEE Internet of Things Journal*, *11*(3), 3812–3819. https://doi.org/10.1109/JIOT.2023.3308564

Ajayi, O. O., Kurien, A., Djouani, K., & Dieng, L. (2025). Artificial intelligence for infrastructure resilience: Transportation systems as a strategic case for policy and practice. *Sustainability*, *17*(20), 8992. https://doi.org/10.3390/su17208992

Alam, M. J. B., Manzano, L. S., Debnath, R., & Ahmed, A. A. (2024). Monitoring slope Movement and soil hydrologic behavior using IoT and AI technologies: A systematic review. *Hydrology*, *11*(8), 111. https://doi.org/10.3390/hydrology11080111

Ankah, M. L. Y., Adjei-Yeboah, S., Ziggah, Y. Y., & Asare, E. N. (2025). Advanced hybrid machine learning models with explainable AI for predicting residual friction angle in clay soils. *Scientific Reports*, *15*(1), 25868. https://doi.org/10.1038/s41598-025-05962-6

Baghbani, A., Choudhury, T., Costa, S., & Reiner, J. (2022). Application of artificial intelligence in geotechnical engineering: A state-of-the-art review. *Earth-Science Reviews*, *228*, 103991. https://doi.org/10.1016/j.earscirev.2022.103991

Bello, O., Holzmann, J., Yaqoob, T., & Teodoriu, C. (2015). Application of artificial intelligence methods in drilling system design and operations: A review of the state of the art. *Journal of Artificial Intelligence and Soft Computing Research*, *5*(2), 121–139. https://doi.org/10.1515/jaiscr-2015-0024

Brinkgreve, R. B. J. (2005). Selection of soil models and parameters for geotechnical engineering application. *Soil Constitutive Models*, 69–98. https://doi.org/10.1061/40771(169)4

Chen, M.-F., Chiu, H.-C., Tseng, K.-S., Yang, Y.-C., Chou, C.-Y., & Jiang, J.-A. (2019). An IoT-based temperature monitoring system for underground cable tunnels. *2019 IEEE PES GTD Grand International Conference and Exposition Asia (GTD Asia)*, 960–964. https://doi.org/10.1109/GTDAsia.2019.8715849

Chien, S., Doubleday, J., Mclaren, D., Tran, D., Tanpipat, V., Chitradon, R., Boonya-aroonnet, S., Thanapakpawin, P., & Mandl, D. (2013). Monitoring flooding in Thailand using earth observing one in a Sensorweb. *IEEE Journal of Selected Topics in Applied Earth Observations and Remote Sensing*, *6*(2), 291–297. https://doi.org/10.1109/JSTARS.2013.2247974

Dui, H., Dong, X., Wu, X., Chen, L., & Bai, G. (2024). IoT-enabled risk warning and maintenance strategy optimization for tunnel-induced ground settlement. *IEEE Internet of Things Journal*, *11*(13), 22966–22981. https://doi.org/10.1109/JIOT.2024.3377440

El-Abbasy, A. A. A. (2025). Artificial intelligence-driven predictive modeling in civil engineering: A comprehensive review. *Journal of Umm Al-Qura University for Engineering and Architecture*. https://doi.org/10.1007/s43995-025-00166-5

Garsole, P. A., Bokil, S., Kumar, V., Pandey, A., & Topare, N. S. (2023). A review of artificial intelligence methods for predicting gravity dam seepage, challenges and way-out. *AQUA — Water Infrastructure, Ecosystems and Society*, *72*(7), 1228–1248. https://doi.org/10.2166/aqua.2023.042

Goodarzi, L., Banihabib, M. E., & Roozbahani, A. (2019). A decision-making model for flood warning system based on ensemble forecasts. *Journal of Hydrology*, *573*, 207–219. https://doi.org/10.1016/j.jhydrol.2019.03.040

Google Flood Forecasting. (2025). https://sites.research.google/gr/floodforecasting/

Hamzah, N., Musa, L. O., Yunus, Muh. Y., Sabarka, M., & Hz, S. R. (2024). Application of internet of things (IoT) for monitoring rainfall and regulating dam gate in hydropower models. In *THE 1st International Multidisciplinary Conference Of Applied Sciences: Applied Sciences for Sustainable Earth, Environment and Management* 18–19 November 2023. Makassar, Indonesia. *AIP Conference Proceedings* (Vol. 3140, No. 1, p. 040024). https://doi.org/10.1063/5.0222604

Harle, S. M. (2024). Advancements and challenges in the application of artificial intelligence in civil engineering: A comprehensive review. *Asian Journal of Civil Engineering*, *25*(1), 1061–1078. https://doi.org/10.1007/s42107-023-00760-9

Hussen Hajjaj, S. S., Hameed Sultan, M. T., Moktar, M. H., & Lee, S. H. (2020). Utilizing the internet of things (IoT) to develop a remotely monitored autonomous floodgate for water management and control. *Water*, *12*(2), 502. https://doi.org/10.3390/w12020502

Hyodo, M., Orense, R. P., Noda, S., Furukawa, S., & Furui, T. (2012). Slope failures in residential land on valley fills in Yamamoto town. *Soils and Foundations*, *52*(5), 975–986. https://doi.org/10.1016/j.sandf.2012.11.015

Jones, A., Kuehnert, J., Fraccaro, P., Meuriot, O., Ishikawa, T., Edwards, B., Stoyanov, N., Remy, S. L., Weldemariam, K., & Assefa, S. (2023). AI for climate impacts: Applications in flood risk. *NPJ Climate and Atmospheric Science*, *6*(1), 63. https://doi.org/10.1038/s41612-023-00388-1

Jong, S. C., Ong, D. E. L., & Oh, E. (2021). State-of-the-art review of geotechnical-driven artificial intelligence techniques in underground soil-structure interaction. *Tunnelling and Underground Space Technology*, *113*, 103946. https://doi.org/10.1016/j.tust.2021.103946

Liang, M.-C., Chen, H.-E., Tfwala, S. S., Lin, Y.-F., & Chen, S.-C. (2023). The application of wireless underground sensor networks to monitor seepage inside an earth dam. *Sensors*, *23*(8), 3795. https://doi.org/10.3390/s23083795

Liang, W., Tadesse, G. A., Ho, D., Fei-Fei, L., Zaharia, M., Zhang, C., & Zou, J. (2022). Advances, challenges and opportunities in creating data for trustworthy AI. *Nature Machine Intelligence*, *4*(8), 669–677. https://doi.org/10.1038/s42256-022-00516-1

Lin, T.-H., Peng, J.-X., Wen, C.-Y., Wang, Y.-T., & Huang, J.-T. (2019). Development of smart stick for soil erosion monitoring. *IGARSS 2019 – 2019 IEEE International Geoscience and Remote Sensing Symposium*, Yokohama, Japan, July 28–August 2, 2019, 9618–9621. https://doi.org/10.1109/IGARSS.2019.8899279

Liu, B., Gao, F., Tian, B., Gong, X., Zhu, H., & Zhao, Y. (2024). Intelligent monitoring system for roller-compacted earth-rock dam construction process. *Journal of Physics: Conference Series*, *2913*(1), 012031. https://doi.org/10.1088/1742-6596/2913/1/012031

Liu, Z., Coleman, N., Patrascu, F. I., Yin, K., Li, X., & Mostafavi, A. (2025). Artificial intelligence for flood risk management: A comprehensive state-of-the-art review and future directions. *International Journal of Disaster Risk Reduction*, *117*, 105110. https://doi.org/10.1016/j.ijdrr.2024.105110

Luo, J., Zhang, Q., Li, L., & Xiang, W. (2019). Monitoring and characterizing the deformation of an earth dam in Guangxi Province, China. *Engineering Geology*, *248*, 50–60. https://doi.org/10.1016/j.enggeo.2018.11.007

Luo, W., Zheng, J., Miao, Y., & Gao, L. (2024). Raspberry Pi-based IoT system for grouting void detection in tunnel construction. *Buildings*, *14*(11), 3349. https://doi.org/10.3390/buildings14113349

Ma, Y., & Guo, G. (2020). Intelligent algorithm of geotechnical test data based on Internet of Things. *Computer Communications*, *158*, 32–38. https://doi.org/10.1016/j.comcom.2020.04.028

Mendes, T. A., Rebolledo, J. F. R., Dos Santos Pereira, S. A., Oliveira, M. V. M. D., & Formiga, K. T. M. (2024). Development of a device for monitoring erosion in the field. *Micromachines*, *15*(7), 880. https://doi.org/10.3390/mi15070880

Menon, U. V., Babu Kumaravelu, V., Vinoth Kumar, C., Rammohan, A., Chinnadurai, S., Venkatesan, R., Hai, H., & Selvaprabhu, P. (2025). AI-powered IoT: A survey on integrating artificial intelligence with IoT for enhanced security, efficiency, and smart applications. *IEEE Access*, *13*, 50296–50339. https://doi.org/10.1109/ACCESS.2025.3551750

Mwanza, J., Mashumba, P., & Telukdarie, A. (2024). A framework for monitoring stability of tailings dams in realtime using digital twin simulation and machine learning. *Procedia Computer Science*, *232*, 2279–2288. https://doi.org/10.1016/j.procs.2024.02.047

Nevo, S., Morin, E., Gerzi Rosenthal, A., Metzger, A., Barshai, C., Weitzner, D., Voloshin, D., Kratzert, F., Elidan, G., Dror, G., Begelman, G., Nearing, G., Shalev, G., Noga, H., Shavitt, I., Yuklea, L., Royz, M., Giladi, N., Peled Levi, N., … Matias, Y. (2022). Flood forecasting with machine learning models in an operational framework. *Hydrology and Earth System Sciences*, *26*(15), 4013–4032. https://doi.org/10.5194/hess-26-4013-2022

Nyokum, T., & Tamut, Y. (2025). Artificial intelligence in civil engineering: Emerging applications and opportunities. *Frontiers in Built Environment*, *11*, 1622873. https://doi.org/10.3389/fbuil.2025.1622873

Omar, A. F., Zainordin, N., Khoo, S. L., & Peng, J. O. H. (2022). Barriers in implementing artificial intelligence (AI) and Internet of Things (IoTs) among Malaysian construction industry. *AIP Conference Proceedings*, 030002. https://doi.org/10.1063/5.0111461

Pham, B. T., Luu, C., Phong, T. V., Nguyen, H. D., Le, H. V., Tran, T. Q., Ta, H. T., & Prakash, I. (2021). Flood risk assessment using hybrid artificial intelligence models integrated with multi-criteria decision analysis in Quang Nam Province, Vietnam. *Journal of Hydrology*, *592*, 125815. https://doi.org/10.1016/j.jhydrol.2020.125815

Pham, B. T., Nguyen, M. D., Dao, D. V., Prakash, I., Ly, H.-B., Le, T.-T., Ho, L. S., Nguyen, K. T., Ngo, T. Q., Hoang, V., Son, L. H., Ngo, H. T. T., Tran, H. T., Do, N. M., Van Le, H., Ho, H. L., & Tien Bui, D. (2019). Development of artificial intelligence models for the prediction of compression coefficient of soil: An application of Monte Carlo sensitivity analysis. *Science of The Total Environment*, *679*, 172–184. https://doi.org/10.1016/j.scitotenv.2019.05.061

Piciullo, L., Abraham, M. T., Drøsdal, I. N., & Paulsen, E. S. (2025). An operational IoT-based slope stability forecast using a digital twin. *Environmental Modelling & Software*, *183*, 106228. https://doi.org/10.1016/j.envsoft.2024.106228

Plevris, V., & Hosamo, H. (2025). Responsible AI in structural engineering: A framework for ethical use. *Frontiers in Built Environment*, *11*, 1612575. https://doi.org/10.3389/fbuil.2025.1612575

Raj, M., & Sengupta, A. (2014). Rain-triggered slope failure of the railway embankment at Malda, India. *Acta Geotechnica*, *9*(5), 789–798. https://doi.org/10.1007/s11440-014-0345-9

Razip, N. A. M., & Daud, N. N. N. (2022). Ground surface and building settlement monitoring using Internet of Things (IoT) approach. *IOP Conference Series: Earth and Environmental Science*, *1091*(1), 012001. https://doi.org/10.1088/1755-1315/1091/1/012001

Rio Tinto. (2022). *Rio Tinto Opens Gudai-Darri, its Most Technologically Advanced Mine* [Online post]. https://www.riotinto.com/en/news/releases/2022/rio-tinto-opens-gudai-darri-its-most-technologically-advanced-mine

Russell, S., & Norvig, P. (2010). *Artificial Intelligence: A Modern Approach* (Third Edition). Prentice Hall.

Santhosh Kumar, V., & Sennimalai Chandrasekaran, S. (2022). Analysis of failure of high slope subjected to rainfall infiltration at Peringavu in Kerala, India. *Engineering Failure Analysis*, *138*, 106423. https://doi.org/10.1016/j.engfailanal.2022.106423

Sarker, I. H. (2021). Machine learning: Algorithms, real-world applications and research directions. *SN Computer Science*, *2*(3), 160. https://doi.org/10.1007/s42979-021-00592-x

Schrotter, G., & Hürzeler, C. (2020). The digital twin of the city of Zurich for urban planning. *PFG – Journal of Photogrammetry, Remote Sensing and Geoinformation Science*, *88*(1), 99–112. https://doi.org/10.1007/s41064-020-00092-2

Setiawan, A., Usman, S., & Qodir, Z. (2025). Implementation of open government through the Jakarta Kini (Jaki) Application on environmental features in the dissemination of public information. *IOP Conference Series: Earth and Environmental Science*, *1475*(1), 012006. https://doi.org/10.1088/1755-1315/1475/1/012006

Shahat, E., Hyun, C. T., & Yeom, C. (2021). City digital twin potentials: A review and research agenda. *Sustainability*, *13*(6), 3386. https://doi.org/10.3390/su13063386

Shao, C., Zheng, S., Xu, Y., Gu, H., Qin, X., & Hu, Y. (2025). A visualization system for dam safety monitoring with application of digital twin platform. *Expert Systems with Applications*, *271*, 126740. https://doi.org/10.1016/j.eswa.2025.126740

Sheil, B. B., Suryasentana, S. K., Mooney, M. A., & Zhu, H. (2020). Machine learning to inform tunnelling operations: Recent advances and future trends. *Proceedings of the Institution of Civil Engineers - Smart Infrastructure and Construction*, *173*(4), 74–95. https://doi.org/10.1680/jsmic.20.00011

Siddula, S. S., Babu, P., & Jain, P. C. (2018). Water level monitoring and management of dams using IoT. *2018 3rd International Conference on Internet of Things: Smart Innovation and Usages (IoT-SIU)*, 1–5. https://doi.org/10.1109/IoT-SIU.2018.8519843

Singh, C. R., & Manoharan, G. (2024). Strengthening resilience: AI and machine learning in emergency decision-making for natural disasters. In D. Satishkumar & M. Sivaraja (Eds.), *Advances in Computational Intelligence and Robotics* (pp. 249–278). IGI Global. https://doi.org/10.4018/979-8-3693-4284-8.ch012

Suryasentana, S., & Shiel, B. (2023). Demystifying the connections between Gaussian Process Regression and Kriging. In *9th International Offshore Site Investigation and Geotechnics (OSIG) Conference*, September 12–14, 2023 at Imperial College: London, 1–8. https://doi.org/10.3723/DAGZ8416

Tsiukhai, T. (2024). *Smart Water Management with IoT: Key Application Areas* [Online post]. https://www.softeq.com/blog/smart-water-management-using-iot-real-world-examples

Waqar, A., Alharbi, L. A., Abdullah Alotaibi, F., Alrasheed, K. A., Khan, A. M., & Almujibah, H. (2024). Challenges of blockchain implementation in construction. *Journal of Engineering*, *2024*(1), 2442345. https://doi.org/10.1155/2024/2442345

Wu, K., Zhang, P., Yuan, J., Qian, X., & Qi, R. (2025). IoT monitoring and evaluating system for the construction quality of foundation pile. *Buildings*, *15*(15), 2660. https://doi.org/10.3390/buildings15152660

Xu, J., & Zhang, Y. (2025). AI-powered digital twin technology for highway system slope stability risk monitoring. *Geotechnics*, *5*(1), 19. https://doi.org/10.3390/geotechnics5010019

Yi, Z., Jinshen, G., & Xu, W. (2019). Intelligent settlement monitoring system of high-speed railway bridge. *Journal of Civil Structural Health Monitoring*, *9*(3), 307–323. https://doi.org/10.1007/s13349-019-00334-x

Yusuf, D. A., Ahmed, A., Sagir, A., Yusuf, A. A., Yakubu, A., Zakari, A. T., Usman, A. M., Nashe, A. S., & Hamma, A. S. (2023). A review of conceptual design and self health monitoring program in a vertical city: A case of Burj Khalifa, U.A.E. *Buildings*, *13*(4), 1049. https://doi.org/10.3390/buildings13041049

Zvarivadza, T., Onifade, M., Dayo-Olupona, O., Said, K. O., Githiria, J. M., Genc, B., & Celik, T. (2024). On the impact of Industrial Internet of Things (IIoT)—Mining sector perspectives. *International Journal of Mining, Reclamation and Environment*, *38*(10), 771–809. https://doi.org/10.1080/17480930.2024.2347131

2 Enhancing Resilience of Geotechnical Structures with AI and IoT

R. S. V. Rashma

2.1 INTRODUCTION

Geotechnical structures play a crucial role in infrastructure development as a permanent or temporary structure, which in turn becomes the backbone of society. The geotechnical structures such as dams, embankments, slopes, and foundations support water management, transportation, energy production, and urban expansion, thereby playing a central role in economic growth and human well-being (Gariano & Guzzetti, 2016; Neuman, 2020). Natural disasters such as earthquakes, floods, and landslides and human-induced activities such as rapid construction and changes in use of land especially in slopes possess significant risks to the safety and performance of these structures (Huang et al., 2021). The history of catastrophic failure of geotechnical structures has led to loss of human life and financial loss along with environmental harm, underscoring the urgent need for resilient and adaptive engineering solutions (Shah et al., 2014).

DOI: 10.1201/9781003714125-2

According to Zhan et al. (2024), traditional geotechnical inspections such as routine inspections, geotechnical surveys, and empirical or analytical modelling are being implemented to assess the performance of structures, which provide valuable insights. However, the conventional performance monitoring systems are often limited in their ability to capture real-time behaviour, address uncertainties, and account for complex soil–structure interactions (Zhan et al., 2024). Technological progressions like enhanced geotechnical and geoenvironmental engineering, usage of satellite monitoring, and predictive AI modelling can boost better understanding patterns of landslide patterns and mitigation, thereby implementing rapid disaster warning arrangements (Vung et al., 2023). Strauss and Wadzuk (2022) reported on the progressive monitoring of stormwater systems using IoT technology for assessing the probability and consequences of reduced performance of storm water facilities due to reduced capacity. With the growing need of infrastructure and climate-related occurrences, it is evident that the conventional practices will not be sufficient to ensure the long-term stability of these structures and associated features.

The earthen dams or embankments as a result of seepage, piping, and erosion can lead to man-made disasters. As per Liang et al. (2023), currently wireless monitoring methods use efficient wireless underground transmission in monitoring the soil moisture content, thereby determining water content inside earthen dams is limited. However, they studied wireless transmission sensors set underground, which can monitor the rising seepage water level which initiates failure behaviour of an earthen dam. Therefore, the study concluded that the awareness of rising water levels within the earthen dams with wireless sensors informs better than conventional water level gauges. Further, they summarized the significance of implementing the same for rapid disaster warning systems in this age of drastic climatic variations that leads to most of the overflowing catastrophic conditions.

These challenges can be further overcome with the advancement of the ongoing digital transformation in engineering. The integration of artificial intelligence (AI) and the Internet of Things (IoT) has emerged as a promising path to enhance resilience of geotechnical engineering structures. IoT devices can deliver continuous monitoring by providing high resolution data on the behaviour of geotechnical structures and the surrounding environments. AI algorithms can process these vast amounts of data to identify the hidden patterns, predict potential failure patterns, and for swift decision-making (Onyelowe et al., 2023). Together, these technologies present opportunities to shift from responsive risk management towards predictive and adaptive frameworks that improve safety, efficiency, and sustainability, which is essential for current and future society.

This chapter provides a bibliometric review of the existing literature to identify current trends, challenges, and opportunities in this rapidly evolving domain on the integration of AI and IoT in strengthening the resilience of geotechnical structures. The discussion highlights the potential benefits and limitations of these technologies along with future directions for research and practice on the performance of existing and new geotechnical structures. Additionally, this chapter aims to review the contributions to a deeper understanding of how digital innovations can transform the way geotechnical infrastructure systems are designed, monitored, managed, and protected for the prevention of loss of life and property by providing early warning.

TABLE 2.1 Research methodology for performing bibliometric analysis to review the impact of evolution of IoT and AI on the enhancement of geotechnical structures

METHOD	*DETAILED STEPS*
1. Document selection by using search string	1. Bibliographic database: Lens 2. Search string: 3. Document type: Topic (Article, Abstract, keyword, Full chapter) 4. Total documents 5. Inclusion criteria: 25 years (2000–2025) 6. Final selection
2. Data extraction using available medium	1. MS Excel 2. VOSviewer software: Bibliometric analysis with maps
3. Analysis and results	1. Database analysis 2. Science mapping

2.2 METHODOLOGY

Bibliometric analysis is conducted to understand trends and developments concerning the impact of evolution of IoT and AI on the enhancement of geotechnical structures. The overall methodology is depicted in Table 2.1, which shows two distinct segments: (i) Data accumulation and adopted search strategy; and (ii) Data Analysis.

2.3 DATA ACCUMULATION AND ADOPTED SEARCH STRATEGY

For the accumulation of data, the scientific literature published in English language using the Lens database (*Lens.Org, Release 9.7*, 2025) (www.lens.org) is used, because of its features such as free accessibility of substantial coverage of scientific articles and easiness in transferring the bibliometric data.

Articles published up to 2025 are incorporated in the study as the list was collected during September/October 2025. The articles were filtered beforehand for accepting for investigation by using filters. The search string "Enhancing resilient geotechnical structures" OR "AI and IOT" for the criteria "Article title, abstract, Keywords" (topic area) of the database: Lens was used, limiting the search criteria only for a span of 25 years from 2000 to 2025. Figure 2.1 shows the implemented exploration of articles. A total of 2076 papers were chosen for further procedure after adopting the above-mentioned strategy. The information from the Lens database was transferred to an Excel file. The exported file comprised of information regarding authors, institutes, countries, and many more information related to chosen articles.

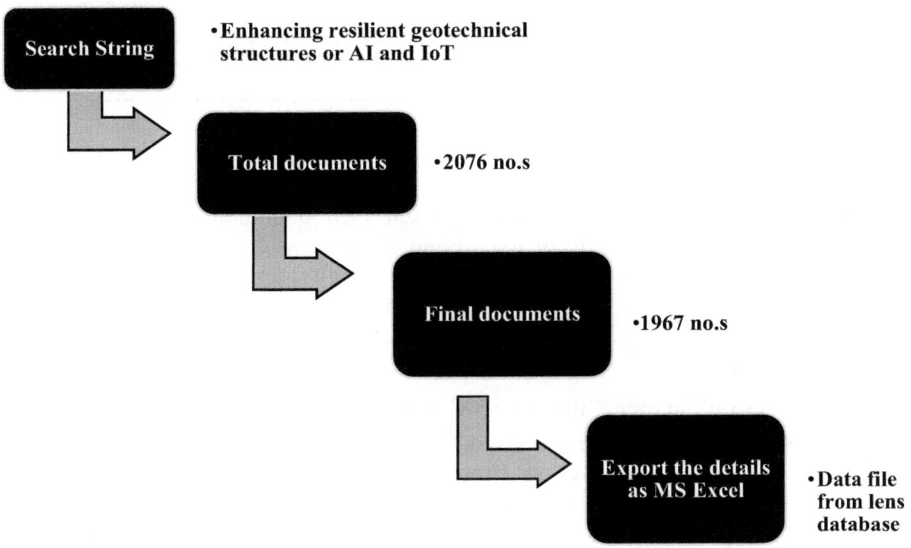

FIGURE 2.1 Search strategy adopted.

2.4 DATA ANALYSIS AND VISUALIZATION

The VOSviewer software (version 1.6.20) was utilized in this study for bibliometric analysis due to its compatibility with multiple operating systems and user-friendly interface (van Eck & Waltman, 2010, 2014, 2017). A total of 1967 research publications were included in the dataset. VOSviewer provides three main modes of visualization: network, overlay, and density to facilitate comprehensive data interpretation. Several researchers have emphasized the superiority of VOSviewer over other bibliometric tools because of its effectiveness in clustering and analysing research outputs (Anjali & Remesan, 2023; Munim et al., 2020). Leveraging VOSviewer's analytical capabilities, this study conducted (i) database investigation across diverse published articles and (ii) mapping to investigate the thematic assembly and progression of the research domain. The database investigation examined scientific production trends, document types, publication years, and institutional contributions related to resilient geotechnical structures and the influence of AI and IoT. The generated maps depicting link strengths, keyword co-occurrences, and co-citation patterns among sources and authors were produced through VOSviewer to represent the intellectual structure of the field. Science mapping further enabled a quantitative classification, visualization, and analysis of all bibliographic entities in the dataset, elucidating their composition and interconnections (Andersen & Swami, 2021).

2.5 RESULTS AND DISCUSSION

2.5.1 Database Analysis

Figure 2.2 shows the distribution of the chosen 1967 documents as per the adopted strategy along with the quantity of issued papers per year. The data depict the growing interest in the field with 1967 documents from 2000 to 2025. There has been peak research from 2015 onwards on this research topic which shows the global significance of this topic. Related to chosen topic, 200 publications were available in 2015 and 370 articles in 2021.

The articles on the topic "Impact of AI and IOT on the enhancement of geotechnical structures" during the 25 years from 2000 to 2025 involve four foremost categories. The search strategy adopted for this study of documents include journal articles, books, report, preprints, conference proceedings, dissertation, book chapters, and others. Figure 2.3 depicts the same with identifying the foremost type as article types (82% of the total documents) followed by reports (6.2%).

The publications on the impact of IoT and AI on geotechnical structure enhancement come across extensive array of research ranges like engineering, geotechnical engineering, geology, environmental sciences, computer sciences, material sciences, composite sciences, geography, etc. as shown in Figure 2.4.

Figure 2.5 shows the ten major fields of research emerging from the examination of all collected documents for 25 years. Major documents (34%) are contributed to "Engineering mainly Geotechnical Engineering and Geology." The categories "environmental science, computer science and material science" constitute the second largest

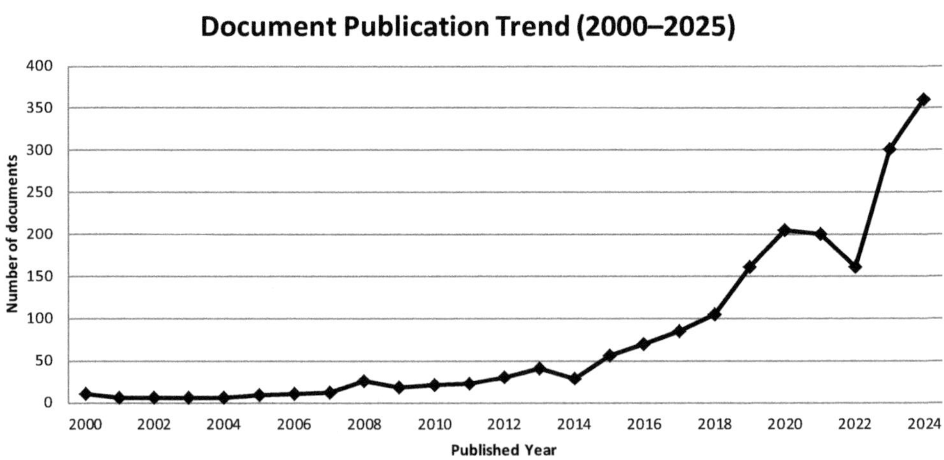

FIGURE 2.2 Growth in scientific articles on the topic "Enhancing resilience of geotechnical structures with IOT and AI" during the last 25 years.

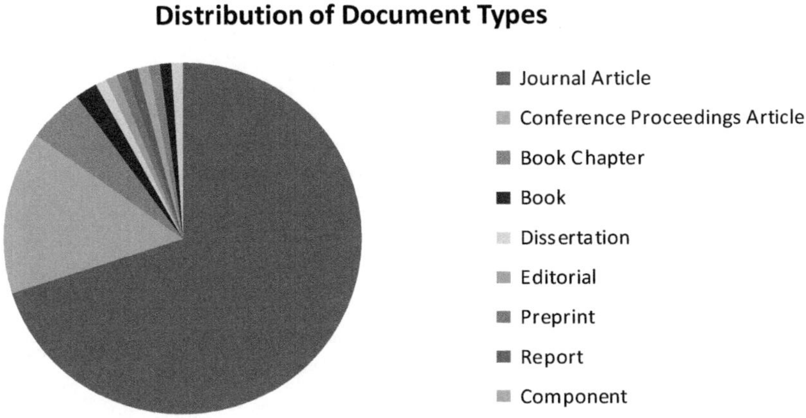

Journal Article	82%
Books	4.5%
Report	6.2%
Preprint	1.6%
Conference Proceedings	1.8%
Dissertation	1.3%
Book Chapter	1.4%
Others	0.56%

FIGURE 2.3 Percentage of documents on the topic "Enhancing resilience of geotechnical structures with IOT and AI" during the last 25 years.

category, accounting for 26% of the total publications. The "composite material" category has a substantial number of articles (20%) and comes third followed by geography (10%), soil water (5%), and mathematics (5%) respectively.

2.5.2 Analysis of Contributions by Most Cited Articles

The citations obtained from the Lens database referring to the subject area are also studied, and the top 10 most cited documents are shown in Figure 2.6. The total number of publications (1967 documents) presents the total number of citations of all the articles as 1200, showing the research still developing. The most cited article corresponds to article "landslides in a changing climate" which includes the review on how climate change affects landslide activity, emphasizing uncertainties in magnitude, extent, and regional impacts. It highlights research gaps in Asia, South America, and Africa; critiques

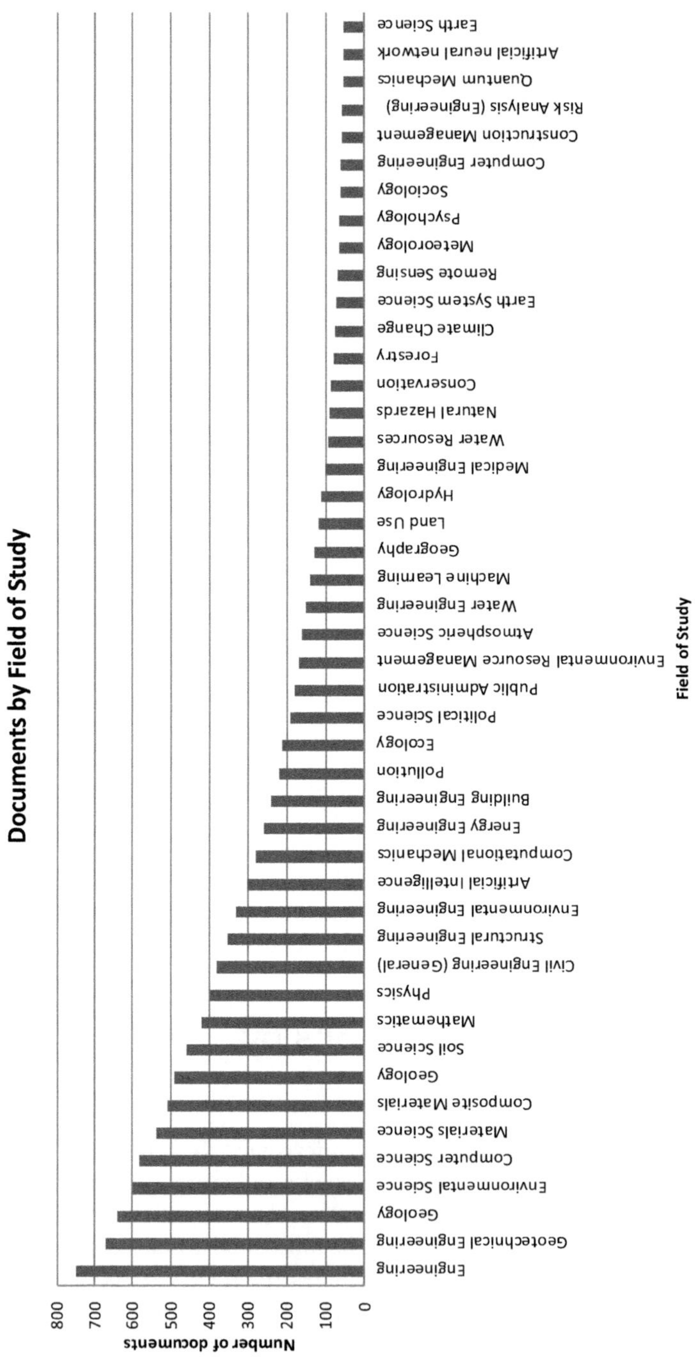

FIGURE 2.4 Major 50 fields of study on the topic "Enhancing resilience of geotechnical structures with IOT and AI" during the last 25 years.

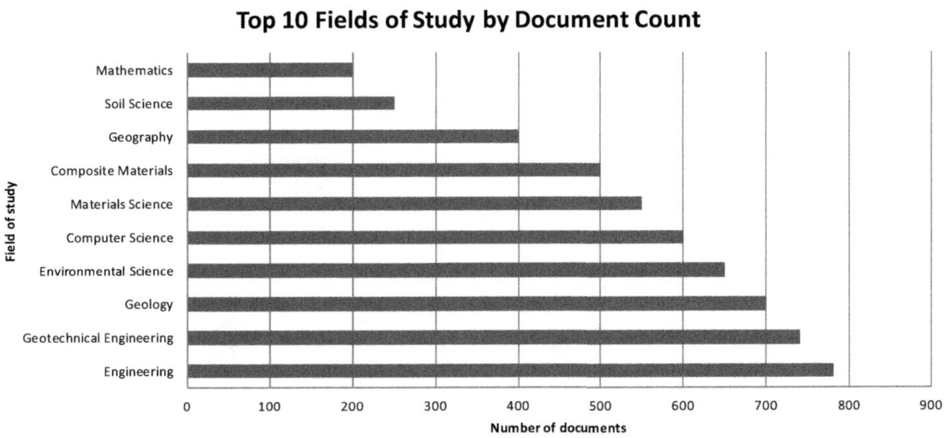

FIGURE 2.5 Distribution of research on the topic "Enhancing resilience of geotechnical structures with IOT and AI" among ten major research areas during the last 25 years.

modelling limitations; and advocates for uncertainty in quantification, ensemble projections, and adaptive risk reduction strategies to better manage future landslide hazards under global warming (Gariano & Guzzetti, 2016). The concept of risk as a product of social processes interacting with the environment, emphasizing coping and adaptive capacities is reported by Cardona et al. (2012). The dimensions and trends of vulnerability and exposure, offering a general overview due to their context-specific nature are then explored. A methodological discussion outlined approaches for identifying and assessing risk, highlighting applications in adaptation planning and the importance of risk perception and communication. Cardona et al. (2012) concluded that risk accumulation and the nature of disasters provide a comprehensive understanding of how risks develop and manifest over time.

Zhang et al. (2021) critically reviewed the application of deep learning (DL) in geotechnical engineering and summarized that DL plays a critical role in analysing the performance of geotechnical infrastructure activities like construction of tunnel, prediction of slope stability, evaluation of susceptibility of the occurrence of landslides, load-carrying capacity, evaluation of pile foundations, etc. They concluded their study stating that the big data and DL are the new trend of AI in geotechnical engineering in the future.

The shear strength of tyre shred–sand mixtures is analysed which points out the connection between sustainability using tyre shreds and the shear strength enhancement of the mixture. It clearly depicts many ways to incorporate sustainable alternatives for improving the performance of the geotechnical structures (Zornberg et al., 2004).

The pattern and link between disaster and next cycle is reported by Cardona et al. (2015) which clearly projects the exposure and vulnerability of the geotechnical infrastructure systems due to change in inadequate water management, changes in land use, urban migration from rural areas, unplanned urban growth, the expansion of informal settlements in zones prone to landslides, and issues associated with proper drainage infrastructure.

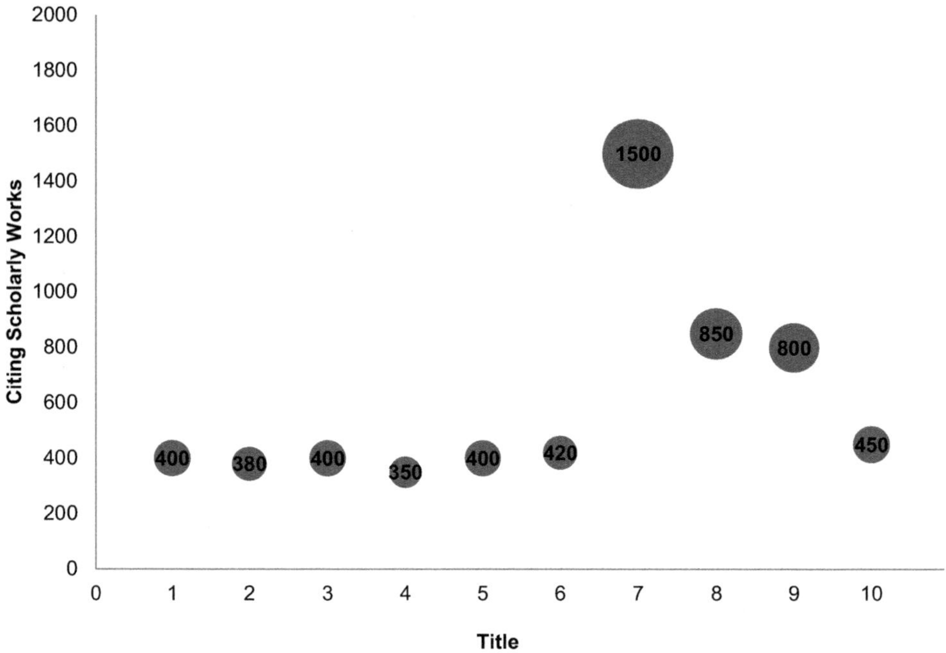

1. Application of deep learning algorithms in geotechnical engineering: a short critical review

2. Behavior of tireshed sand mixtures

3. Determinants of risk: Exposure and vulnerability

4. Ecological mitigation of hillslope instability: ten key issues facing researchers and practitioners

5. Effect of confining pressure on ballast degradation and deformation under cyclic triaxial loading

6. Fundamentals of soil stabilization

7. Landslides in a changing climate

8. Recent changes in the fire regime across the North American boreal regional Spatial and temporal patterns of burning across Canada and Alaska

9. Soil-Erosion and Runoff Prevention by Plant Covers: A Review

10. The Hindu Kush Himalaya Assessment

FIGURE 2.6 Top ten articles which are cited on the subject over the last 25 years.

The devastating effect of landslides occurred due to slope instability, and the effects on infrastructure and communities are reported by Ge et al. (2024), Petley (2012), and Stokes et al. (2014). A sustainable alternative to civil engineering solutions when protecting against shallow landslides and soil erosion by using vegetation is proposed by Stokes et al. (2014) in soil fields where natural soil protection systems by plants or shrubs are subjected to the exploitation in the name of cultivation of crops, built environment activities, overconsumption of tress, etc. They further emphasized the collection of data through field investigation, open-source records, modelling, awareness, and acceptance of the sustainable alternatives in slope renovation projects predominantly in the geotechnical societies. Additionally, the study conducted by Durán Zuazo and Rodríguez Pleguezuelo (2008) concluded that for highly erodible soils, vegetation enhances biodiversity for steeply sloped areas and maintains crucial interrelation with soil properties. Therefore, the requirement of database and modelling for the betterment of civil engineering structures have been clearly documented a decade ago to link between the sustainable needs and for enhancing the resilience of civil engineering structures.

Lackenby et al. (2007) found that the traditional railway foundations or substructures have been subjected to heavy loads due to faster and heavier trains, thereby increasing the frequent and expensive maintenance cycles due the lack of substructure re-engineering. This indicates that with the advancement of train technologies like high-speed railway lines, railway substructures require re-engineering, and AI data modelling and IoT implementation on existing railway substructures can provide databases to develop systems which can be implemented in the future. This AI advancement can result in better track stability which is crucial for a high-speed railway network with improved rail tracks with degradation and deformation resistance along with enhanced resilience characteristics.

The low compressive strength and excessive settlement of soft clays result in lower carrying capacity because of the water content which could result in catastrophic costs to civil and geotechnical infrastructures. Therefore, when an infrastructure is planned on such soils, large amount of existing data on the behaviour of drying, swelling, shrinkage, and cracks in clayey soil and how it significantly impacts the soil's behaviour needs to be researched well (Firoozi et al. 2017). This is possible with the ongoing AI advancement which can open wider research in the field of IoT instrumentation.

Gariano and Guzzetti (2016) concluded that the climatic changes greatly distress the structure, strength, and stability of natural and man-made slopes, which will eventually lead to landslides. Further, they reported complications on the evaluation of the climate impacts on landslides in terms of lack of clarity on the type, extent, magnitude, and direction of the changes in the stability conditions of the slopes, and on the location, abundance, and frequency of the landslides. The effects of the warming climate on landslide risk, especially population risk, were reported as difficult to quantify. However, they recommended slope stability models and climate modelling for further risk assessment and for predicting different variables (Gariano & Guzzetti, 2016).

Innovative approaches to integrating technology in areas susceptible to landslides due to ongoing climate changes such as improved engineering, data analytics, and resilient infrastructure along with advanced modelling and predictive analysis are required (Alcántara-Ayala, 2025; Vung et al., 2023).

Alcántara-Ayala (2025) clearly depicts the necessity for robust, resilient, and rapid disaster warning infrastructure systems with the intensity of extreme climatic events such as increased rainfall, floods, etc. The study also emphasized on the opportunities adapted on a global scale to boost rapid disaster warning facilities and calamity awareness by simplifying the discussion of evidence, technology, and necessity of building these systems. Additionally, ecofriendly and sustainable resolutions, identical to environment rebuilding, is suggested to even out susceptible regions and lessen landslide hazards in a sustainable way. Further, the importance of resident data and community-grounded tactics in designing frameworks and detailed involvement for addressing the complex, interconnected challenges of landslide risk in a changing world is emphasized. The study was concluded with firming governance agendas to synchronize procedures and implement worldwide collaboration in addressing challenges related to landslide mitigation globally.

Along with the climate warming, changes in the fire regime characteristics and the impact of the fire on the environment and society are captured by Kasischke and Turetsky (2006) and Pickering et al. (2025). The study conducted by Pickering et al. (2025) compared present climatic conditions with warmer drier future and hotter wetter future with different models and simulation tools. They summarized that the fire range and strength will surge during the upcoming environments, irrespective of prearranged effort and arrangements. Their study suggested that a boosted understanding of the intricacy of destruction will lead to a better understanding of the cost efficiency of diverse management tactics under the present and impending weathers. Therefore, the advancement of AI and other relevant IoT technologies will be useful for further researching on fire resistance of geotechnical structures.

Wester et al. (2019) in their book assessed the link between mountains, climate change, sustainability, and people in the Himalayan region, which is known for a series of long-standing and innovative challenges moving forward, with climate change, globalization, movement of people, conflict, and environmental degradation. They prepared monitoring and assessment reports, to deal with major issues such as climate change, biodiversity, energy, cryosphere (frozen water), food safety, air contamination, calamity and resilience, adaptation, and migration. This initial evaluation discourses the three pillars of sustainability and will aid as a foundation for decision-making to maintain the environment and social welfare in the region.

2.5.3 Mapping the Co-occurrence of Author Keywords, Author Co-citation Patterns, and Top Journals

The keyword co-occurrence network map identifies the most frequently used keywords represented as nodes in the study area. Results from the VOSviewer analysis reveal that the Lens dataset includes 4390 keywords, with 24 keywords appearing at least ten times, as depicted in Figure 2.7.

Based on the number of nodes, cluster 1 represents the largest research area focusing on climate change, followed by clusters 2 and 3 related to sustainability and resilience, and cluster 4 on machine learning, as shown in Figure 2.7. Other clusters indicate civil engineering; especially geotechnical field and other improvements connected to the present topic. This shows clear research between machine learning, climate conditions, geotechnical engineering, and sustainability with all the 11 clusters shown in Figure 2.7.

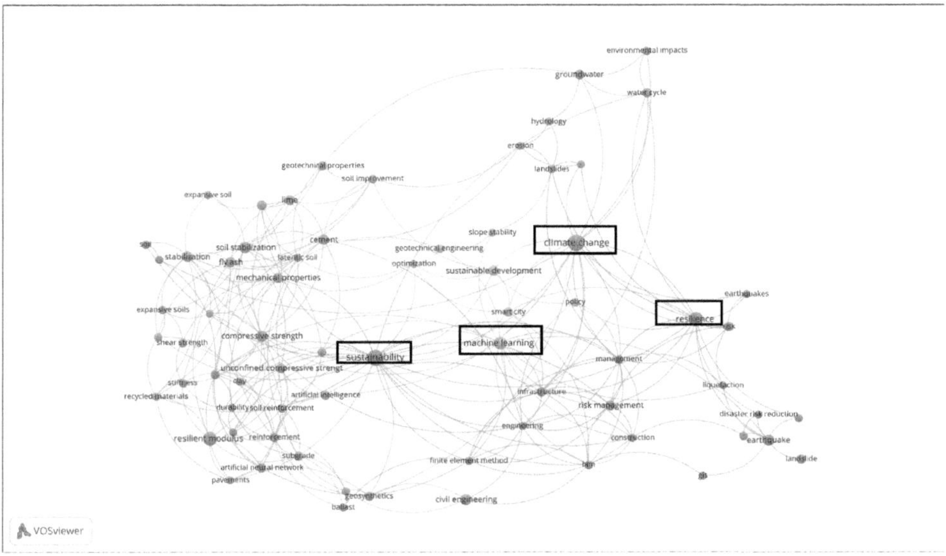

FIGURE 2.7 Mapping the co-occurrence of author keywords, author co-citation patterns among 70 keywords during the last 25 years.

Phoon and Zhang (2023) proposed data-centric geotechnics with three fundamentals: data centricity, fit for (and transform) preparation, and geotechnical focus while exploring machine learning in geotechnical engineering. They emphasized on the data-driven site characterization for better decision-making for crucial projects. Further, the challenges are identified as crude data, site recognition, still advancing technology, and upcoming evolutions like Digital Twin (DT).

Figure 2.8 illustrates that authors grouped within same-coloured cluster are strongly interconnected through research collaborations or citation relationships within the selected topic. In total, 6458 authors contributed to studies related to this theme, of whom 150 authors met the minimum threshold of three citations. The author co-citation network shown in Figure 2.8 consists of 12 clusters, 98 nodes, and 249 links, with a total link strength of 464. This network enables the identification of influential authors who underpin the intellectual structure of the field, as reflected in the Lens database.

The topmost authors whose work has been highlighted in the database on the chosen topic is tabulated in Table 2.2. This table also shows that the utmost significant studies absorbed the field investigation on geotechnical applications using instrumentation and data interpretation (Indraratna et al., 2019, 2023; Riyad et al., 2021). The above reports indicate the significance of sustainability and geotechnical infrastructure resilience using different techniques and/or materials across different geotechnical applications.

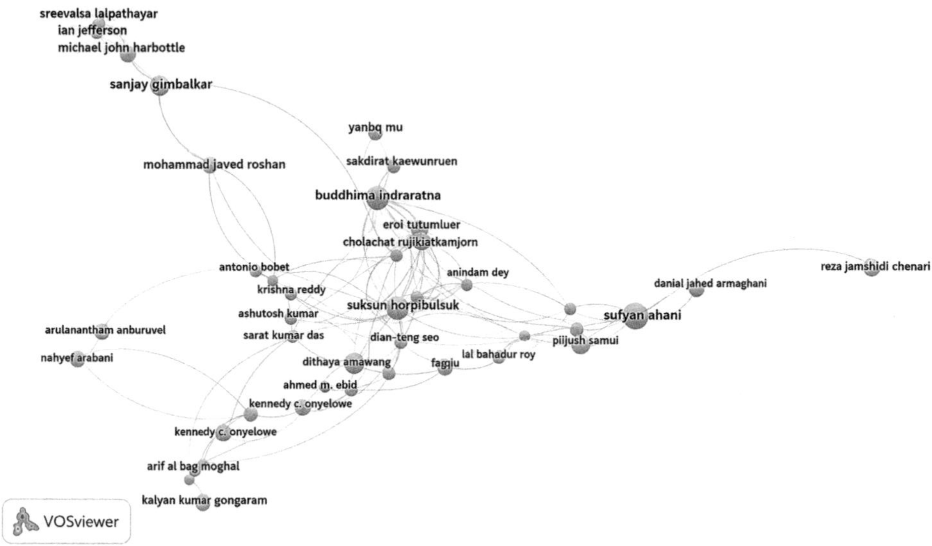

FIGURE 2.8 Author's co-citation bibliometric map constructed based on authors who have received a minimum of three citations over the past 25 years.

TABLE 2.2 Top 10 primary authors who have contributed to the selected field of research

SELECTED	*AUTHOR*	*DOCUMENTS*	*CITATIONS*	*TOTAL LINK STRENGTH*
1	Buddhima Indraratna	15	998	90
2	Cholachat Rujikiatkamjorn	9	367	73
3	Ngoc Trung Ngo	4	255	50
4	Fernanda Bessa Ferreira	4	55	29
5	Trung Ngo	4	56	29
6	Yujie Qi	3	74	29
7	Amin Soltani	5	148	28
8	An Deng	4	144	28
9	Arul Arulrajah	9	334	24
10	Suksun Horpibulsuk	10	324	24

Table 2.3 shows the ten most appropriate co-cited scientific articles regarding the impact of AI and IoT in enhancing geotechnical structures focused on Geotechnical Engineering and Sustainability such as *Geotechnical and Geological Engineering*, *Indian Geotechnical Journal*, and *Sustainability*. The 1043 citations from the source "Sustainability" indicates the strong connection on need of enhancement of the safety and monitoring of geotechnical structures which is directly impacting humanity. This also suggests usage of IoT/AI for advanced monitoring as well as prediction in regions recently impacted by landslides and other natural calamities.

TABLE 2.3 Ten topmost sources that refer to the selected field of research over the last 25 years

SELECTED	*SOURCE*	*DOCUMENTS*	*CITATIONS*	*TOTAL LINK STRENGTH*
1	*International Journal of Pavement Research and Technology*	43	536	54
2	*Geotechnical and Geological Engineering*	60	548	52
3	*Indian Geotechnical Journal*	67	274	51
4	*Transportation Infrastructure Geotechnology*	50	217	49
5	*Sustainability*	50	1043	42
6	*Scientific Reports*	55	285	39
7	*Innovative Infrastructure Solutions*	32	260	36
8	*International Journal of Geosynthetics and Ground Engineering*	37	229	36
9	*Materials (Basel, Switzerland)*	33	755	36
10	*Bulletin of Engineering Geology and the Environment*	30	358	35

2.5.4 Bibliometric Trends of Geotechnical Infrastructure Resilience with AI and IoT

Keyword concurrence study is implemented to unearth the core link among research areas of selected topic. Figure 2.9 displays the keyword co-occurrence map with the topmost keywords of at least five occurrences, among the full 4740 keywords in the list assumed for the study. The top 70 keywords were categorized into seven clusters with clusters 1 to 7 focusing on main keywords such as sustainability, resilience, climate change, machine learning, geotechnical properties, and risk management. Geotechnical infrastructure-based bibliometric investigation has shown that "landslides," "groundwater," and "erosion" are the intensive area of exploration in linking with the influence of climate change. Even though extensive research has been published on "climate change," "sustainability," "resilient," etc., most of the studies did not emphasize the use of BIM/DT/Sophisticated developments so far, implying that research focusing on these keywords, particularly in the context of AI and IoT, has not been documented as a research work in progress.

Research contributions and gaps identified by various researchers in the domain of AI and IoT to enhance the resilience of geotechnical structures are summarized in Table 2.4.

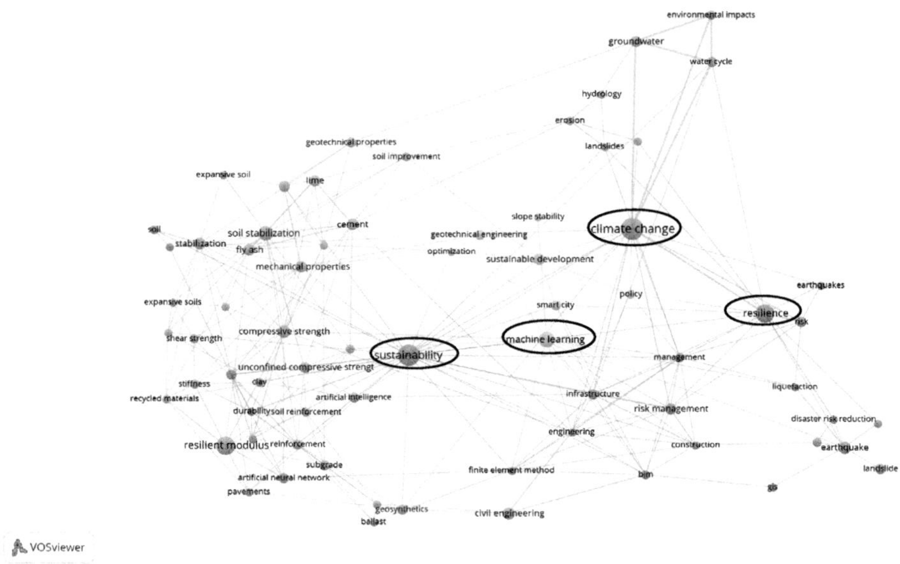

FIGURE 2.9 Keyword co-occurrence visualization on the connection between climate change, resilience, sustainability, and machine learning.

TABLE 2.4 Research contributions in the field of study

AUTHOR (YEAR)	*FINDINGS*	*IDENTIFIED GAPS*
Firoozi and Firoozi (2024)	IoT and AI integration for smart geotechnics to enhance infrastructure resilience.	Limited real-world deployment and interoperability standards for IoT in geotechnical systems.
Hemdan and Al-Atroush (2024)	Soil image recognition using IoT-based and hybrid deep learning for geotechnical applications.	Lack of large-scale datasets and real-time adaptability for diverse soil conditions.
Onyelowe et al. (2023)	Review AI optimization procedures and applications in geotechnical engineering.	Need for hybrid AI models merging physics-based and data-driven methods for reliability.
Phoon and Zhang (2023)	Discusses future of machine learning in geotechnics, emphasizing risk assessment.	Insufficient frameworks for uncertainty quantification and explainability in ML models.
Xu et al. (2024)	Multi-fidelity DeepONet for real-time settlement prediction during tunnelling.	Requires validation under varying geological conditions and integration with IoT sensors.
Latif et al. (2023)	Digital twin-driven framework for TBM performance forecast using ML.	Limited integration with IoT for real-time monitoring and predictive maintenance.
Pregnolato et al. (2022)	Conceptualized Civil Engineering 4.0 with digital twins for infrastructure resilience.	Lack of standardized protocols for data exchange between IoT devices and digital twins.

(Continued)

TABLE 2.4 (*Continued*) Research contributions in the field of study

AUTHOR (YEAR)	*FINDINGS*	*IDENTIFIED GAPS*
Chen et al. (2024)	Utility of large language models in geotechnical education and problem-solving.	Application in real-time decision-making for resilience enhancement remains unexplored.
Tuzelbayev et al. (2025)	Interpretable ensemble ML for liquefaction risk prediction.	Integration with IoT-based early warning systems is missing.
Ge et al. (2024)	GAN-based approach for landslide displacement prediction using augmented data.	Limited field validation and integration with IoT sensors for continuous monitoring.
Campos Montero et al. (2025)	SchemaGAN for geotechnical subsurface schematization.	Scalability and adaptability to heterogeneous soil conditions remain untested.
Febrianto et al. (2022)	Digital twinning of self-sensing structures using statistical FEM.	Lack of real-time IoT integration for predictive maintenance and resilience.
Ninić et al. (2024)	Instantaneous evaluation of tunnelling-induced losses inside BIM framework.	Limited AI-driven predictive capabilities and IoT sensor fusion for proactive risk mitigation.
Xu et al. (2025)	Multimodal framework mixing several LLM representatives for intelligent geotechnical design.	Practical implementation and integration with IoT for automated resilience planning are lacking.
Zhan et al. (2024)	State-of-the-art review on seismic assessment considering soil–structure interaction.	Need for AI-driven predictive models and IoT-based real-time seismic monitoring.
Strauss and Wadzuk (2022)	Predictive maintenance of stormwater infrastructure using IoT.	Application to geotechnical assets and integration with AI for resilience optimization is missing.

Firoozi and Firoozi (2024) highlighted the significant advantages of advanced AI and IoT technologies in strengthening the resilience and sustainability of critical infrastructure, including bridges, tunnels, and foundations. These advantages encompass enhanced operational efficiency, improved safety, and a reduced environmental footprint. The study also examined key challenges such as data security, system integration, and scalability, and proposed future research directions and policy measures to address these issues. Beyond theoretical frameworks, the authors employed an extensive literature review and in-depth case studies to advocate for the broader adoption of smart geotechnics, emphasizing its vital role in promoting sustainable and resilient infrastructure within the context of smart cities.

Sheil et al. (2026) highlighted the recent AI application such as intelligent site investigation, soil behaviour modelling, geotechnical design optimization by making use of predictive models and streamlined operations. The intricate advantages of AI in addressing various complex interaction between soils and soil infrastructure, ground water, etc. has been addressed too. The key challenges are also identified as data scarcity which hinders training machine learning due to complex interactions in infrastructure projects.

The application of IoT and AI in identifying soil characteristics is reported by Hemdan and Al-Atroush (2024), which aids in upgraded site characterization and supports informed decision-making in engineering projects, especially for critical infrastructure projects. They proposed soil classification based on the vast soil images and proposed a framework that could provide competent decision-making for present geotechnical engineering applications.

The bibliometric study indicates that there is a well-established link between climate change, resilience, sustainability, and machine learning. In Figure 2.9, the density visualization and the colour of the nodes indicate the interconnections among subject categories, whereas the node size reflects the frequency of occurrence of the respective items. As shown in Figure 2.9, it is noteworthy that only a limited number of subject categories overlap with others, which shows the impact of advanced technologies such as AI and IoT on resilient geotechnical structures contributing to climate resilience. Strauss and Wadzuk (2022) concluded that the alerts on predictive maintenance using IoT help to manage stormwater systems effectively through maintenance alerts and to identify issues related to resilience of stormwater networks.

Febrianto et al. (2022) demonstrated the application of statistical finite element method in developing a DT of a self-sensing structure and an instrumented steel railway bridge of 27.34 m length located along the West Coast Mainline near Staffordshire in the UK, considered as a case study. The study concluded with the reliability of DT in predicting the system response and predictions. The study further highlighted the application of DT for long-term structural health monitoring and assessment in locations and under loading conditions where measurement data are unavailable, with optimization of sensor placement and other relevant analyses (Febrianto et al., 2022).

Campos Montero et al. (2025) introduced schemaGAN, a conditional Generative Adversarial Network (GAN) to create geotechnical subsurface schematizations from site investigation statistics. This new scheme can utilize multifaceted data from cone penetration tests in a quick timeframe to generate subsurface profiles. The accuracy and superior performance of the novel method have been confirmed in two real case studies for known CPT data. Further, these advancements will help in solving geotechnical problems which need subsurface schematizations especially crucial infrastructure systems. Campos Montero et al. (2025) also reported significant obstacles in DL applications, particularly related to the availability of data and the challenge in consolidating data from the disparate nature of geotechnical investigations. Thus, the instrumentation of geotechnical infrastructure systems and the need of data for addressing future development are highlighted. Further, the advantage of AI and IoT would result in generating realistic and accurate schematizations with minimum input information.

It is reported that the data unavailability of on-site measurement has been an ongoing challenge to develop ML models. Therefore, to make use of scarce data to generate high-fidelity synthetic landslide data, the study by Ge et al. (2024) proposed a data augmentation framework that employs GANs, an advanced form of contemporary generative AI, to enhance predictive accuracy. This development plays a crucial role in enhancing the predictive capability of advanced machine learning models for landslide displacement, especially in scenarios where data availability is limited. Therefore, the proposed framework has further scope to develop with the usage of AI in geo-hazard risk management and to enhance early warning systems through synthetic data generation for diverse applications (Ge et al., 2024).

J. Xu et al. (2022) compiled a comprehensive slope failure database and a Bayesian machine learning (BML)-based method was established. The slope failure time (SFT) was based on data-driven phenomenological models and observational uncertainties drawn in SFT forecast, through which the probabilistic distribution of the SFT can be obtained using BML-based method.

Finally, the findings suggest sustained interdisciplinary investigation into the long-term impacts of machine learning with advanced technology and sustainability, including climate change-related disasters happening across the globe. In the Architecture, Engineering and Construction (AEC) sector, more research is needed to accelerate the digital transformation concept of DTs which can fundamentally transform the design, construction, and maintenance of infrastructure assets. Pregnolato et al. (2022) established a sequential five-step process to guide the creation of DTs for existing infrastructure assets within existing review and the workflow processes are identified as: (i) data and need acquisition, (ii) digital modelling, (iii) dynamic data transmission, (iv) data/model integration, (v) and operation. A similar procedural framework was executed for the Clifton Suspension Bridge (Bristol, UK) and it aimed to facilitate predictive maintenance by analysing patterns of component degradation, ensuring that essential repairs or mitigation actions could be undertaken ahead of potential asset failure, which is equivalent to Level 3. The authors concluded that the current Level 3 model could be enhanced with machine learning and AI to make it DT with autonomous decision-making (Level 4). Further, the same workflow could be adopted for other infrastructure projects in a longer term and with other advancements (Pregnolato et al., 2022).

The study conducted by Chen et al. (2024) revealed that the effectiveness of large language models (LLMs) in handling core geotechnical concepts and problem-solving tasks depends on the subject matter, the level of complexity involved, and the prompting approach used such as zero-shot learning, chain-of-thought (CoT) prompting, and custom instructional prompting. Further the study underlines the critical role of human judgement in reviewing and validating outputs, particularly when addressing complex, higher-level analytical tasks especially for geotechnical concepts. The study further highlights the potential and limitations of current AI tools in its approach to geotechnical problem-solving. Therefore, a balanced integration of AI while stressing the fundamental role of human interpretation and insight alongside evolving digital technologies is reported (Chen et al., 2024). Tuzelbayev et al. (2025) formulated a comprehensive ML methodology for liquefaction risk prediction, a critical factor in evaluating seismic hazards. This is one of the crucial components in geotechnical engineering and shows the scope for further research and implementation to enhance the overall accuracy and effectiveness of liquefaction prediction models. Ninić et al. (2024) reported that the influence of tunnelling activities on surrounding structures can be effectively analysed using BIM through the creation of risk maps and deformation-scale visual outputs. This would help in understanding the design across all stakeholders to collaborate effectively.

They also emphasized on fully parametric design model and real-time predictions to enable the assessment and visualization of tunnelling-induced damages along with the damage risk group of constructions affected by tunnelling. However, this also shows the research to be redirected to technologies such as BIM and DT and to resilient infrastructure with a focus on climate resilience.

For automated geotechnical site planning and quick geological site characterization, an LLM-based agent is proposed by Qian and Shi (2025). A progressive collaboration driven by LLMs and humans is required for a site characterization to be quick and reliable which is crucial for infrastructure projects. This clearly depicts that site characterization will improve with evolvement of ongoing AI and IoT with more data availability.

C. Xu et al. (2024) reinstated the significance of ground settlement prediction during tunnelling operations which remains crucial in geotechnical infrastructure. Therefore, numerical simulation models for prediction based on physics-driven, and ML techniques based on data-driven approach are used to learn links between influencing factors and the settlement. The study proposed a framework with ground settlement patterns from finite element simulations and real engineering monitoring data. Subsurface profiling data have been modelled and generative models have been used by many researchers (C. Xu et al., 2024; H.-R. Xu et al., 2025; J. Xu et al., 2022), which are crucial for the design of infrastructure geotechnical structures or data related to these structures.

The rapid development of underground infrastructure such as tunnelling activities generates multifaceted and diverse figures, which makes it difficult to envisage the implementation of a project. Recent studies also are conducted to lessen the hazard connected with the cost and schedule of a tunnelling project by concurrently envisaging and supervising the routine functioning of Tunnel Boring Machines using DT and machine learning algorithms (Latif et al., 2023). This shows the application of AI and IoT for tunnelling projects in minimizing risk to different stakeholders even during the planning and implementation phase.

The complex challenges associated with implementing AI to geotechnics is identified as interpretability of models, generalization applied for specific conditions, and integration of uncertain components. Therefore, connecting AI with traditional models of geotechnical base and launching targets becomes critical. This ensures the proper orientation of AI models with traditional engineering concepts, removing potential bias, and encouraging use of AI with the element of trust and reliability within the geotechnical engineers across various types of infrastructure systems where priorities differ. The paper also summarizes the technological advancements in implementing AI along with human oversight using LLMs, advanced models, generative models, BIM and DT, predictive warning systems, etc. However, the paper points to the implementation of IoT devices and use of other traditional systems across the globe to collect vast amount of high-quality and reliable data. The paper further considers it vital and essential to continue a multi-disciplinary alliance among AI and geotechnical researchers to align with established engineering principles and to accomplish progressions. Furthermore, ethical considerations, predominantly addressing bias and liability with the use of synthetic data, feature the importance of accountable AI practice within legal contexts.

2.6 CONCLUSION

This chapter has reviewed the potential of AI and IoT in advancing geotechnical resilience through bibliometric analysis, emphasizing how these tools can shift the field from reactive to predictive and adaptive management. The emergence of AI and IoT technologies provides transformative opportunities to overcome these challenges. The bibliometric study indicates that there is a well-established link between climate change, resilience, sustainability, and machine learning. Therefore, by enabling continuous, real-time data collection and predictive analytics, AI and IoT integration can significantly improve the early detection of risks, enhance decision-making, and support proactive resilience strategies for critical geotechnical structures. However, data integration, algorithm transparency, cost-effectiveness, and the scalability of smart monitoring systems are challenges which need further research and development. Moreover, cross-disciplinary collaboration between various stakeholders such as geotechnical engineers, computer scientists, and policymakers is essential to translate these technological advancements into robust and viable solutions which could be implemented.

The integration of AI and IoT in monitoring, analysing and predicting the behaviour of complex soil and geotechnical engineering structures presents significant opportunities for advancement. The development of DTs, where real-time sensor data and AI analytics create virtual models of geotechnical structures to support predictive maintenance and early warning, is the future. Further, research should also advance IoT and AI by incorporating improved transparency, ethical standards, along with user trust especially among engineers, together with high-performance computing speed and connectivity to enable swift data processing in remote or hazard-prone locations. Further, addressing climate resilience remains essential, as AI- and IoT-enabled systems can enhance adaptive responses to floods, landslides, and extreme weather. In parallel, sustainability must be prioritized through cost-effective, energy-efficient monitoring systems that encourage broader adoption. Finally, global standards and governance frameworks will be needed to ensure interoperability, ethical data use, and reliability. These directions can drive a paradigm shift towards smarter, more resilient geotechnical infrastructure.

AUTHOR CONTRIBUTIONS

The author used AI tools, specifically ChatGPT-4 (OpenAI) and Copilot (Microsoft/OpenAI), to support language editing and improve readability in selected sections of the manuscript. These tools were used solely to enhance clarity and did not contribute to the study design, data collection, analysis, interpretation of findings, or generation of original scholarly content. The author has carefully reviewed, verified, and approved the final manuscript and accepts full responsibility for its content.

CONFLICT OF INTEREST

The author states that there is no conflict of interest.

FUNDING

No funding was received for conducting this research.

DATA AVAILABILITY

The datasets generated and/or analysed during the current research are available from the corresponding author on reasonable request.

REFERENCES

Alcántara-Ayala, I. (2025). Landslides in a changing world. *Landslides*, *22*(9), 2851–2865. https://doi.org/10.1007/s10346-024-02451-1

Andersen, N., & Swami, V. (2021). Science mapping research on body image: A bibliometric review of publications in body image, 2004–2020. *Body Image*, *38*, 106–119. https://doi.org/10.1016/j.bodyim.2021.03.015

Anjali, K., & Remesan, R. (2023). Exploring the last 50 years of Indian research on the impact of coal mining using bibliometric analysis with an overview of water-related impacts. *Environmental Science and Pollution Research*, *30*(6), 16449–16463. https://doi.org/10.1007/s11356-022-23381-2

Campos Montero, F. A., Zuada Coelho, B., Smyrniou, E., Taormina, R., & Vardon, P. J. (2025). SchemaGAN: A conditional Generative Adversarial Network for geotechnical subsurface schematisation. *Computers and Geotechnics*, *183*, 107177. https://doi.org/10.1016/j.compgeo.2025.107177

Cardona, O. D., Birkmann, J., & Fordham, M. (2015). *Determinants of risk: Exposure and vulnerability Maarten van Aalst Red Cross Red Crescent Climate Centre*. https://www.researchgate.net/publication/244062037

Cardona, O. D., van Aalst, M., Birkmann, J., Fordham, M., McGregor, G. R., Perez, R., Pulwarty, R. S., Schipper, E. L. F., Sinh, B. T., Décamps, H., Keim, M., Davis, I., Ebi, K. L., Lavell, A., Mechler, R., Murray, V., Pelling, M., Pohl, J., Smith, A.-O., & Thomalla, F. (2012). *Determinants of risk: Exposure and vulnerability* (pp. 65–108). Cambridge University Press. https://doi.org/10.1017/cbo9781139177245.005

Chen, L., Tophel, A., Hettiyadura, U., & Kodikara, J. (2024). An investigation into the utility of large language models in geotechnical education and problem solving. *Geotechnics*, *4*(2), 470–498. https://doi.org/10.3390/geotechnics4020026

Durán Zuazo, V. H., & Rodríguez Pleguezuelo, C. R. (2008). Soil-erosion and runoff prevention by plant covers. A review. *Agronomy for Sustainable Development*, *28*(1), 65–86. https://doi.org/10.1051/agro:2007062

Febrianto, E., Butler, L., Girolami, M., & Cirak, F. (2022). Digital twinning of self-sensing structures using the statistical finite element method. *Data-Centric Engineering*, *3*, e31. https://doi.org/10.1017/dce.2022.28

Firoozi, A. A., & Firoozi, A. A. (2024). Smart geotechnics: Enhancing infrastructure resilience with IoT and AI. *Journal of Civil Engineering and Urbanism*, *14*(1), 89–101. https://doi.org/10.54203/jceu.2024.7

Firoozi, Ali Akbar, Guney Olgun, C., Firoozi, Ali Asghar, & Baghini, M. S. (2017). Fundamentals of soil stabilization. *International Journal of Geo-Engineering*, *8*(1), 26. https://doi.org/10.1186/s40703-017-0064-9

Gariano, S. L., & Guzzetti, F. (2016). Landslides in a changing climate. *Earth-Science Reviews*, *162*, 227–252. https://doi.org/10.1016/j.earscirev.2016.08.011

Ge, Q., Li, J., Lacasse, S., Sun, H., & Liu, Z. (2024). Data-augmented landslide displacement prediction using generative adversarial network. *Journal of Rock Mechanics and Geotechnical Engineering*, *16*(10), 4017–4033. https://doi.org/10.1016/j.jrmge.2024.01.003

Hemdan, E. E.-D., & Al-Atroush, M. E. (2024). An efficient IoT-based soil image recognition system using hybrid deep learning for smart geotechnical and geological engineering applications. *Multimedia Tools and Applications*, *83*(25), 66591–66612. https://doi.org/10.1007/s11042-024-18230-y

Huang, Y., Xu, C., Zhang, X., Xue, C., & Wang, S. (2021). An updated database and spatial distribution of landslides triggered by the Milin, Tibet Mw6.4 earthquake of 18 November 2017. *Journal of Earth Science*, *32*(5), 1069–1078. https://doi.org/10.1007/s12583-021-1433-z

Indraratna, B., Malisetty, R. S., Nair, L., & Rujikiatkamjorn, C. (2023). Instrumentation and data interpretation in transportation geotechnics. *Indian Geotechnical Journal*, *54*(1), 40–62. https://doi.org/10.1007/s40098-023-00765-w

Indraratna, B., Qi, Y., Ngo, T., Rujikiatkamjorn, C., Neville, T., Ferreira, F. B., & Shahkolahi, A. (2019). Use of geogrids and recycled rubber in railroad infrastructure for enhanced performance. *Geosciences*, *9*(1), 30. https://doi.org/10.3390/geosciences9010030

Kasischke, E. S., & Turetsky, M. R. (2006). Recent changes in the fire regime across the North American boreal region—Spatial and temporal patterns of burning across Canada and Alaska. *Geophysical Research Letters*, *33*(9). https://doi.org/10.1029/2006GL025677

Lackenby, J., Indraratna, B., McDowell, G., & Christie, D. (2007). Effect of confining pressure on ballast degradation and deformation under cyclic triaxial loading. *Géotechnique*, *57*(6), 527–536. https://doi.org/10.1680/geot.2007.57.6.527

Latif, K., Sharafat, A., & Seo, J. (2023). Digital twin-driven framework for TBM performance prediction, visualization, and monitoring through machine learning. *Applied Sciences*, *13*(20), 11435. https://doi.org/10.3390/app132011435

Lens.org, *Release 9.7*. (2025). https://www.lens.org/lens/

Liang, M.-C., Chen, H.-E., Tfwala, S. S., Lin, Y.-F., & Chen, S.-C. (2023). The application of wireless underground sensor networks to monitor seepage inside an earth dam. *Sensors*, *23*(8), 3795. https://doi.org/10.3390/s23083795

Munim, Z. H., Dushenko, M., Jimenez, V. J., Shakil, M. H., & Imset, M. (2020). Big data and artificial intelligence in the maritime industry: A bibliometric review and future research directions. *Maritime Policy and Management*, 577–597. https://doi.org/10.1080/03088839.2020.1788731

Neuman, M. (2020). Infrastructure is key to make cities sustainable. *Sustainability*, *12*(20), 8308. https://doi.org/10.3390/su12208308

Ninić, J., Gamra, A., & Ghiassi, B. (2024). Real-time assessment of tunnelling-induced damage to structures within the building information modelling framework. *Underground Space*, *14*, 99–117. https://doi.org/10.1016/j.undsp.2023.05.010

Onyelowe, K. C., Mojtahedi, F. F., Ebid, A. M., Rezaei, A., Osinubi, K. J., Eberemu, A. O., Salahudeen, B., Gadzama, E. W., Rezazadeh, D., Jahangir, H., Yohanna, P., Onyia, M. E., Jalal, F. E., Iqbal, M., Ikpa, C., Obianyo, I. I., & Rehman, Z. U. (2023). Selected AI optimization techniques and applications in geotechnical engineering. *Cogent Engineering*, *10*(1). https://doi.org/10.1080/23311916.2022.2153419

Petley, D. (2012). Global patterns of loss of life from landslides. *Geology*, *40*(10), 927–930. https://doi.org/10.1130/G33217.1

Phoon, K.-K., & Zhang, W. (2023). Future of machine learning in geotechnics. *Georisk: Assessment and Management of Risk for Engineered Systems and Geohazards*, *17*(1), 7–22. https://doi.org/10.1080/17499518.2022.2087884

Pickering, B. J., Kultaev, D., Holyland, B., Ababei, D., & Penman, T. D. (2025). The changing risk of fire to human and environmental assets under climate induced altered fire regimes in south-east Australia. *International Journal of Disaster Risk Reduction*, *127*, 105668. https://doi.org/10.1016/j.ijdrr.2025.105668

Pregnolato, M., Gunner, S., Voyagaki, E., De Risi, R., Carhart, N., Gavriel, G., Tully, P., Tryfonas, T., Macdonald, J., & Taylor, C. (2022). Towards civil engineering 4.0: Concept, workflow and application of digital twins for existing infrastructure. *Automation in Construction*, *141*, 104421. https://doi.org/10.1016/j.autcon.2022.104421

Qian, Z., & Shi, C. (2025). Large language model-empowered paradigm for automated geotechnical site planning and geological characterization. *Automation in Construction*, *173*, 106103. https://doi.org/10.1016/j.autcon.2025.106103

Riyad, A. S. M., Ferreira, F. B., Indraratna, B., & Ngo, T. (2021). A critical review on the performance of pile-supported rail embankments under cyclic loading: Numerical modeling approach. *Sustainability*, *13*(5), 2509. https://doi.org/10.3390/su13052509

Shah, J., Jefferson, I., & Hunt, D. (2014). Resilience assessment for geotechnical infrastructure assets. *Infrastructure Asset Management*, *1*(4), 95–104. https://doi.org/10.1680/iasma.14.00007

Stokes, A., Douglas, G. B., Fourcaud, T., Giadrossich, F., Gillies, C., Hubble, T., Kim, J. H., Loades, K. W., Mao, Z., McIvor, I. R., Mickovski, S. B., Mitchell, S., Osman, N., Phillips, C., Poesen, J., Polster, D., Preti, F., Raymond, P., Rey, F., … Walker, L. R. (2014). Ecological mitigation of hillslope instability: Ten key issues facing researchers and practitioners. *Plant and Soil*, *377*(1–2), 1–23. https://doi.org/10.1007/s11104-014-2044-6

Strauss, M., & Wadzuk, B. (2022). Predictive maintenance of stormwater infrastructure using internet-of-things technology. *Journal of Environmental Engineering*, *148*(2). https://doi.org/10.1061/(ASCE)EE.1943-7870.0001972

Tuzelbayev, D., Moon, S.-W., Lee, M., Abdialim, S., Aremu, E. A., Satyanaga, A., & Kim, J. (2025). Interpretable ensemble machine learning for liquefaction risk prediction. *Infrastructures*, *10*(11), 304. https://doi.org/10.3390/infrastructures10110304

van Eck, N. J., & Waltman, L. (2010). Software survey: VOSviewer, a computer program for bibliometric mapping. *Scientometrics*, *84*(2), 523–538. https://doi.org/10.1007/s11192-009-0146-3

van Eck, N. J., & Waltman, L. (2014). Visualizing bibliometric networks. In Ding, Y., Rousseau, R., Wolfram, D. (Eds.). *Measuring Scholarly Impact* (pp. 285–320). Springer International Publishing. https://doi.org/10.1007/978-3-319-10377-8_13

van Eck, N. J., & Waltman, L. (2017). Citation-based clustering of publications using CitNetExplorer and VOSviewer. *Scientometrics*, *111*(2), 1053–1070. https://doi.org/10.1007/s11192-017-2300-7

Vung, D. Van, Tran, T. V., Duc Ha, N., & Huy Duong, N. (2023). Advancements, challenges, and future directions in rainfall-induced landslide prediction: A comprehensive review. *Journal of Engineering and Technological Sciences*, *55*(4), 466–477. https://doi.org/10.5614/j.eng.technol.sci.2023.55.4.9

Wester, P., Mishra, A., Mukherji, A., & Shrestha, A. B. (Eds.). (2019). *The Hindu Kush Himalaya Assessment*. Springer International Publishing. https://doi.org/10.1007/978-3-319-92288-1

Xu, C., Cao, B. T., Yuan, Y., & Meschke, G. (2024). A multi-fidelity deep operator network (DeepONet) for fusing simulation and monitoring data: Application to real-time settlement prediction during tunnel construction. *Engineering Applications of Artificial Intelligence*, *133*, 108156. https://doi.org/10.1016/j.engappai.2024.108156

Xu, H.-R., Zhang, N., Yin, Z.-Y., & Njock, P. G. A. (2025). Multimodal framework integrating multiple large language model agents for intelligent geotechnical design. *Automation in Construction*, *176*, 106257. https://doi.org/10.1016/j.autcon.2025.106257

Xu, J., Wang, Y., & Zhang, L. (2022). Fusion of geotechnical and geophysical data for 2D subsurface site characterization using multi-source Bayesian compressive sampling. *Canadian Geotechnical Journal*, *59*(10), 1756–1773. https://doi.org/10.1139/cgj-2021-0323

Zhan, P., Xue, S., Li, X., Sun, G., & Ma, R. (2024). Seismic assessment of large-span spatial structures considering soil–structure interaction (SSI): A state-of-the-art review. *Buildings*, *14*(4), 1174. https://doi.org/10.3390/buildings14041174

Zhang, W., Li, H., Li, Y., Liu, H., Chen, Y., & Ding, X. (2021). Application of deep learning algorithms in geotechnical engineering: A short critical review. *Artificial Intelligence Review*, *54*(8), 5633–5673. https://doi.org/10.1007/s10462-021-09967-1

Zornberg, J. G., Cabral, A. R., & Viratjandr, C. (2004). Behaviour of tire shred sand mixtures. *Canadian Geotechnical Journal*, *41*(2), 227–241. https://doi.org/10.1139/t03-086

AI-Based Soil Classification and Characterization 3

Techniques and Case Studies

Ifeyinwa Ijeoma Obianyo, Frank Nnebe, Gloria Nkeiruka Matthew, Abubakar Dayyabu, Abdulhameed Danjuma Mambo, and Azikiwe Peter Onwualu

3.1 INTRODUCTION

Characterization and classification of soil in geotechnical engineering is a key requirement during execution of any construction project because there is need to always assess the suitability of soils for different construction projects. When engineers characterize and classify soil accurately, potential hazards such as building collapse can be averted by using the predicted soil behavior data to design and build stable structures. The conventional methods used for soil classification and characterization include laboratory testing, manual sampling, and visual evaluation. According to Aydin et al. (2023), conventional methods of characterizing and classifying soils are mostly

DOI: 10.1201/9781003714125-3

time-consuming, labor-intensive, and prone to error. This is because these methods do not take into consideration the complex and dynamic nature of soils, which may result in inaccurate or unsatisfactory soil assessment results. There is need to develop more accurate and efficient methods for classifying and characterizing soils for large-scale projects especially in locations with limited access to laboratory.

In recent years, researchers have been exploring more cost-effective, accurate, and efficient methods of classifying and characterizing soils for construction projects due to the disadvantages associated with the conventional methods. Artificial intelligence (AI) is a potential tool that can be used to improve classification and characterization. Machine learning (ML) and deep learning (DL) are two major forms of AI that have the capacity of analyzing large datasets, identifying patterns, and producing accurate predictions (Abrar et al., 2024). Using these techniques to characterize and classify soils will play a vital role in improving soil assessment because of its accuracy, cost-effectiveness, and efficiency. Different AI-based methods have been used to conduct a wide range of soil characterization and classification activities including identifying anomalies in soil, predicting soil parameters, and identifying soil types (İnci et al., 2024; Samptkumari Bandagar, 2022; Wadoux, 2025).

In classifying and classifying soil, several subfields of AI have indicated its potential to produce accurate results. The ML algorithms have the potential to identify the patterns and relationships in data for the purpose of making predictions (Kang et al., 2022). Conversely, complex patterns and relationships that may not be identified by ML algorithms can easily be identified using DL algorithms with improved prediction accuracy (Alzubaidi et al., 2021). A variety of data, including sensor data, environmental parameters, and soil quality, can be analyzed using both the ML and DL algorithm techniques. Studies that have explored the potential applications of AI in soil classification exist in literature. These studies found that AL has the capacity of improving the accuracy and efficiency of soil assessments (Ameer Abdulridha AjmiAlali, 2024; Wadoux, 2025).

Despite the potential benefits of using AI as a technique for soil classification and characterization, there are limitations to AI-based soil classification and characterization, which require thorough research for accurate assessment. Previous studies have highlighted the importance of selecting appropriate AI model and algorithm, gathering quality data as well as proper training of model for different applications. Classification and characterization of soil can be carried out using AI models and techniques such as support vector machines (SVMs), decision trees (DTs), random forest (RF), and neural networks, and each of these models and methods has advantages and disadvantages. The best model and approach to be used in each case scenario depends on the data properties and applications of the models. To analyze continuous data, SVMs and neural networks are two ML algorithms that perform better than others (Nguyen et al., 2022; Uddin et al., 2019), while RF and DTs work better for data classification (Mutale et al., 2024). In order to efficiently and accurately classify and characterize soils, the benefits and drawbacks of different AI models and algorithms must be evaluated.

In this chapter, a thorough overview of AI-based techniques for soil classification and characterization was carried out. Different AI-based techniques for characterizing and classifying soil were examined using case studies. The potential applications and limitations of these techniques were highlighted. Soil characterization and classification using ML and DL methods were covered in this chapter. This chapter provides insights in enhancing the methods of assessing the characteristics and behavior of soils for accurate and efficient soil classification and characterization using AI-based methods.

3.2 METHODOLOGY

3.2.1 Data Collection and Preprocessing

To gather information for this review study, a thorough search of the literature on AI-based methods for soil classification and characterization was conducted. This involves obtaining articles, research papers, and other pertinent documents from journals, conferences, and scholarly databases. Preprocessing ensures that studies are relevant and contribute significantly to the review by screening and selecting them based on predefined inclusion and exclusion criteria.

3.2.2 AI-based Techniques Used in the Study (ML and DL)

The review study focuses on the use of various AI-based methods, including ML and DL, as reported in the latest literature. ML techniques can make use of algorithms such as DTs, RF, SVMs, and others. Two neural network types that can be used in DL techniques are convolutional neural networks (CNNs) and recurrent neural networks (RNNs) (Alzubaidi et al., 2021). The study looks at the benefits, drawbacks, and possible uses of soil characterization and classification techniques.

3.2.3 Model Development and Training

The methods, results, and conclusions of previous research that created and trained AI models for soil classification and characterization are integrated and evaluated in this review. This work covers a variety of model construction strategies, including feature selection, data preparation, model architecture, and training techniques.

3.2.4 Performance Evaluation Metrics

As shown in Figure 3.1, the review study focuses on performance evaluation metrics used in the literature to evaluate the efficacy of AI-based models for soil characterization and classification. Common metrics that can be employed include the F1 score,

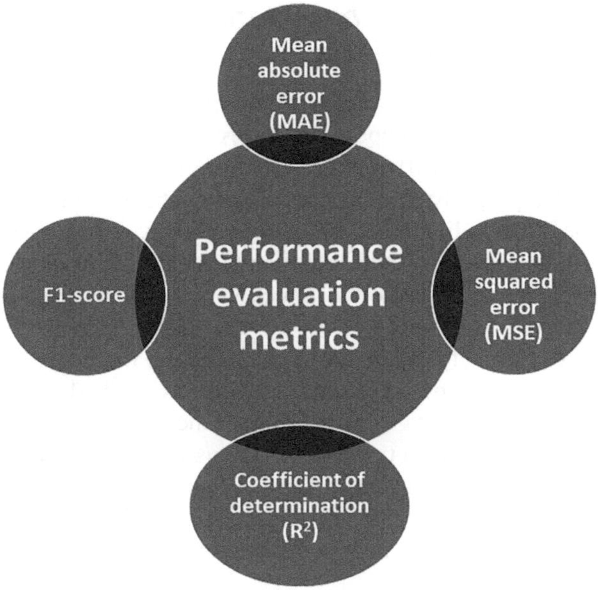

FIGURE 3.1 Performance evaluation metrics of AI-based models for soil characterization and classification.

coefficient of determination (R^2), mean absolute error (MAE), and mean squared error (MSE) (Gómez-Miranda et al., 2025; Osisanwo et al., 2017; Ratke et al., 2024). The study evaluates the value of comparing the performance of various AI models using these criteria and discusses how the results may affect future research and real-world applications.

3.3 RESULTS AND DISCUSSION

3.3.1 Overview of Case Studies

3.3.1.1 Case study 1: soil classification for foundation design

Soil classification and foundation analysis have been made easier by AI techniques like ML, artificial neural network (ANN), and DL. Using genetic algorithm (GA)-based ANN models, input datasets comprising compaction parameters of coarse-grained and fine-grained soils resulted in estimations of the ultimate bearing capacity at +80% accuracy rates, signifying its high efficiency, speed, and accuracy (Abrar et al., 2024; Ameer Abdulridha AjmiAlali, 2024). These estimations can be further enhanced by incorporating sub-clustered data division techniques, as a series of ML algorithms tested with 10-fold cross-validation resulted in high accuracy rates in soil classification.

3.3.1.2 Case study 2: soil characterization for tunneling projects

Tunneling projects are subsurface constructions that require tunnel boring machines (TBMs) to excavate and build support structures. The performance of TBMs is very difficult to predict due to the different geological formations and geotechnical factors, but when constructed properly, there is a reduction in the amount of caving in a tunnel, disturbance to the surrounding area, and workers compared to conventional methods (Singh et al., 2024). However, with the inclusion of AI, the accurate prediction of TBM's performance, lower capital cost, shorter tunneling excavation time, and management of tunneling projects can be improved.

AI research on a tunnel project aimed at predicting the penetration rate of TBMs utilizing two neural network-based models. AI methods resulted in an ANN model with one hidden layer and seven neurons, which showed the highest level of capability at 0.94 accuracy in predicting TBM penetration rate (Ding et al., 2024). Other ML techniques, including GBM, AdaBoost, HMM, and Deep Networks, achieved a high accuracy of 0.92 and an F1 score of 0.90 to enhance tunnel construction's reliability and environmental sustainability (Ding et al., 2024; Ghader et al., 2023; Ghiasi & Koushki, 2020).

3.3.1.3 Case study 3: soil classification and characterization for landslide risk assessment

A landslide is a downward movement of slope materials that could be triggered by rainfall, humans, or seismic activity. The early identification of the possible occurrence of landslide through landslide risk assessment helps to mitigate the risk and ensure continuity of the ecosystem infrastructure and human lives. Landslide patterns are usually predicted using the Soil Water Index (SWI) and the slope failure scale using meteorological data that are robust for efficient risk assessment and mitigation (Zhu et al., 2025). One key advantage of using AI-powered tool for landslide risk assessment is its ability to predict the area of high risk for landslides as well as developing a landslide hazard map to help mitigate and prevent the occurrence of landslides (Zhu et al., 2025).

Different ML algorithms for landslide risk assessment have been explored. On each database, the different algorithms, namely RF, logistic regression (LR), boosting methods (BM), SVM, and CNN, were evaluated (Ehsan et al., 2025). Based on the assessment, the CNN achieved the highest prediction accuracy at 92.5%, followed by BM, RF, LR, and SVM accordingly (Akin & Sayil, 2025; Ehsan et al., 2025; Kudaibergenov et al., 2025).

Table 3.1 presents the performance of AI models for landslide susceptibility based on the testing data using various techniques for prediction. The AI models assessed include the Naive Bayes (NB) model, Quadratic Discriminant Analysis (QDA) model, Linear Discriminant Analysis (LDA), DT model, SVM model, and RF classification. Class 0 and class 1 refer to the dominant class and minority class, respectively, adopted to an optimal linear combination of predictions. In comparison of AI models assessed for landslide susceptibility, the SVM model showed the overall best performance, achieving an accuracy of over 80% in both accuracy and recall parameters (Gómez-Miranda et al., 2025). LDA and DT models displayed acceptable results of about 70% accuracy,

TABLE 3.1 AI Models' performance of landslide susceptibility

ALGORITHM	*ACCURACY*	*CLASS*	*PRECISION*	*F1 SCORE*
Naive Bayes	0.5005	0	1.00	0.00
		1	0.50	1.00
Linear Discriminant Analysis	0.7478	0	0.80	0.67
		1	0.71	0.83
Quadratic Discriminant Analysis	0.5000	0	0.00	0.00
		1	0.50	1.00
Decision Tree	0.6370	0	0.58	1.00
		1	1.00	0.27
Random Forest	0.5770	0	0.54	1.00
		1	1.00	0.15
Support Vector Machine	0.8450	0	0.82	0.89
		1	0.88	0.80

Source: Gómez-Miranda et al. (2025).

making it viable for use, while the NB, QDA, and RF models resulted in very poor results of about 50% accuracy, making it unviable for landslide susceptibility prediction using AI unless modified with boosters.

3.3.2 Application of AI-Based Techniques in Soil Classification and Characterization

3.3.2.1 ML techniques for soil classification

ML is a branch of AI that focuses on the development of algorithms and statistical models that enable computers to perform specific tasks using patterns found in training data, and it is currently the most popular domain of AI in soil science (Minasny et al., 2024; Mutale et al., 2024). It is separated into supervised and unsupervised algorithms. In supervised ML, the algorithms are trained on labeled data to predict outcomes for new, unseen data based on the calibrated relationships. Common supervised algorithms are neural networks and RF. In unsupervised ML, the algorithms are not trained on labeled data, but they identify patterns and structures directly from data, such as with clustering and association algorithms. K-means and hierarchical clustering are popular unsupervised techniques.

Innovative methods for automating soil categorization have emerged as ML has grown in popularity as a solution to challenges in geotechnical engineering. Multinomial Logistic Regression (MLR), Gaussian Naive Bayes (GNB), RF, Extreme Gradient Boosting (XGBoost), and Artificial Neural Network-Multilayer Perceptron (ANN-MLP) are among the ML algorithms that are most effective at classifying soils; their accuracy ranges from 87% to 93%, with the NB model trailing at 77.6% (Osisanwo et al., 2017).

In geotechnical engineering, ANNs are used for slope stability analysis, whereas SVMs are used for foundation design (Teodosio et al., 2025). For example, researchers at the University of Cambridge used ANNs to predict the stability of slopes in a copper mine in Chile. After training the ANNs on a set of mine slope stability data, the researchers used them to predict the mine slope stability (Parra et al., 2023). The results showed that the ANNs could accurately predict slope stability, reducing the risk of slope failure. Another useful application of AI and ML in geotechnical engineering is the use of neural networks to predict the behavior of rock and soil. Neural networks are one type of ML system that can be trained on large datasets to predict new data. In geotechnical engineering, neural networks can predict the behavior of rock and soil under a variety of conditions, such as changes in moisture content, strain, and stress (Pace et al., 2025). The performance of existing structures over time can be evaluated using this data, and more effective retaining structures and foundations can be created. In geotechnical engineering, soil settlement beneath foundations is predicted using neural networks (Maralapalle et al., 2026a). Because it can result in differential settling and structural damage, settlement can be a major concern when designing foundations. By analyzing data from field measurements and laboratory experiments, neural networks can be trained to predict soil settling under different conditions. This information can be used to build more effective foundations and track the performance of existing structures over time.

3.3.2.2 DL models for soil characterization

Recently, CNNs, a subset of ML, have drawn a lot of interest because of their exceptional performance in tasks like speech and picture recognition. This method uses several layers, each of which merely modifies the data, to learn data representations. However, DL methods often require large amounts of data to train models efficiently; insufficient data may lead to overfitting. Furthermore, it can be difficult to optimize CNN parameters for accurate prediction or classification. Interpreting CNN results presents more challenges in soil research. DL, a branch of ML, has garnered a lot of attention since the introduction of CNNs, which have shown impressive performance in tasks like image and speech recognition. This approach makes use of hierarchical data representations, where the input data is transformed nonlinearly by multiple layers composed of neurons. However, a significant challenge in DL is the need for large, high-quality datasets to properly train models because smaller datasets may lead to overfitting and reduce the model's capacity for generalization.

Multiple photographs of soil can be used to support and identify soils using a Deep ML Algorithm. Using their classification, information about soils is filtered and grouped. Accurate identification of soil types enables engineers to make informed decisions regarding foundation design, slope stability, and overall project feasibility. It provides high accuracy in detecting various soil types from photos. This high accuracy shows that the model can successfully categorize soil samples, which can greatly help civil engineers make decisions.

3.3.2.3 Hybrid approaches for soil classification and characterization

In order to address complicated, data-rich problems in civil engineering, hybrid techniques combine the best aspects of ML and DL into a multi-stage data pipeline that mixes models from several frameworks. They perform well with structured, quantifiable data, such as soil grain size, material compressive strength, or traffic counts, but poorly with unstructured data, such as pictures or sensor time series. From this unstructured data, DL models—in particular, CNNs—automatically and reliably extract complicated characteristics (e.g., finding cracks in bridge photos or recognizing soil texture). The rich, high-level data that is learned by the simpler ML model (the "brain") is fed into the DL model, which acts as a feature extractor (the "eyes") in a hybrid system. This leverages DL's superior pattern recognition on critical engineering tasks, such as structural health monitoring or predictive maintenance, while preserving the better interpretability and generalization of conventional ML techniques. It will be more expensive and time-consuming to classify soil using the conventional method. In order to address this, the hybrid approach processed easily accessible but complex sensor data (such as Cone Penetration Test with pore pressure (CPTu) logs) and even borehole log images using a DL model (such as a CNN or MLP). The DL model automatically extracts rich, non-linear features. For the final classification of the soil, these features are then fed into a traditional ML algorithm (such as RF or XGBoost). Additionally, hybrid models can help manage the inherent heterogeneity and lack of data in subsurface investigations. Previously dependent on subjective visual inspection, the DL model functions as an automated feature extractor, learning the subtle textural and color patterns that differentiate fine-grained from coarse-grained soils (Maralapalle et al., 2026b). The ML model makes the final, trustworthy classification using these features. Scientists can quickly characterize soils in remote or data-poor regions using only photographic or basic geophysical inputs, enhancing the quality and speed of site investigations for stability and material property prediction, and it creates an objective, digital classification standard for large datasets. Compared to manual methods, this fusion produced classification accuracies exceeding 90%, which was reliable, multipurpose, significantly faster, and less expensive.

3.3.3 Strengths and Limitations of AI-Based Techniques

AI-based soil classification is strong because it is quick, reliable, and has more analytical capacity than conventional laboratory testing. AI models, particularly hybrid ML/DL systems, can quickly process a variety of data, including CPT, images, and sensor readings, significantly cutting down on the time and cost associated with site characterization. By identifying intricate, non-linear patterns that empirical methods frequently overlook, they minimize human error and provide objective classifications. This ability allows engineers to forecast important soil characteristics, such as shear strength, from sparse field data, leading to faster, safer, and more dependable foundation and slope stability designs.

There are significant limitations to the widespread use of AI in soil characterization, primarily related to model interpretability and data issues. AI models frequently exhibit overfitting or data bias due to the inherently sparse, non-standardized, and expensive nature of geotechnical data. This results in poor performance when applied to new geologically distinct sites. In addition, highly accurate DL models frequently lack the transparency and physical foundation that engineers require to support crucial design decisions (such as foundation sizing) (Maralapalle et al., 2026c). This lack of reliability restricts AI's role in critical design workflows and makes it challenging to integrate AI results with established soil mechanics principles and regulatory requirements.

3.3.4 Comparison of AI-Based Techniques with Traditional Methods

The rapid development of AI has altered soil classification and properties. Soil evaluation has long been based on conventional techniques that mainly rely on laboratory testing, professional interpretation, and manual sampling. The intricate relationships between soil properties may not be captured by these time-consuming and labor-intensive techniques. However, methods based on AI, like ML and DL, have demonstrated significant promise for improving the accuracy, efficiency, and reliability of soil classification and characterization. The AI-based algorithms in Tables 3.2 and 3.3 are compared to conventional methods for characterizing and classifying soil. The advantages and disadvantages of AI-based techniques in contrast to traditional ones are highlighted in this comparison, along with details regarding their accuracy, uses, and future development potential.

For soil classification and characterization, a comparison of ML, DL, and conventional approaches reveals clear benefits and drawbacks for each approach (Alzubaidi et al., 2021). Despite being well-established and comprehensible, traditional methods can occasionally be labor-intensive and time-consuming. Although ML techniques can easily handle massive datasets, their use may require a high degree of expertise (Mehrishal et al., 2025). DL models can successfully identify complex patterns in

TABLE 3.2 Comparison of AI-based techniques with traditional methods in soil classification

METHOD	*ACCURACY*	*EFFICIENCY*	*DATA REQUIREMENTS*	*EXPERTISE REQUIRED*
Traditional methods	60%–80%	Low	Limited dataset	High
ML	80%–90%	High	Large dataset	Moderate
DL	90%–95%	High	Large dataset	High

Source: Aydın et al. (2023); Nguyen et al. (2022).

TABLE 3.3 Comparison of AI-based techniques with traditional methods in soil characterization

METHOD	*PRECISION*	*RECALL*	*F1 SCORE*	*INTERPRETABILITY*
Traditional methods	High	Moderate	Moderate	High
ML	High	High	High	Moderate
DL	Very high	Very high	Very high	Low

Source: Awais et al. (2023); İnci et al. (2024); Naeimi et al. (2024).

soil data, despite the fact that they are computationally demanding and require large datasets. Despite these limitations, AI-based techniques have several applications in soil research, such as real-time monitoring of soil properties like moisture, organic matter, and nutrient levels (Bharti et al., 2025; Minasny et al., 2024). In contrast to DL models' capacity to recognize complex patterns in soil data, which permits accurate classification and prediction, ML techniques such as RF and SVMs can make incredibly accurate predictions about the properties of soil.

3.3.5 Implications of AI-Based Techniques for Geotechnical Engineering

For figuring out complicated soil mechanics, improving constitutive modeling, and dealing with spatial variability, AI is a powerful tool. The goal is to improve our comprehension of geomaterials. It offers instantly applicable solutions to enhance project efficiency, risk management, and design optimization over the course of an infrastructure project's whole lifecycle, as shown in Figure 3.2. AI is excellent at quickly automating data transcription from core images and borehole logs, relieving engineers of time-consuming duties. Important uses consist of:

Foundation Design Optimization: By examining sizable datasets of CPT and Standard Penetration Test results, AI models accurately forecast bearing capacity and settlement. This enables rapid comparison of shallow and deep

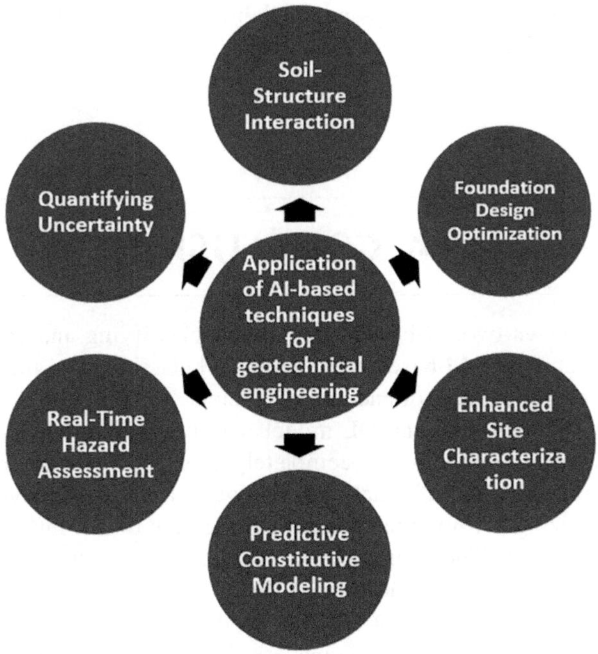

FIGURE 3.2 Application of AI-based techniques for geotechnical engineering.

foundations and optimization of design parameters, leading to substantial material and cost savings. **Real-Time Hazard Assessment**: By integrating real-time sensor data from construction sites, such as pore pressure and displacement, AI provides predictive early warnings for potential hazards like slope failure or tunnel face instability. To avoid catastrophic delays and enhance on-site safety, proactive monitoring is essential.

Soil–Structure Interaction: Advanced ML generates surrogate models that quickly and accurately simulate complex soil–structure interaction, assisting design engineers in evaluating the performance of earth-retaining systems and tunnels. This is in contrast to traditional, time-consuming Finite Element Analysis.

Predictive Constitutive Modeling: AI algorithms, especially neural networks, are being used to create data-driven constitutive models that more accurately capture the highly non-linear, stress-dependent behavior of rock and soil than traditional, physics-based models. By learning from intricate laboratory tests like triaxial and shear box, these models can predict how the soil will react under a variety of loading conditions.

Enhanced Site Characterization: AI models successfully overcome spatial variability by combining diverse data sources, including remote sensing, geophysics, and sparse borehole data, to interpolate soil properties in 3D. This enables scientists to produce subsurface profiles that are more precise, continuous, and detailed.

Quantifying Uncertainty: ML offers a strong statistical framework for managing and quantifying intrinsic geotechnical uncertainty. By converting raw data into practical risk maps and reliability indices that can be utilized to create new geotechnical theories and standards, scientists use AI to assess the probability of soil liquefaction or slope instability.

3.4 CONCLUSION

This chapter discusses two AI-based methods for classifying and characterizing soil: ML and DL. By applying AI-based methods to soil classification and characterization, soil assessment can be made more accurate, efficient, and reliable. By identifying complex patterns in the data, ML and DL models can precisely predict and describe soil. These methods have the potential to completely alter how soils are tested due to their effectiveness, precision, and affordability. Although an AI-based system for classifying and characterizing soil may have benefits, there are disadvantages that need to be taken into consideration. The complexity of soil systems and the models' interpretability are just two of the many considerations that must be made when creating AI algorithms. Furthermore, the quality of the data used to train the models determines how reliable and accurate AI-based techniques are. By using AI-based soil categorization and characterization, incorporating domain expertise into AI models, and creating more comprehensible and interpretable AI models, we may be able to manage uncertainty and unpredictability in soil data in the future. Researchers will also need to look into how well AI-based techniques work in actual projects in order to assess their effectiveness and performance in geotechnical engineering. In the near future, geotechnical engineering practice and construction are expected to be significantly impacted by AI-based soil classification and characterization. By offering more accurate and dependable soil assessments, AI-based techniques can aid in better decision-making. This could result in more affordable solutions, increased safety, and a decreased overall risk. AI-based methods are expected to become more prevalent in geotechnical engineering practice since they allow engineers to design and build more reliable, sustainable, and safe infrastructure.

REFERENCES

Abrar, M., Haque, M. M., Goswami, P., Ishtiaq, A., Abrar, M. M., Haque, M. M., Goswami, P., Rahman, M. A., Rishad, J. K., Hassan, M. T., Islam, J., & Asif, K. A. (2024, December). A comprehensive study of soil classification and identification using deep ML algorithm based on soil images. *International Conference on Advances in Civil Engineering (ICACE2024).* https://www.researchgate.net/publication/387057101

Akin, O., & Sayil, N. (2025). Soil characterization in landslide-prone areas using ground shear strain based on active and passive source surface wave methods. *Pure and Applied Geophysics, 182*(4), 1579–1600. https://doi.org/10.1007/s00024-025-03696-0

Alzubaidi, L., Zhang, J., Humaidi, A. J., Al-Dujaili, A., Duan, Y., Al-Shamma, O., Santamaría, J., Fadhel, M. A., Al-Amidie, M., & Farhan, L. (2021). Review of DL: concepts, CNN architectures, challenges, applications, future directions. *Journal of Big Data, 8*(1). https://doi.org/10.1186/s40537-021-00444-8

Ameer Abdulridha AjmiAlali. (2024). Applying AI techniques to analyze soil data and improve foundation design in construction in Iraq. *Jurnal Kendali Teknik Dan Sains, 2*(3), 141–157. https://doi.org/10.59581/jkts-widyakarya.v2i3.3769

Awais, M., Naqvi, S. M. Z. A., Zhang, H., Li, L., Zhang, W., Awwad, F. A., Ismail, E. A. A., Khan, M. I., Raghavan, V., & Hu, J. (2023). AI and ML for soil analysis: an assessment of sustainable agricultural practices. *Bioresources and Bioprocessing, 10*(1). Springer. https://doi.org/10.1186/s40643-023-00710-y

Aydın, Y., Işıkdağ, Ü., Bekdaş, G., Nigdeli, S. M., & Geem, Z. W. (2023). Use of ML techniques in soil classification. *Sustainability, 15*(3). https://doi.org/10.3390/su15032374

Bharti, B., Bharti, V., Negi, N., Mohan, J., Arya, R., & Kumar, A. (2025). Role of AI technologies in soil health monitoring system. *International Journal of Advanced Biochemistry Research, 9*(4), 1158–1162. https://doi.org/10.33545/26174693.2025.v9.i4n.4262

Ding, Z., Tu, W., & Li, X. (2024). Enhancing ground loss rate prediction in soft-soil shield tunneling: a synergistic approach of peck back analysis and eXtreme Gradient Boosting and Bayesian optimization. *Scientific Reports, 14*(1). https://doi.org/10.1038/s41598-024-73025-3

Ehsan, M., Anees, M. T., Bakar, A. F. B. A., & Ahmed, A. (2025). A review of geological and triggering factors influencing landslide susceptibility: artificial intelligence-based trends in mapping and prediction. *International Journal of Environmental Science and Technology.* Springer Nature. https://doi.org/10.1007/s13762-025-06741-6

Ghader, S., Barbara, S.-M., Javankhoshdel, S., Michael, M., & Heiner, K. (2023). *Future Challenges in Tunnel Stability Analysis Using AIand ML* (pp. 317–324). https://doi.org/10.2991/978-94-6463-258-3_33

Ghiasi, V., & Koushki, M. (2020). Numerical and artificial neural network analyses of ground surface settlement of tunnel in saturated soil. *SN Applied Sciences*, *2*(5). https://doi.org/10.1007/s42452-020-2742-z

Gómez-Miranda, I. N., Restrepo-Estrada, C., Builes-Jaramillo, A., & Albuquerque, J. P. de. (2025). Advanced AI techniques for landslide susceptibility mapping and spatial prediction: a case study in Medellín, Colombia. *Applied Computing and Geosciences, 25.* https://doi.org/10.1016/j.acags.2025.100226

İnci, Y., Bilgili, A. V., Gündoğan, R., Gözükara, G., Karadağ, K., & Tenekeci, M. E. (2024). ML-based classification of soil parent materials using elemental concentration and vis-NIR data. *Sensors, 24*(16). https://doi.org/10.3390/s24165126

Kang, T. H., Choi, S. W., Lee, C., & Chang, S. H. (2022). Soil classification by ML using a tunnel boring machine's operating parameters. *Applied Sciences, 12*(22). https://doi.org/10.3390/app122211480

Kudaibergenov, M., Nurakynov, S., Iskakov, B., Iskaliyeva, G., Maksum, Y., Orynbassarova, E., Akhmetov, B., & Sydyk, N. (2025). Application of artificial intelligence in landslide susceptibility assessment: Review of recent progress. *Remote Sensing, 17*(1). Multidisciplinary Digital Publishing Institute (MDPI). https://doi.org/10.3390/rs17010034

Maralapalle, V., Muktinutalapati, J., Ansari, A., & Chithambar Ganesh, A. (2026a). Geospatial analysis of soil quality: GIS-based mapping for sustainable land management. In S. C. Pal, U. Chatterjee, & D. Ruidas (Eds.), *Soil Degradation, Desertification and Restoration. Sustainability Solutions*, vol. 1. Springer. https://doi.org/10.1007/978-3-031-92277-0_11

Maralapalle, V., Muktinutalapati, J., Ansari, A., & Chithambar Ganesh, A. (2026b). Soil quality restoration zone mapping using GIS-based machine learning techniques. In S. C. Pal, U. Chatterjee, & D. Ruidas (Eds.), *Soil Degradation, Desertification and Restoration. Sustainability Solutions*, vol. 1. Springer. https://doi.org/10.1007/978-3-031-92277-0_15

Maralapalle, V., Muktinutalapati, J., Tammineni, G., Krishna, M. V. N., & Marlapalle, S. (2026c). Smart cities, smarter solutions: AI for urban transformation. In R. C. Poonia, D. Thakur, & I. Thakur (Eds.), *Intelligent Systems for Sustainable Infrastructure: AI Solutions Shaping a Green Future. Engineering Cyber-Physical Systems and Critical Infrastructures*, vol. 17. Springer. https://doi.org/10.1007/978-3-031-95240-1_1

Mehrishal, S., Leem, J., Kim, J., Shao, Y., Kang, I. S., & Song, J. J. (2025). Tunnel rapid AI classification (TRaiC): An open-source code for 360° tunnel face mapping, discontinuity analysis, and RAG-LLM-powered geo-engineering reporting. *Remote Sensing*, *17*(16). https://doi.org/10.3390/rs17162891

Minasny, B., Bandai, T., Ghezzehei, T. A., Huang, Y. C., Ma, Y., McBratney, A. B., Ng, W., Norouzi, S., Padarian, J., Rudiyanto, Sharififar, A., Styc, Q., & Widyastuti, M. (2024). Soil science-informed ML. *Geoderma, 452.* Elsevier B.V. https://doi.org/10.1016/j.geoderma.2024.117094

Mutale, B., Withanage, N. C., Mishra, P. K., Shen, J., Abdelrahman, K., & Fnais, M. S. (2024). A performance evaluation of random forest, artificial neural network, and support vector ML algorithms to predict spatio-temporal land use-land cover dynamics: a case from Lusaka and Colombo. *Frontiers in Environmental Science*, *12*. https://doi.org/10.3389/fenvs.2024.1431645

Naeimi, M., Daggupati, P., & Biswas, A. (2024). Image-based soil characterization: a review on smartphone applications. *Computers and Electronics in Agriculture, 227.* Elsevier B.V. https://doi.org/10.1016/j.compag.2024.109502

Nguyen, M. D., Costache, R., Sy, A. H., Ahmadzadeh, H., Van Le, H., Prakash, I., & Pham, B. T. (2022). Novel approach for soil classification using ML methods. *Bulletin of Engineering Geology and the Environment*, *81*(11). https://doi.org/10.1007/s10064-022-02967-7

Osisanwo F.Y., Akinsola J.E.T., Awodele O, Hinmikaiye J. O, & Olakanmi O. (2017). Supervised ML algorithms: classification and comparison. *International Journal of Computer Trends and Technology*, *48*(3). https://doi.org/10.14445/22312803/IJCTT-V48P126

Pace, R., Schiano Di Cola, V., Monti, M. M., Affinito, A., Cuomo, S., Loreto, F., & Ruocco, M. (2025). AI in soil microbiome analysis: a potential application in predicting and enhancing soil health—a review. *Discover Applied Sciences, 7*(2). Springer Nature. https://doi.org/10.1007/s42452-024-06381-4

Parra, F., González, J., Chacón, M., & Marín, M. (2023). Modeling and evaluation of the susceptibility to landslide events using ML algorithms in the province of Chañaral, Atacama region, Chile. *Natural Hazards and Earth System Sciences*. https://doi.org/10.5194/nhess-2023-72

Ratke, R. F., Viana, P. R. N., Teodoro, L. P. R., Baio, F. H. R., Teodoro, P. E., Santana, D. C., Santos, C. E. da S., Zuffo, A. M., & Aguilera, J. G. (2024). Multispectral sensors and ML as modern tools for nutrient content prediction in soil. *AgriEngineering*, *6*(4), 4384–4394. https://doi.org/10.3390/agriengineering6040248

Samptkumari Bandagar, S. M. (2022). Crop and soil monitoring by AI technique. *International Journal for Research Trends and Innovation*, *7*(7), 2456–3315. www.ijrti.org

Singh, J., Singh, P., Ravi, V., Kumar, S., Diwakar, M., Al Mazroa, A., & Gupta, I. (2024). Advancing tunnel construction reliability with automated AI under geotechnical and aleatoric uncertainties. *The Open Civil Engineering Journal*, *18*(1). https://doi.org/10.2174/0118741495343684240910115546

Teodosio, B., Wasantha, P. L. P., Yaghoubi, E., Guerrieri, M., van Staden, R., & Fragomeni, S. (2025). Application of AI in reactive soil research: a scientometric analysis. *Geotechnical and Geological Engineering, 43*(4). Springer Science and Business Media Deutschland GmbH. https://doi.org/10.1007/s10706-025-03097-z

Uddin, S., Khan, A., Hossain, M. E., & Moni, M. A. (2019). Comparing different supervised ML algorithms for disease prediction. *BMC Medical Informatics and Decision Making*, *19*(1). https://doi.org/10.1186/s12911-019-1004-8

Wadoux, A. M. J. C. (2025). AIin soil science. *European Journal of Soil Science, 76*(2). John Wiley and Sons Inc. https://doi.org/10.1111/ejss.70080

Zhu, Y., Wang, B., Zhang, Y., & Sun, Z. (2025). Identification and classification method of landslide pattern in the soil water index-based early warning system. *Scientific Reports*, *15*(1). https://doi.org/10.1038/s41598-025-13348-x

Harnessing Artificial Intelligence Techniques for Prediction of Compaction Parameters of Soil

4

Utkarsh, Pradeep Kumar Jain, Kaustav Chatterjee, and Sudhir Bhatt

4.1 INTRODUCTION

Soil compaction is one of the basic processes in geotechnical engineering and it is substantial in the construction of different types of infrastructures, including highways, embankments, and earth dams (Proctor, 1933). The main aim of compacting soil is to enhance the engineering qualities of soil by minimizing air pores and increasing the

DOI: 10.1201/9781003714125-4

density of soil leading to an increase in the strength, stiffness, and stability of soil. Two more important parameters that describe the compaction behavior of soils are the optimum moisture content (OMC) and maximum dry density (MDD), which are crucial to ensuring that the required compaction degree is attained on the field and is generally 95%–98% of the MDD (Tatsuoka and Correia, 2018; Chen et al., 2019).

Best practices OMC and MDD are characterized by laboratory tests, including the standard Proctor test (ASTM D698) and the Modified Proctor test (ASTM D1557). These tests are however time-consuming, labor-intensive, and demand enormous efforts and resources (Ardakani and Kordnaeij, 2019; Ren et al., 2015). In order to surpass these shortcomings, scholars have established empirical and semi-empirical correlations to predict compaction parameters using simple physical characteristics of the soil, including grain size distribution and Atterberg limits (Al-Khafaji, 1993; Blotz et al., 1998; Sridharan and Nagaraj, 2005; Farooq et al., 2016). Although these correlations can give fast estimations, in many cases, they are limited in their applicability and cannot be applied to a broad range of soils because they are grounded on select datasets and simplified assumptions (Gunaydin, 2009; Nagaraj et al., 2015).

Artificial Neural Networks (ANNs) have also become an influential tool to make predictions of complex non-linear relationships in other engineering disciplines such as geotechnical engineering in recent years (Sinha and Wang, 2008). ANNs are capable of learning without any assumptions on the underlying data distribution and thus can be used in making predictions of the compaction properties of soils (Gunaydin, 2009). The use of machine learning, including ANNs, as the alternative to the traditional empirical models in predicting the soil compaction parameters, has presented some promise because they help with the incorporation of non-linear correlations between soil properties and compaction parameters, resulting in more accurate predictions (Ardakani and Kordnaeij, 2019).

This study is based on a comprehensive experimental study about the compaction properties of a wide series of soils (including 5 natural soils and 73 artificial soils) based on 24 different soil types based on the Indian Standard Soil Classification System (ISSCS). Artificial soils were made by a mixture of different proportions of natural soils, bentonite clay, and marble powder to depict a wide range of soil gradations and plasticity properties. Experimental database is very diversified in the properties of soils, and the gravel content varies in the range 0–71, sand content 0–97, fines content (FC) 1–100, liquid limit (LL) 0–79, and plasticity index (PI) 0–43. Compaction parameters also show a great variability with OMC ranging between 7.0% and 23.0% and MDD ranging between 15.11 and 21.39 kN/m^3. This is a very powerful data source of predicting multi-dimensional and precise model of soil compaction parameters.

In order to increase predictive abilities of machine learning models, experimental data is combined with advanced machine learning methods in the scope of the current study. The state-of-the-art models such as Random Forest (RF), Gradient Boosting (GB), Extreme Gradient Boosting (XGBoost), Light Gradient Boosting Machine (LightGBM), long short-term memory (LSTM), Convolutional Neural Network (CNN), and an Ensemble model are used to forecast OMC and MDD. These models have been implemented successfully in many geotechnical engineering works (Samui and Sitharam, 2011; Kurnaz and Kaya, 2020; Wang and Yin, 2020) and it is likely that they will give the correct prediction of soil compaction parameters, when trained on the overall experimental database.

The primary objectives of this study are: (i) to develop robust and versatile machine learning and deep learning models for predicting OMC and MDD of soils, incorporating a wide range of soil types, plasticity ranges, and compaction energies; (ii) to validate the developed models using a comprehensive experimental database and assess their performance using various statistical metrics; (iii) to compare the performance of the

FIGURE 4.1 Methodology of work process.

proposed models with existing empirical and machine learning models in the literature; and (iv) to investigate the relative importance of input variables on the prediction of soil compaction parameters through sensitivity analysis. The development of accurate prediction models is not the end of the scope of this study. Its goal is to offer a practical tool to the geotechnical engineers to effectively estimate the compaction parameters of soil at a rapid and accurate rate and eliminate the tendencies to use time-consuming laboratory tests. The proposed models are expected to find applications in various geotechnical engineering projects, improving design efficiency and optimizing construction processes, ultimately saving time and resources. Figure 4.1 shows the methodology of the research.

4.2 MATERIALS AND METHODS

4.2.1 Soils

The present study utilized a diverse range of soils, including both natural and artificial soils, to develop a comprehensive database for ANN modeling. Five natural soils were collected from various locations in Jabalpur, Madhya Pradesh, India, representing different soil types and characteristics. These soils were air-dried and all the deleterious materials or clods were removed before further testing.

In order to increase the database and to increase the possible range of soil properties, 73 artificial soils were made by mixing the natural soils, commercially available bentonite clay, and powdered marble in various ratios. The synthetic soils were also aimed at reflecting different soil classification groups and subgroups so that there was a wide range of data on which the model would be developed.

4.2.2 Laboratory Tests

The laboratory tests were done on all the natural and artificial soils to ascertain their basic properties and compaction parameters. Relevant Indian Standard (IS) codes that were used in the following tests include particle size analysis (IS: 2720 Part 4–1985), LL and plastic limit tests (IS: 2720 Part 5–1985), specific gravity (IS: 2720 Part 3/Sec 1–1980), and standard Proctor compaction test (IS: 2720 Part 7–1980). These tests provided the basis on which the soils were categorized as per the ISSCS and to provide input data to ANN modeling. Figure 4.1 can help in understanding the entire process of methodology.

4.2.3 Database Generation

The laboratory tests were conducted and the results combined to establish a complete database that consisted of ANN modeling. The database entries include the following input variables: Gravel content (%), Sand content (%), FC (%), LL (%), and PI (%) percentage. Compaction parameters are the output variables, which include OMC (%) and maximum dry unit weight, MDUW (kN/m^3). The data in the database includes a great variety of soil properties, and the content of gravel is between 0 and 71, sand between 0 and 97, fines between 1 and 100, LL between 0 and 79, and PI between 0 and 43. The compaction parameters also reveal a large variation with OMC and MDUW ranging between 7.0% and 23.0% and 15.11–21.39 kN/m^3, respectively. Figure 4.2a shows the overall pair-wise correlation and distribution plots of all input and output variables of the experimental database. This analysis provides critical insights into the dataset characteristics and modeling requirements:

Data Quality and Coverage Assessment: The diagonal probability density plots reveal well-distributed variables across their respective ranges, with no significant data gaps or clustering. The gravel content shows a slight right-skewed distribution (0%–71%), while sand content exhibits a more uniform distribution (0%–97%), ensuring comprehensive coverage of gradation types. OMC and MDUW display near-normal distributions, indicating balanced representation of compaction characteristics.

Variable Interdependencies: The off-diagonal scatter plots reveal several important correlations:

i. Strong negative correlation ($r \approx -0.75$) between fines and sand content, confirming their complementary relationship in soil gradation
ii. Moderate positive correlation ($r \approx 0.65$) between LL and PI, as expected from Atterberg limits theory
iii. Weak to moderate correlations between particle size fractions and compaction parameters, suggesting complex multi-variable dependencies

Non-linear Relationship Validation: The scatter plots clearly demonstrate non-linear, heteroscedastic relationships between most input–output variable pairs. The absence of clear linear trends, particularly in OMC vs. FC and MDUW vs. LL plots, confirms that simple linear regression would be inadequate for accurate prediction.

Modeling Approach Justification: The complex, multi-dimensional correlation structure and evident non-linearity validate the selection of advanced machine learning algorithms (ANN, XGBoost, Ensemble models) over traditional empirical correlations. The data complexity necessitates models capable of capturing higher-order interactions and non-linear mappings between soil properties and compaction behavior.

The correlation heatmap matrix of Pearson correlation coefficients between all the input and output variables will be provided in Figure 4.2b. The color scale has a range of deep blue (–1.0), which means strongly negative correlation, to deep red (+1.0), which means strong positive correlation, with white being the null case. The correlation table indicates that the relationships between variables are weak and most correlation coefficients fall in the range of –0.15 to +0.19: (a) Particle Size Relationships: The gravel content is weakly associated with the sand content ($r = -0.087$) and weakly interdependent with the

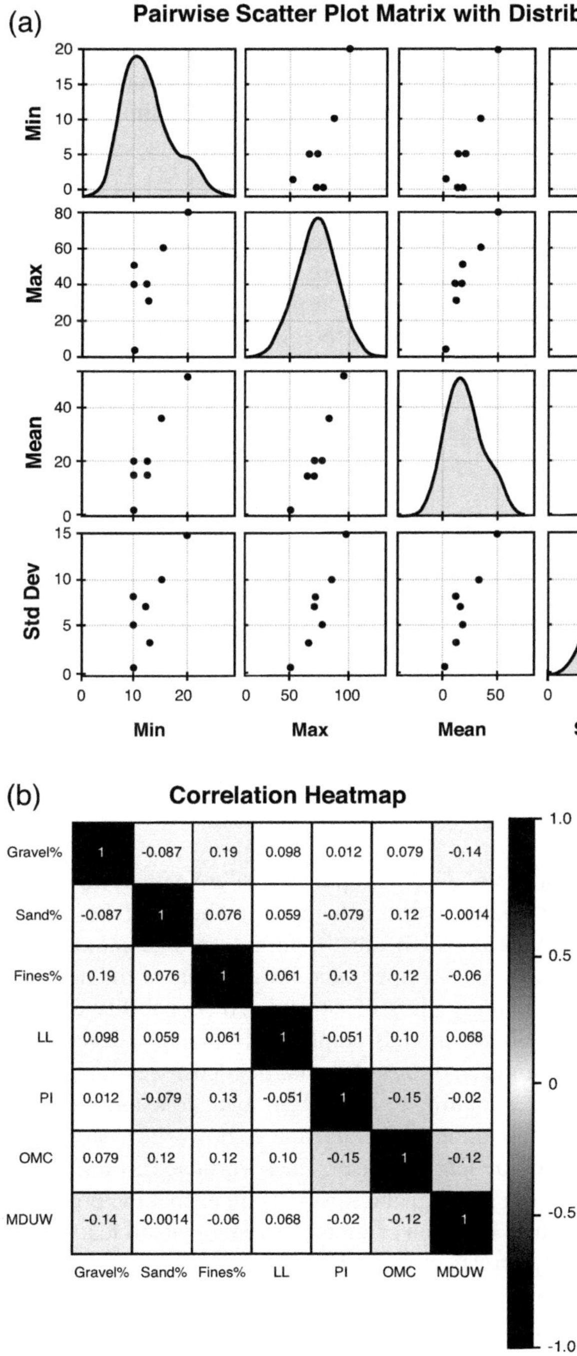

FIGURE 4.2 (a) Input–output variable relationships. (b) Correlation heatmap.

FC ($r = 0.19$): existing (b) as a subset of a complex system of interrelationships between gradation measurements and not as simple complementary relationships. (b) Plasticity Characteristics: LL and PI show a rather weak negative relationship ($r = -0.051$), which indicates the rather independent behavior of these Atterberg limits in this heterogeneous soil database. (c) Correlations between Compaction Parameters: OMC exhibits weak correlations with all of the input variables (|human|)and MDUW, similarly weak correlations with all input variables (|human|). (d) Modeling Implications: The low-order (most of the correlations are less than 0.2) linear correlations suggest that the simple linear regression models could not be used to predict the soil compaction parameters. This confirms the need to have complex machine learning computer programs that could represent non-linear and multi-variable interactions that are not visible in linear correlation analysis. These low values of correlation indicate that the association of the soil properties and the behavior of compaction are very complicated and demand advanced modeling methods to attain precise predictions.

4.2.4 Data Preprocessing

The experimental dataset, comprising 78 soil samples with a wide range of properties, was randomly divided into training (90%) and testing (10%) subsets. To ensure better convergence during training, data normalization was performed to scale the inputs and outputs between –1 and +1 using the following equation (Baughman and Liu, 1995):

$$n_s = \sum_{i=1}^{R} (w_{si}) \cdot p_i + b_s$$

where n_s = total input of the node, p_i = input factor "p" with component p_i (i =1 to R), w_{si} = weight factors associated with each input to the node (w_{11}, w_{12}, ..., and w_{si}), b_s = node's internal threshold which are the minimum and maximum values of the variable, and High and Low are the upper and lower limits of the normalization range.

Figure 4.3a shows violation plots of normalized probability density distributions of the input variables of all the models employed in machine learning. The shapes of the violin are the shape representations of the kernel density estimation that are created with the broader part corresponding to more significant concentration of data at certain normalized values. Normalization between 0 and 1 shows the systematic preprocessing of data necessary in machine learning uses. Each variable has the same characteristic distribution shape following scaling, and this fact proves that the preprocessing pipeline did not destroy relationships between original data but made all features subjected to the algorithm treated equally, irrespective of their original scales of measurement.

The violin plots provide extensive coverage of data in the entire range of the normalized (0–1) input variables where no substantial gaps or missing regions of values are observed. This is validated by the continuous density distributions which ensure there is good sampling coverage in the parameter space; thus, there is strong model training with no data-sparse areas which would invalidate prediction accuracy. The various patterns of distribution confirm the adequacy of the dataset as representative

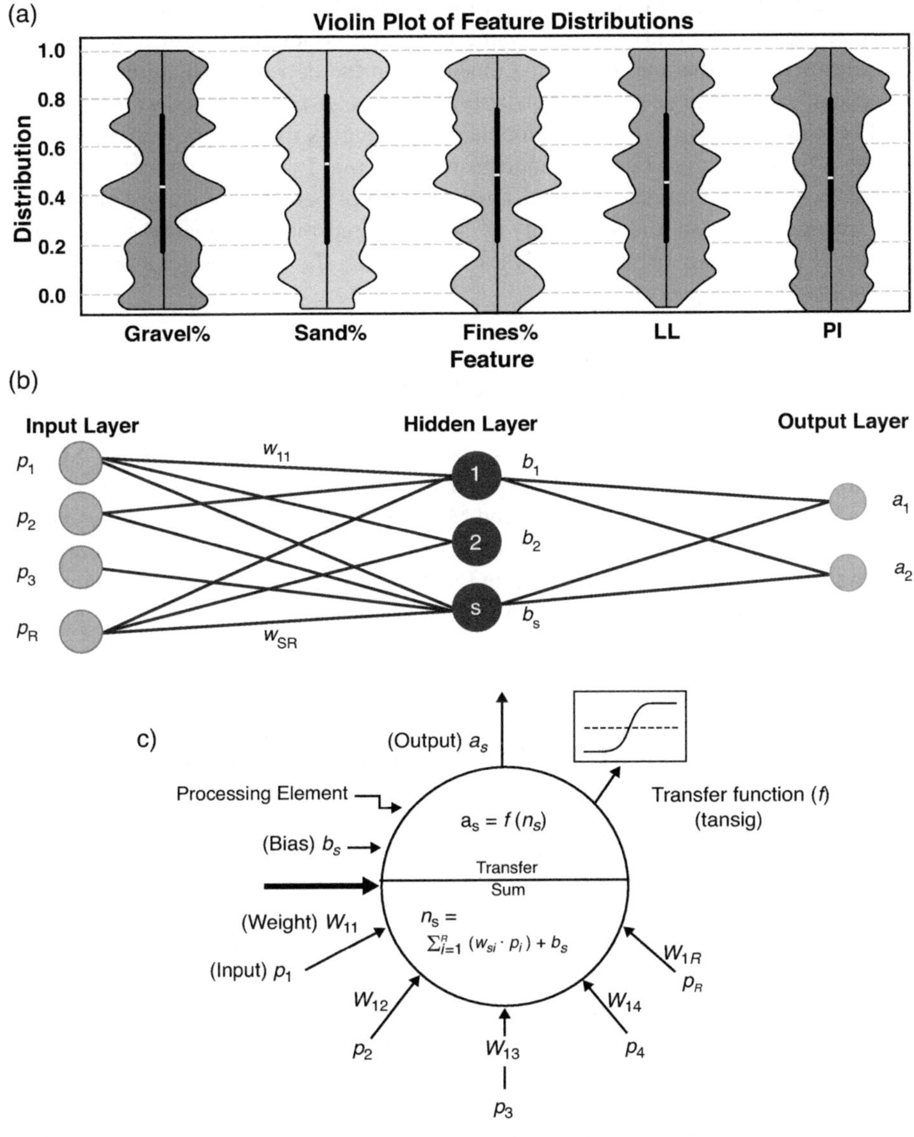

FIGURE 4.3 (a) Violin plot of feature distribution. (b) Structure of a typical neural network. (c) A single neuron.

of actual geotechnical conditions: (i) Gravel content exhibits natural bimodal behavior with respect to gravelly and fine-grained population of soils. (ii) The content of sand has a balanced distribution throughout the range, which is a representation of different gradations. (iii) When the content of fine concentrates is in high levels, it indicates the usual engineering soil conditions. (iv) Appropriate representation of low to high plasticity soils is demonstrated by plasticity parameters (LL and PI).

Potential Modeling Issues: Distribution Characteristics: (i) Bimodal distributions (especially gravel content) can have some modeling issues because it may need some algorithmic model that can take into consideration the non-linear feature relationships as opposed to assuming a normal distribution. (ii) The distribution of the content of fines and PIs is skewed, which calls on models with strengths in non-symmetric data distributions. (iii) Multi-modal density patterns indicate that Ensemble approaches could be especially helpful in modeling the different pattern of behavior of soils. The quality of experimental database and suitability of sophisticated machine learning methods are confirmed by these distribution analyses in depicting the intricate relationships in the behavior of soil compaction.

4.2.5 Development of Models

Several state-of-the-art machine learning models including RF, GB, XGBoost, LightGBM, Ensemble model, and deep learning models including ANN, LSTM, and CNN were employed to predict OMC and MDUW. The input variables for these models were the same as those used for the ANN models, i.e., gravel content, sand content, FC, LL, and PI. The models were trained and tested using the experimental database, and their performance was evaluated using various statistical metrics.

4.3 DEEP LEARNING MODELS

4.3.1 ANN Algorithms

In the present study, five distinct ANN algorithms were employed to develop predictive models for estimating soils' OMC and MDUW. These algorithms include Levenberg-Marquardt backpropagation (LMB), Gradient Descent with Momentum backpropagation (GDM), Scaled Conjugate Gradient backpropagation (SCG), Radial Basis Function (RBF), and Generalized Regression Neural Network (GRNN) (Fattah et al., 2009; Zell, 1994). The arrangement of neurons within the network is structured in layers, with connections established such that neurons in one layer receive inputs from the preceding layer and transmit outputs to the subsequent layer. External inputs are introduced at the initial layer, while system outputs are obtained at the final layer. The layers situated between the input and output layers are referred to as hidden layers. Figure 4.3b illustrates the architecture of a conventional neural network.

LMB, GDM, and SCG are backpropagation-based algorithms that iteratively adjust the network weights and biases to minimize the error between predicted and target outputs. RBF and GRNN are radial basis networks with a hidden layer of radial units and a unique linear output layer (Broomhead and Lowe, 1988; Poggio and Girosi, 1990). The whole process of a single neuron can be better understood by Figure 4.3c.

For the backpropagation algorithms, various network architectures were explored by varying the number of hidden layers and neurons. The optimal architecture was chosen by evaluating performance metrics, including the coefficient of determination (R^2) and mean squared error (MSE). For the radial basis networks, different spread constant values (0.1, 0.5, and 1.0) were investigated to identify the best-performing model.

4.3.2 Convolutional Neural Network

CNNs are one of the popular deep learning networks popularly used in computer vision. The CNN architecture is composed of convolutional layer, maxpooling layer, and dense layer. The convolutional layer performs the job of extracting different features from the input data. The maxpooling layer down-samples the input data to reduce dimensionality and computational load, and the dense layer facilitates in prediction of the output. Figure 4.4a shows a schematic representation of the CNN architecture.

4.3.3 LSTM Network

LSTM Network is one of the popular deep learning models found, and its application is in natural language processing and time series forecasting. The LSTM architecture consists of three different gates namely: forget gate, input gate, and the output gate. The forget gate of the LSTM architecture performs the task of eliminating useless information from the memory, the input gate encompasses information into the memory cell, and the output gate outputs the information from the current memory cell into the next cell. Figure 4.4b show a schematic representation of the LSTM cell.

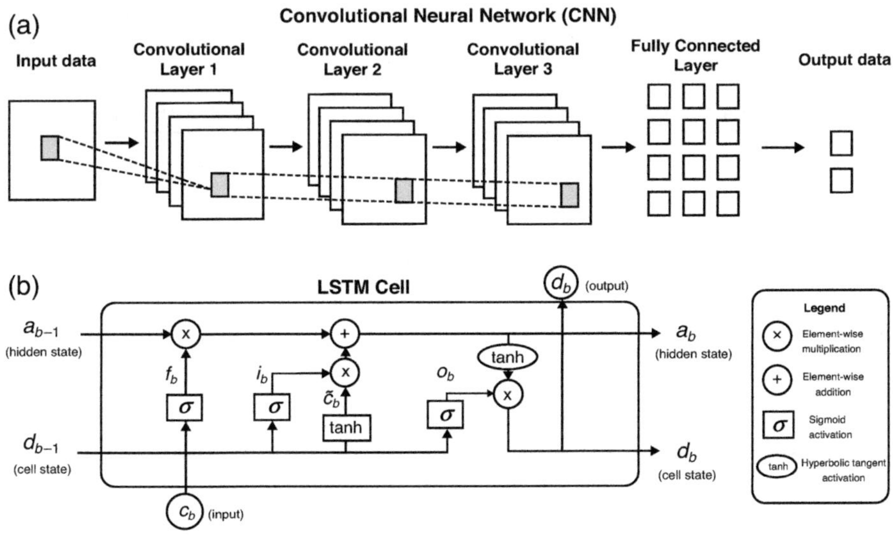

FIGURE 4.4 (a) Schematic representation of CNN architecture (Ansari et al., 2025). (b) Schematic representation of the LSTM cell (Chatterjee et al., 2025).

4.4 MODEL DEVELOPMENT AND PERFORMANCE EVALUATION

For each ANN algorithm, three models were developed by varying the input variables. The first model (M-3) utilized FC, LL, and PI as inputs. The second model (M-4) included sand content (S) in addition to the M-3 inputs, while the third model (M-5) further incorporated gravel content (G). OMC and MDUW were the output variables for all models.

The backpropagation models (LMB, GDM, and SCG) were trained using the tan-sigmoid transfer function in the hidden layers and the purelin transfer function in the output layer. The number of hidden layers and neurons was varied to identify the optimal network architecture for each model. The radial basis networks (RBF and GRNN) were designed using the "newrb" and "newgrnn" functions in MATLAB, respectively (MATLAB-R2010b, Help).

The CNN and the LSTM models were developed in Python using the tensorflow package. The input to the CNN and the LSTM models is FC, LL, plastic limit, PI, and percentage passing through different sieves and the output from the deep learning models is OMC and MDD. Final models were selected after the operation of hyperparameter tuning. Hyperparameter tuning was performed by varying the batch size, activation function, and different components of the LSTM and the CNN network. Figure 4.5a and b shows the MAE of the model over different epochs during the process of model development.

4.4.1 Performance Evaluation

The efficacy of the created machine learning and deep learning models was assessed utilizing statistical metrics, including the coefficient of determination (R^2) and MSE,

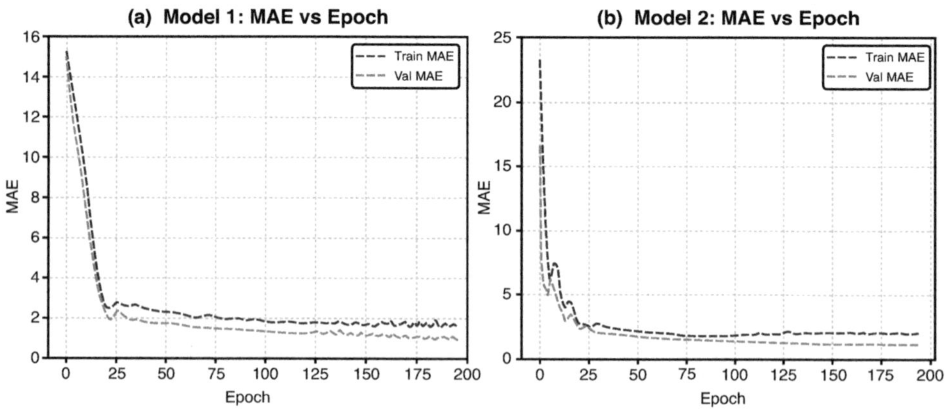

FIGURE 4.5 MAE during training and validation phase of model development (a) LSTM model. (b) CNN model.

comparing the anticipated and actual values of OMC and MDUW. The metrics were computed for both the training and testing datasets to evaluate the models' accuracy and generalization capacity.

Each algorithm's best-performing model was identified based on the highest R^2 and lowest MSE values. The optimal models from each algorithm were then compared to determine the overall "key model" that exhibited the best performance in predicting OMC and MDUW.

4.4.2 Model Validation

As an additional confirmation of the strength and generalizability of the created key model, a literature-based independent data (Gunaydin, 2009) was employed. The variables of the validation set as input into the key model were run, and the predicted values of OMC and MDUW were compared with the measured ones using the R^2 and MSE indicators. The objective of this validation process was to determine that the model is effective in the prediction of the compaction parameters of the soils that were not studied in the experimental data on which the model was developed.

4.5 RESULTS AND DISCUSSION

4.5.1 Experimental Results of Soil Properties and Compaction Parameters

The laboratory tests carried out on experimental results on the natural and artificial soils gave a holistic insight into the characteristics of the soils and their compaction characteristics. In the present study, the soils that were tested include 24 types of soils as used in the ISSCS, which are GW, GP, GC, GM, GM-GC, GW-GM, GP-GM, GW-GC, GP-GC, SW, SP, SM, SC, SW-SC, SW-SM, SP-SC, SP-SM, ML, MI, MH, CL, CI, CH, and CL-ML.

The particle size analysis showed that the soils possessed a broad span of gradation properties, whereby the gravel content was between 0 and 71 for sand, and the fine content was between 1 and 100. The test of Atterberg limits indicated that the LLs of the soils were 0%–79% and the PI was 0%–43%. The soils had specific gravity of between 2.61 and 2.69.

The standard Proctor parameters were also found to be in a wide range where OMC was between 7.0% and 23.0% and MDUW was between 15.11 and 21.39 kN/m^3. These findings not only indicate the heterogeneity of the samples of soil that were employed in this research, but it also forms a strong basis on which ANN and machine learning models can be developed.

4.5.2 Performance of Machine Learning and Deep Learning Models

The evaluation of the developed machine learning and deep learning models was conducted using three statistical metrics: coefficient of determination (R^2), root mean square error (RMSE), and mean absolute error (MAE). The results of the model performance evaluation are presented in Table 4.1.

Figure 4.6a presents a comprehensive performance comparison of all developed machine learning models using a bubble chart format, where the x-axis represents training R^2 values and the y-axis represents testing R^2 values. The bubble size indicates the relative magnitude of prediction errors. The Ensemble model demonstrates

TABLE 4.1 Performance evaluation of machine learning and deep learning models

	TRAINING DATASET			*TESTING DATASET*		
MODEL	*R^2*	*RMSE*	*MAE*	*R^2*	*RMSE*	*MAE*
ANN	0.9883	0.7084	0.4311	0.9676	1.1445	0.7140
Random Forest (RF)	0.9824	0.8665	0.6051	0.9459	1.4795	0.9706
Gradient Boosting (GB)	0.9985	0.2543	0.1779	0.9604	1.2660	0.7947
XGBoost	0.9929	0.5500	0.3767	0.9671	1.1537	0.7357
LightGBM	0.9883	0.7084	0.4311	0.9676	1.1445	0.7140
Ensemble model	0.9917	0.5943	0.3787	0.9808	0.8808	0.6344
CNN	0.9411	0.9412	0.8984	0.8603	1.8545	1.0931
LSTM	0.8399	1.3491	0.9107	0.8603	1.6300	0.9333

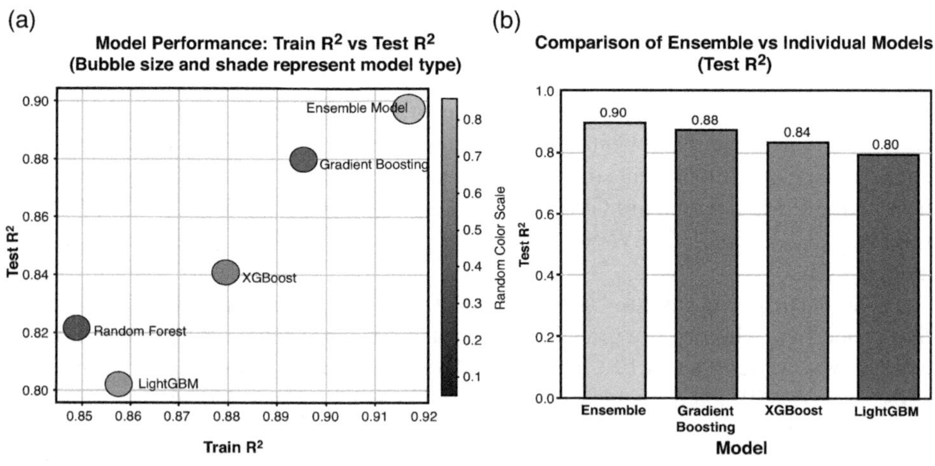

FIGURE 4.6 (a) Bubble chart of machine learning model performance. (b) Comparison of Ensemble with individual models.

superior performance with the highest testing R^2 (0.90) and training R^2 (0.92), positioned in the upper-right quadrant indicating optimal balance between accuracy and generalization. Models closer to the diagonal line (where training R^2 = testing R^2) exhibit better generalization with minimal overfitting. The GB model shows strong training performance (R^2 = 0.90) but slightly lower testing performance (R^2 = 0.88), while RF and LightGBM demonstrate more conservative but stable performance across both datasets. The diagonal positioning shows the bias–variance tradeoff, Bubble size represents model complexity or error magnitude, upper-right positioning indicates optimal performance, and distance from the ideal diagonal line indicates overfitting tendency. The results indicate that all models, including the ANN, exhibited strong performance in both training and testing datasets, achieving R^2 values exceeding 0.94 along with comparatively low RMSE and MAE values. The Ensemble model demonstrated superior overall performance, achieving the highest R^2 values along with the lowest RMSE and MAE values across both the training and testing datasets.

The ANN model performed well, with R^2 values of 0.9883 and 0.9676 in the training and testing datasets, respectively. It also had a relatively low RMSE and MAE, which shows that it can forecast the compaction parameters correctly. The ANN model was similar to the LightGBM model in its performance and there is an indication that both models are effective in showing the complex association between soil properties and compaction parameters.

The GB model exhibited the highest levels of performance in the training dataset with the largest R^2 value and least RMSE and MAE values. Its performance in the testing dataset was however a bit poor compared to the other models. Both the XGBoost and the RF models were also good performers, but XGBoost had a slightly higher R^2 values and lower RMSE and MAE values than RF.

Figure 4.6b shows the high performance of the Ensemble model over the individual machine learning algorithms in terms of testing dataset R^2 values. The Ensemble model achieves the highest prediction accuracy with R^2 = 0.90, representing a significant improvement over the best individual model (XGBoost, R^2 = 0.87).

Performance Analysis: The Ensemble model outperforms individual models by 3.4% compared to XGBoost, 5.9% compared to GB (R^2 = 0.85), and 4.7% compared to LightGBM (R^2 = 0.86). This improvement demonstrates the effectiveness of combining multiple algorithms to leverage their individual strengths while compensating for their respective weaknesses.

Ensemble Learning Benefits: The superior performance validates the Ensemble learning approach, where the final prediction is derived from weighted combinations of individual model outputs. This technique reduces prediction variance and improves generalization by:

i. Mitigating individual model biases
ii. Capturing diverse feature patterns through different algorithmic approaches
iii. Reducing overfitting through model averaging
iv. Enhancing robustness across varying soil conditions

The consistent improvement across all model comparisons confirms that Ensemble methods are particularly effective for complex geotechnical problems where multiple non-linear relationships exist between soil properties and compaction parameters.

4.5.3 Comparison with Existing Models

The results of the developed ANN and machine learning models were compared to the existing empirical and machine learning models found in the literature. The findings indicated that the proposed models were better than the existing models in prediction accuracy and generalizability.

The semi-empirical and empirical correlations that existed before among the previous researchers were limited and their prediction ability was less than the ANN and machine learning models created during this chapter. This was because these correlations were made using certain datasets and simplistic assumptions that cannot be applied to a large variety of soils.

The machine learning models that had been developed by previous researchers through ANNs yielded promising results but had less prediction accuracy than those that were developed in this research. It is possible to explain this by the bigger and more varied data of the current research and by the fact that the more sophisticated machine learning models, including XGBoost, LightGBM, and the Ensemble model, have been used.

4.5.4 Sensitivity Analysis

To test the relative significance of the input variables on the prediction of soil compaction parameters, a sensitivity analysis has been done. The review showed that both OMC and MDUW were predicted by the most significant variables, which consisted of FC and LL. This is owed to the fact that these variables play a direct role in the water-holding capacity as well as the extent of compaction done by the soil.

The percentage of gravel and sand affected the prediction of compaction parameters moderately and its significance also depended on the type of soil and gradation. PI also exerted the least effect on prediction of the compaction parameters and this could be explained by the fact that it is interdependent with the LL and does not contribute to more information than the other input variables. The results of the sensitivity analysis influence the soil compaction properties and can be used in the choice of input variables to use in future research and model building.

In short, this study suggests that ANN and machine learning models can be effectively used to predict soil compaction parameters. The ANN and the developed models were very efficient in the training and test sets and were better than the pre-existing empirical and machine learning models. The ANN model had a similar performance with the LightGBM model, which suggests that it has the capacity to represent the non-linear association among soil characteristics and compaction factors. The sensitivity analysis gave information on the relative significance of the input variables that can help in the interpretation of soil compaction behavior and formulation of better prediction models.

4.6 CONCLUSIONS

The study provides a detailed model of how to predict the soil compaction parameters through the state-of-the-art artificial intelligence (AI) methods, which provides an effective alternative to the traditional methods of laboratory tests that require a lot of time. The research employed a large experimental dataset of 78 soil samples (5 natural and 73 artificially prepared) in 24 different soil classifications using the ISSCS which has a very good coverage in a wide range of geotechnical circumstances.

Seven developed machine learning models, including ANNs (which have 5 variants: LMB, GDM, SCG, RBF, and GRNN), RF, GB, XGBoost, LightGBM, LSTM, CNN, and a novel Ensemble model, were carefully tested in predicting OMC and MDD based on five critical input parameters, namely, gravel content, sand content, FC, LL, and PI. The backpropagation approach and radial basis network approach were used to develop these models, and optimal architecture was established by massive hyperparameter tuning.

Ensemble model had better predictive power as it has $R^2 = 0.9808$, RMSE = 0.8808, and MAE = 0.6344 on the test data, which is much higher than the individual models. This is an increase of 3.4% more than XGBoost ($R^2 = 0.9671$) and 5.9% more than GB ($R^2 = 0.9604$). The model can predict the parameters of compaction accurately and explains 98.08% of the test data variance with an excellent ability to generalize.

The sensitivity analysis showed that the most important parameters are FC and LL on both the OMC and MDD predictions with secondary contribution by the gravel and sand content. The PI had the least impact, implying that the properties of particle size distributions are more important to the correct estimation of the compaction parameters than plasticity properties.

This has important practical implications of this research. The AI model that has been developed can save 80%–90% of laboratory testing time while ensuring accuracy in prediction, which can be used in critical infrastructure endeavors. This efficiency saving is especially useful with large-scale highway construction where the properties of soils are very different along the highway alignment, and this efficiency saving can save a lot of time and resources in geotechnical investigations.

The empirical correlations show that compared to the existing empirical correlations, the machine learning method offers a better level of accuracy and increased applicability to a variety of soil types. The Ensemble learning approach is useful in reducing the bias in the individual models, and it also captures complicated non-linear interactions that are not well represented in the other forms of empirical methods.

This chapter confirms that AI algorithms, specifically the Ensemble methods, can provide a quality, quick, and economical solution to the soil compaction parameter prediction. The framework offers geotechnical engineers with an effective tool that balances efficiency of computations and accuracy of predictions, as well as informative decision-making in the pavement design and quality control in the construction quality. The research can be developed into future directions by adding more soil parameters to the framework, applying deep learning architecture in real time in the field, and transfer learning methods to adapt to other geographical areas.

REFERENCES

Al-Khafaji, A. N. (1993). Estimation of soil compaction parameters by means of Atterberg limits. *Quarterly Journal of Engineering Geology and Hydrogeology*, 26(4), 359–368.

Ansari, F., Chatterjee, K., Li, J. Q., Wang, K., & Golalipour, A. (2025). Multi-object pavement surface feature detection with CNN and transformer deep learning architecture. In Hasan Ozer, PhD and Quintin Watkins, P.E. (Eds.), *Airfield and Highway Pavements 2025* (pp. 350–359). Glendale, AZ: American Society of Civil Engineers (ASCE).

Ardakani, A., & Kordnaeij, A. (2019). Soil compaction parameters prediction using GMDH-type neural network and genetic algorithm. *European Journal of Environmental and Civil Engineering*, 23(4), 449–462.

Baughman, D., & Liu, Y. (1995). *Neural Networks in Bioprocessing and Chemical Engineering*. San Diego, CA: Academic Press.. ISBN 0120830302.

Blotz, L. R., Benson, C. H., & Boutwell, G. P. (1998). Estimating optimum water content and maximum dry unit weight for compacted clays. *Journal of Geotechnical and Geoenvironmental Engineering*, 124(9), 907–912.

Broomhead, D. S., & Lowe, D. (1988). Multivariable functional interpolation and adaptive networks. *Complex Systems*, 2(3), 321–355.

Chatterjee, K., Li, J. Q., Ansari, F., Munna, M. R., Parajulee, K., & Schwennesen, J. (2026). Hybrid LSTM-Transformer Models for Profiling Highway–Railway Grade Crossings. *Journal of Transportation Engineering, Part A: Systems, 152*(2), 04025138.

Chen, R. P., Wang, H. L., Hong, P. Y., Cui, Y. J., Qi, S., & Cheng, W. (2019). Effects of degree of compaction and fines content of the subgrade bottom layer on moisture migration in the substructure of high-speed railways. *Proceedings of the Institution of Mechanical Engineers, Part F: Journal of Rail and Rapid Transit*, 232(5), 1197–1210.

Farooq, K., Khalid, U., & Mujtaba, H. (2016). Prediction of compaction characteristics of fine-grained soils using consistency limits. *Arabian Journal for Science and Engineering*, 41(4), 1319–1328.

Fattah, M. A., Ren, F., & Gowayed, Y. (2009). GA, MR, FFNN, PNN and GMM based models for automatic text summarization. *Computer Speech & Language*, 23(1), 126–144.

Gunaydin, O. (2009). Estimation of soil compaction parameters by using statistical analyses and artificial neural networks. *Environmental Geology*, 57(1), 203–215.

Kurnaz, T. F., & Kaya, Y. (2020). The performance comparison of the soft computing methods on the prediction of soil compaction parameters. *Arabian Journal of Geosciences*, 13(7), 1–13.

MATLAB-R2010b Help. (2010). *Neural Network Toolbox*. Natick, MA: The MathWorks, Inc.

Nagaraj, H. B., Reesha, B., Sravan, M. V., & Suresh, M. R. (2015). Correlation of compaction characteristics of natural soils with modified plastic limit. *Transportation Geotechnics*, 2, 65–77.

Poggio, T., & Girosi, F. (1990). Networks for approximation and learning. *Proceedings of the IEEE*, 78(9), 1481–1497.

Proctor, R. R. (1933). Fundamental principles of soil compaction. *Engineering News-Record*, 111, 245–248.

Ren, X. C., Lai, Y. M., Zhang, F. Y., & Hu, K. (2015). Test method for determination of optimum moisture content of soil and maximum dry density. *KSCE Journal of Civil Engineering*, 19(7), 2061–2066.

Samui, P., & Sitharam, T. G. (2011). Machine learning modelling for predicting soil liquefaction susceptibility. *Natural Hazards and Earth System Sciences*, 11(1), 1–9.

Sinha, S. K., & Wang, M. C. (2008). Artificial neural network prediction models for soil compaction and permeability. *Geotechnical and Geological Engineering*, 26(1), 47–64.

Sridharan, A., & Nagaraj, H. B. (2005). Plastic limit and compaction characteristics of fine grained soils. *Proceedings of the Institution of Civil Engineers-Ground Improvement*, 9(1), 17–22.
Tatsuoka, F., & Correia, A. G. (2018). Importance of controlling the degree of saturation in soil compaction linked to soil structure design. *Transportation Geotechnics*, 17, 3–23.
Wang, H. L., & Yin, Z. Y. (2020). High performance prediction of soil compaction parameters using multi expression programming. *Engineering Geology*, 276, 105758.
Zell, A. (1994). *Simulation Neuronaler Netze* (Vol. 1, No. 5.3). Bonn: Addison-Wesley.

5 Application of Data-Driven Techniques for Prediction of Soil Water Characteristic Curve

A State-of-the-Art Review

Mohak Desai and Kaustav Chatterjee

5.1 INTRODUCTION

Soil suction is a very important factor in geotechnical engineering and unsaturated soil mechanics, affecting slope stability, behavior of expansive soils, and soil load bearing capacity. Soil suction is the capacity of soil to hold the water due to capillary and adsorptive forces. Soil suction can be bifurcated into two types, matric suction and osmotic suction. Matric suction is created by capillary forces due to soil's pore structures,

DOI: 10.1201/9781003714125-5

depending on the distribution and size of the pores. Osmotic suction arises due to the availability of dissolved salts in soil water, affecting the movement of water in soil.

Established techniques to evaluate soil suction include tensiometers, pressure plate apparatus, filter paper method, and psychrometers. Tensiometer is one of the commonly used tools which can measure matric suction directly, but it can also be used for measuring low suction (below 100 kPa) range. Pressure plate apparatus can be employed for measuring higher suction values. In this method, soil samples are subjected to controlled air pressure to calculate water retention characteristics. Filter paper technique is another straightforward technique where equilibrium moisture content of filter paper in contact with soil is used to calculate suction. Dew point hygrometers and psychrometers can calculate total suction by evaluating the relative humidity of the soil atmosphere.

These traditional methods to evaluate soil suction are widely used but also have some limitations such as time-consuming experiments, manual data acquisition, and limited accuracy because of sensitivity of the instruments, environment variability, and human errors in some cases. Machine learning (ML) can alleviate these problems by analyzing the large datasets obtained from different sources like remote sensing, sensor network, and weather data. Different ML models were employed by researchers, practitioners, and geotechnical engineers to measure soil suction and to gain insight into the complicated relationship between soil properties and suction behavior.

This study aims to introduce a comprehensive summary of the application of state-of-the art data-driven techniques for measuring soil suction and related properties. The research commenced with selection of keywords (data-driven techniques, soil suction, unsaturated soil mechanics, ML), followed by literature acquisition from different databases such as Science Direct, Google Scholar, and some reputed journals. After literature acquisition, the contents of different literatures were summarized in the body of the paper and in a table. Moreover, this chapter provides a brief insight into some of the popular ML models. Figure 5.1 presents an overview of the methodology used in this research.

The literature collected in this study was analyzed in the three different categories namely: (i) ML algorithm employed for predicting parameters related to unsaturated properties of soil, (ii) number of papers published in different years, and (iii) objectives of the papers. Analyzing the ML algorithms employed by different researchers provided the information about the most efficient ML algorithms for predicting unsaturated properties of soil. Figure 5.2a shows the bar plot of commonly used algorithms and their frequency. Artificial neural network (ANN) was the most popular algorithm followed by random forest (RF) algorithm. The reason for higher application of ANN by researchers was due to its ability to capture the nonlinear relationship in the data. RF was frequently used by researchers because it is an ensemble-based algorithm and can perform nonlinear modeling without overfitting. Figure 5.2b shows the number of papers published in different years. Analyzing the number of papers published during different years revealed that ML has become a popular tool among engineers and researchers in the last five years, with 15, 6, 6, 6, and 6 papers published in 2025, 2024, 2023, 2022, and 2021, respectively. Moreover, analyzing the paper with respect to research objective reveals that majority of the research covers soil water characteristic curve (SWCC) which is considered as the principal factor affecting relationship between soil suction and water content followed by soil water retention curve (SWRC) and universal water characteristic (UWC).

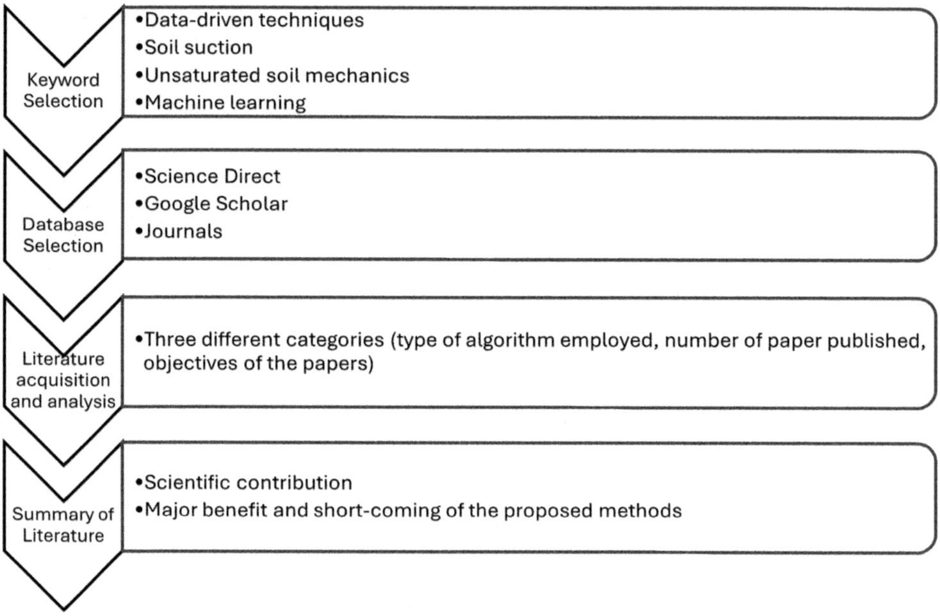

FIGURE 5.1 Methodology of research.

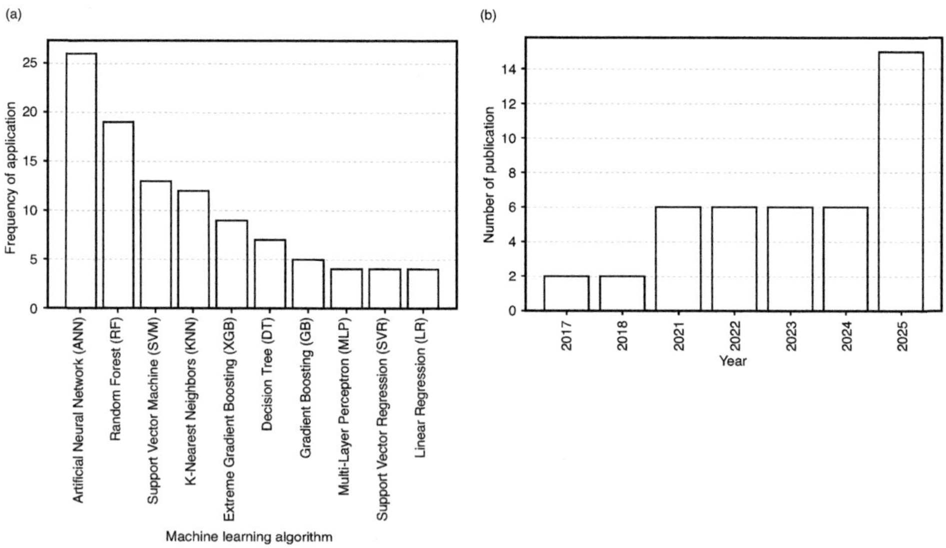

FIGURE 5.2 (a) Frequency of application of ML algorithms. (b) Numbers of papers published each year.

5.2 STATE-OF-THE-ART ML ALGORITHMS

5.2.1 Decision Tree

Decision tree (DT) is the Boolean-based data classification and regression technique, performs automatic decisions by starting at the root (top level) and moving down through a series of Boolean-based questions until it finally reaches a final answer at the leaf. This tree-based model automatically takes the decision of which and how many questions to be asked and where to split the data to reach the final leaf based on "YES/NO" condition. Figure 5.3 illustrates the schematic representation of DT algorithm.

5.2.2 Support Vector Machine

Support Vector Machine (SVM) is a supervised ML algorithm utilized for both data classification and regression. It works by defining the optimal hyperplane that maximally splits data points of different classes, creating the maximum possible gap between two classes of data to minimize the classification errors. Figure 5.4 shows a schematic representation of the SVM model. Sometimes, it is not possible to split data linearly, in those cases kernel transformation techniques can be used to separate the nonlinear data into higher dimensions.

5.2.3 Random Forest

RF (Chatterjee et al., 2024; Parajulee et al., 2025; Rana et al., 2025) model is a group of multiple DTs to generate the best accuracy and avoid overfitting. It provides the output by combining the results from numbers of DTs and by considering the output which is common from majority of the DTs.

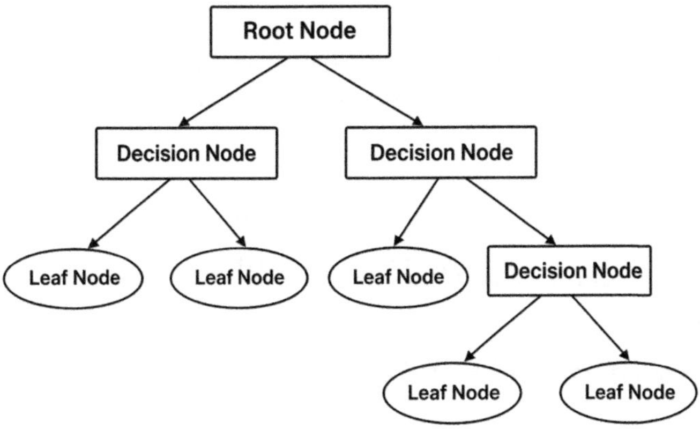

FIGURE 5.3 Schematic representation of DT.

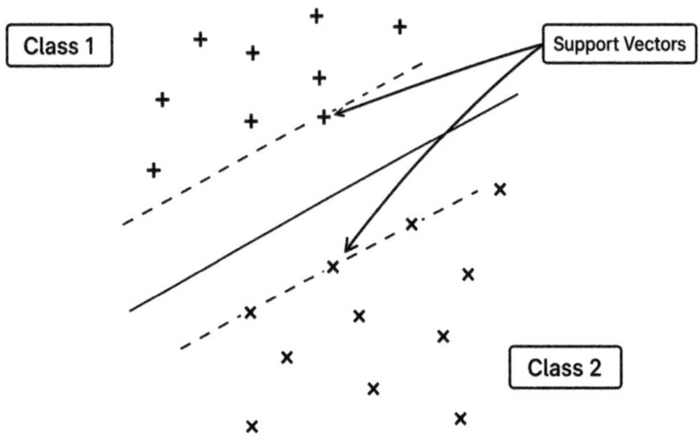

FIGURE 5.4 Schematic representation of SVM.

5.2.4 K-Nearest Neighbors

K-nearest neighbor (KNN) is the type of ML model which separates objects based on its nearest neighbor's class in the dataset. The KNN model assumes that objects near each other are alike. In this method, first the value of K is chosen, which is the number of neighbors. Then the K number of nearest points to the new point is chosen. The label or value of a new point is determined based on the most common value among the neighbors.

5.2.5 Gradient Boosting

Gradient Boosting (GB; Chatterjee et al., 2024; Parajulee et al., 2025; Rana et al., 2025), just like AdaBoost method, is an ensemble technique that inflates estimation accuracy by combining multiple weak learners. In this model, each new model tries to correct the mistakes created by the previous model. GB creates models by reducing a loss function using gradient descent, offering greater flexibility and accuracy for complex, nonlinear problems. The overall prediction is obtained by adding all appropriately weighted individual models. GB is usually opted in geotechnical engineering because of its capability to capture relationships between soil properties, especially when predicting soil suction or SWCC parameters.

5.3 APPLICATION OF ML

5.3.1 Estimation of SWCC

Johari et al. (2006) conducted a study in which SWCC was predicted using Genetic Programming (GP) describing how soil suction varies different water content in an unsaturated soil. The author used soil parameters like initial water content, void ratio, clay, and silt content and GP automatically gave mathematical relationships without hypothesizing any predefined functional form. The GP model obtained high prediction accuracy (R^2 = 0.94 for training dataset and 0.93 for validation) which outperformed traditional empirical methods like those of Fredlund et al. (1997). By removing the need for high effortful laboratory tests, the given ML model provides a quick, cost-effective, and reliable option to estimate SWCC. The main advantage of this method is that it achieved high accuracy ($R^{2=}$0.93) and it works well on unseen data which shows its good adaptability for various soil types. The only limitations of this model are its complexity because of its lengthy expressions and model accuracy depending on the quality of the training dataset.

Nikhil et al. (2016) conducted a study on ANN-based approach for prediction of SWCC parameters for Korean weathered soil. For this study the authors trained an ANN by using soil properties like void ratio, in-situ water content, sand percentage, and fines percentage. These parameters were linked with the fitting variables of the Fredlund and Xing model that describes the nonlinear interaction between volumetric water content and matric suction. By combining ML and soil index properties, suction behavior of unsaturated soil can be determined. For this study, the ANN model was developed with two hidden layers with eight and four neurons, which achieved the highest accuracy of R^2 = 0.99 on training dataset and R^2 = 0.89 on testing dataset. The outcome shows the effectiveness of ANN model for capturing the nonlinear dependence of suction on soil structure and moisture conditions. Additionally, the inclusion of in-situ characteristics significantly enhanced model realism which made the predictions more applicable to natural slopes compared to historical models that relied only on laboratory-prepared samples. Limitations of these models are the complex black box type nature of ANN model, data dependency, and limited parameter range, whereas advantages of the model are high accuracy, ability to create nonlinear relationships between soil properties and suction, and ability to integrate with analytical models.

Zainal and Fadhil (2018) developed an ANN-based model to estimate the SWCC (relationship between matric suction and water content in unsaturated soils) using soil parameters such as air entry point and the residual degree of saturation. The ANN model was trained on eight experimental SWCC datasets and tested against two datasets. The outcome showed that ANNs can successfully capture the nonlinear behavior of SWCC with high accuracy (R^2 = 0.97). This method removes the need to estimate fitting parameters from traditional methods such as Brooks–Corey, van Genuchten (vG), or Fredlund–Xing, instead providing a direct method for curve generation. Limitations of this model are requirement of soil-specific parameters, overfitting risk, and data dependency on quality and quantity of training datasets. The advantages of the ANN include high predictive accuracy, direct prediction of SWCC without need of fitting parameters, and its ability to estimate the nonlinear relationship between suction and water content.

Ramos-Rivera et al. (2021) conducted a study to estimate SWCC based on soil type and best fitting regression. The author used KNN ML algorithm to estimate SWCC parameters by employing basic soil index properties and soil type information. This model enables fast computation of SWCC which is very important for measuring soil suction in an unsaturated soil. The KNN achieved the accuracy of R^2 = 0.82–0.88 which is better than traditional methods. The main advantage of this model is that it can represent SWCC of soil with dual porosity. This is very difficult to measure with traditional techniques. Another advantage of this model is that it reduces tedious laboratory tests, and it is completely data driven and flexible. It captures nonlinear relationships between soil properties and SWCC parameters. Limitation of this method is its moderate accuracy compared to ensemble methods like RF and its data dependency on the quality and representativeness of the training dataset. Additionally, KNN is a slow learner method. So, prediction time may increase in case of a large dataset.

Yang et al. (2021) conducted a study on estimation of SWCC using different ML algorithms. The authors studied how soil suction can be estimated by using ML in a faster and accurate manner as measuring suction and water content, which leads to calculating SWCC, is a time-consuming and costly process. The soil parameters considered in this study are particle size, void ratio, and density (whatever), and seven different ML models are used. RF model was found to be the most effective model ($R^2 = 0.35$), and the R^2 of the other models are 0.21 for SVM, 0.22 for KNN, and lowest –10 for DT. This approach can help the researcher understand the reliability of the output, which will allow better understanding of soil behavior and will eliminate the need for expensive and time-consuming soil testing while improving slope stability and design of the foundations. The major advantage of using ML to determine SWCC is that it reduces time and expense of laboratory work.

Li et al. (2022a) performed a study on shallow depth soil database to predict unsaturated soil properties (especially SWCC and soil suction) employing two ML algorithms RF and ANN. This study concluded that RF model outperformed ANN model in terms of prediction accuracy and computational efficiency. The main advantage of this model is that it enables large-scale seepage and stability analyses without requirement of costly laboratory tests, which shows practical usage of ML in geotechnical engineering and unsaturated soil modeling. Limitations of this model are its less interpretability and longer training time.

Javid (2023) performed a study on variation of soil suction and applications of remote sensing and statistical techniques in estimating the behavior of unsaturated soil. The main purpose of this study was to create a data-driven model to evaluate soil suction and water diffusivity which plays an important role in moisture movement, swelling, and shrinkage of subgrade soils. To discuss the complicated relation between soil and climate, this study used ML techniques (clustering, nonlinear least squares regression, and ridge regression) to estimate vG diffusivity parameters and equilibrium soil suction. The fusion of ML and remote sensing gives a powerful framework to estimate unsaturated soil behavior. The main advantage of this ML model is that it is less expensive and less time-consuming compared to traditional laboratory methods. Another benefit of this model is that in addition to soil properties using climatic parameters like humidity, rainfall, and temperature, the model can provide a realistic behavior of soil suction and provides high accuracy ($R^2 = 0.80$ with Ridge Regression model). The limitations of this model are that it is highly dependent on the quality of input data. Also, the created models are site specific. So, it may need modification for the soil of other regions and different types of soil.

Pham et al. (2023) presented an ML-based method for predicting SWCC by using 2748 data points across 11 soil textures. This study compared seven ML based on pedo-transfer functions (ML-PTFs). Additionally, this study elaborated practical applications of Extreme GB (XGB)-PTF with Bayesian optimization to calculate field matric suction. The researchers compared seven ML models which are KNN, SVM, DT,

Neural network, RF, GB, and XGB. Of these models, XGB showed the best accuracy of $R^2 = 0.972$, RMSE = 0.028. The advantages include less overfitting; strong generalization of ensemble methods like RF, GB, XGB; and high accuracy of predicting SWCC and less laboratory work, whereas limitations include low accuracy of single models like SVM, NN, and KNN; less interpretability because of black box type nature of the model; and model's heavy dependency on dataset size and diversity.

Nazem et al. (2023) conducted a study on prediction of SWCC that defines the relationship between soil suction and water content by using optimized ML approaches. ML approaches used in this study include Multilayer Perceptron ANN (MLPANN), XGB, RF, and Support Vector Regression (SVR). Input parameters were initial void ratio, initial gravimetric water content, clay content, and silt content. Among all ML models, XGB achieved the highest accuracy of $R^2 = 0.97$ for training dataset and $R^2 = 0.90$ for testing dataset which was proven to be the biggest advantage. The limitation of this study is that it was conducted only on a small set of samples.

Bakhshi et al. (2023) performed a study to evaluate SWCC with different ML algorithms including GB, AdaBoost, DT, RF, ANN, SVM, KNN, and linear regression. The different input parameters for this study are organic carbon, bulk density, and micro-porosity derived from image analysis, and the outputs from the model are estimated SWCC and water content and different matric suction. RF model achieved the highest value for coefficient of determination R^2 nearly equal to 0.91 with minimal RMSE and MAE. KNN model showed moderate accuracy with $R^2 \approx 0.80$, whereas other ML models like SVM and DT performed reasonably but with lower value of R^2. This study proves that ensemble methods like RF have an ability to exceed the results compared to single models when soil suction parameters are calculated. The main limitations are its heavy dependency on the quality and size of the training dataset and its possibilities of overfitting training data.

dos Santos Pereira (2024) predicted SWCC of tropical bimodal soil using a GB model. Tropical soil offers unique challenges because of its structure that influences soil suction. To eliminate this difficulty, the author presented an ML approach by using a GB model to estimate the degree of saturation at any suction value. Input parameters for this model were soil index properties like particle size distribution of sand, silt and clay, liquid limit, and plasticity index. This model achieved high accuracy of $R^2 = 0.95$ for training dataset and $R^2 = 0.90$ for testing dataset, which outperforms traditional methods and ANN especially while predicting air entry value. It was concluded that GB can capture a nonlinear relationship between soil suction and water retention. The primary benefit of this model is its capability to provide a nonlinear relationship between soil suction and water retention with high accuracy. Limitations of this study are its less accuracy at high degree of saturation and database constraints. One important point to be noted about this study is that training dataset was limited to tropical bimodal soils because of which model generalization to other soil samples can be limited.

Yang et al. (2024) estimated SWCC employing ML model considering multiple soil and environmental aspects such as soil particle size, density, plasticity index, temperature, void ratio, salinity, and stress history. The Bayesian Regularization Neural Network (BRNN) model was trained with more than 1000 experimental datasets and can predict the relationship between soil suction and water content under complex conditions, reducing the requirement for experimental studies. The scientific contribution of this study lies in the fact that this study considered multiple factors like temperature, salinity, and deformation for developing prediction models.

Sharma et al. (2025) predicted SWCC of an unsaturated soil employing ML including RF and ANN. The study concluded that the RF model showed the most reliable prediction with minimum error followed by the ANN model. The main advantage of this study is that tested ML models offer a cost-effective and fast alternative to predict SWCC. Limitations of this study include data dependency, risk of overfitting, the model predictions may not generalize well for soil from other regions, and lack of validation using field data.

Lamichhane et al. (2025) predicted soil moisture by employing remote sensing and ML. ML models possess strong capacity to capture a nonlinear relationship between RS-derived features and soil moisture dynamics. This study highlighted the most important features for soil suction prediction like NDVI, SAR backscatter, elevation, and soil texture. In general, ML models like RF, SVM, ANN are proven to be robust tools for soil suction measurement, which are scalable and better data-driven alternatives compared to conventional models. RF model is reported to have the highest accuracy of R^2=0.95. The primary benefit of this technique is that it covers a wide range of soil types and provides performance metrics across multiple models, whereas the major limitation of this study includes limited focus on physical modeling of suction and transferability issues for different environmental conditions.

He et al. (2025) conducted a study on indirect models for SWCC parameters to minimize prediction uncertainty with ML. As direct measurement of SWCC parameters is expensive and time consuming, indirect probabilistic models were used by using easily measurable properties of soil such as particle size distribution and porosity. The author used ANNs, predictor-dependent error scaling, and enriched PSD features to estimate SWCC. The negative log-likelihood (NLL) value decreased from ~1.36 and 1.33 (joint model) to 1.18 and 1.20 on the test set which shows that the predictions are reliable. The advantage of this model is that it reduces testing efforts and improved accuracy, which ensures reliability and generalizability of the models. Bayesian inference assumes specific statistical distributions, which may not always represent actual soil behavior which is the major limitation of this method. In addition, the model relies on the training dataset.

Wang et al. (2025) developed a framework to estimate matric suction in compacted fine-grained soils employing ML algorithms such as multi-gene genetic programming (MGGP) and particle swarm optimized SVR (PSO-SVR). A parameter called effective degree of aggregation was created using ML to take soil structure variations done by compaction and initial water content. It was concluded that proposed ML-assisted framework can simulate the nonlinear relationship between soil properties, compaction characteristics, and suction behavior. PSO-SVR achieved the highest accuracy of $R^2 = 0.95$ with RMSE nearly equal to 0.15. The advantages of this method are its high predictive accuracy, reduction of laboratory efforts, and creation of degree of aggregation, whereas limitations include data quality dependency of the model, computationally expensive PSO-SVR, and the model being range specific (0–1500 kPa). Projection beyond this limit might be unreliable.

5.3.2 Projection of SWRC

Jain et al. (2004) carried out research on soil water retention data employing ANN. The study elaborates how ANN predicts water holding capacity of soil, an important factor affecting soil suction. The RMSE for this model was very low (0.002–0.006 for different soils) which demonstrates accuracy of the model. The advantages of this model are its high accuracy, ability to capture nonlinear relationships between suction and water content, and ability to reproduce drying, wetting, and scanning curves. The limitations of this model are its sufficient and well-distributed data requirement, and its incapability to generalize well to soils apart from training dataset.

Lamorski et al. (2017) conducted a study to estimate the main wetting branch of SWRC (that shows the relationship between soil suction and volumetric water content) derived from its main drying branch by employing ANN model. The ANN models were trained on soil hydraulic property datasets to estimate SWRC parameters. The accuracy of the ML model is measured in terms of RMSE which was noted as 0.021 m^3/m^3 for volumetric water content predictions. The volumetric water content ranges from 0 to nearly 0.5 m^3/m^3 for most soils and RMSE of 0.021 represents an error of nearly 4%–5% of the total range. So, in terms of percentage, the accuracy of ANN model can be estimated as 95%–96%, which shows its major advantage of high predictive accuracy. Some advantages of this model include its low cost and its ability to capture the nonlinear relationship between soil properties and hydraulic behavior, including hysteresis effects. Model's limitations include its dependency on training dataset, less interpretability, and its need for retraining when applying to new regions.

Albuquerque et al. (2022) undertook a study to predict SWRC based on physical characterization parameters by use of ML. The author used a dataset with 794 water retention and suction points from 51 soils and the study used parameters like sand, silt, clay, gravel content, porosity, and plasticity index as predictors. Various ML models like RF, GB, SVM, ANN were used to predict the accuracy, but the extremely randomized trees (Extra Trees) model achieved the highest accuracy (training R^2 = 0.99 and testing R^2 = 0.90), which successfully predicted volumetric water content as a function of suction. In addition, the major advantage of this model is that it estimates SWRC points that can be fitted to continuous analytical models like vG or Costa & Cavalcante for use in numerical simulations. Another advantage is that it reduced dependency on expensive and time-taking laboratory tests for suction measurement. The limitations of this model are data dependency on the size and quality of training dataset, model complexity, and risk of overfitting.

dos Santos Pereira et al. (2024) executed research to estimate SWRC with the use of datasets from 794 soil samples with basic physical parameters like sand, silt, clay, gravel content, porosity, and plasticity index. The authors used multiple ML models like RF, DTs, Extremely Randomized Trees, SVM, KNNs, Logistic Regression, and MLP. By linking ML outputs with functions like vG, Cavalcante & Zornberg, and Costa & Cavalcante, it is clear that ML models can estimate soil suction behavior without time-consuming laboratory tests. The study showed that ML models achieved high accuracy of R^2 = 0.84–0.99, with ensemble models and ANN. Additionally, fusion of ML models and analytical SWRC functions enables robust modeling of water retention. The major advantages of this study are high accuracy of ML models, model's ability to capture complicated relation between soil properties and water retention, and model's use in geotechnical, agricultural, and hydrological modeling, whereas limitations are dependent on size and diversity of training dataset, less interpretability, and model's need for continuous update of data.

Abdallah (2022) performed a study and compared two ML approaches, which are pointwise and continuous, to predict SWRC by using ANN and by using physical soil parameters like porosity and clay/silt/sand fractions. The pointwise method estimates water content at a specific suction value, whereas the continuous method predicts the vG equation parameters to form the curve. It was concluded that the pointwise model achieved better performance (RMSE = 0.027) compared to the continuous model (RMSE = 0.326). But the continuous model gives consistent curve shape as well as better theoretical compliance. Some of the advantages of this study are high accuracy of pointwise prediction, curve consistency in continuous prediction, and global applicability. The limitations are that the performance is limited on clayey soil as well as no hybrid approach was experimented that combines both approaches.

Cisty et al. (2021) conducted a study to evaluate water retention curves by employing regression and ML algorithms. For this, two ML methodologies were used, which are MLR and SVM. The authors created PTFs based on easily measurable properties of soil like particle size distribution and dry bulk density. The authors concluded that while MLR gives a straightforward modeling approach, SVM outperforms it in terms of predictive accuracy. The use of SVM with kernel-based learning and structural risk minimization allowed for better generalization to new data. Predictive accuracy of the SVM model was reported to be 0.878–0.925 for correlation coefficients and 0.831–0.894 for MLR. The study concluded the potential of ML techniques to enhance hydrological modeling by giving efficient and accurate alternatives. The advantages of these models are high predictive accuracy, nonlinear modeling capacity, and robust generalization. Limitations include complexity of SVL and MLR models and its computational cost.

5.3.3 Estimation of Hydraulic Conductivity

Bashar et al. (2025) conducted a study by utilizing SWCC parameters like saturated volumetric water content, residual volumetric water content, air entry suction, and residual suction in ANN model to predict the unsaturated hydraulic conductivity, which is closely related to soil suction. By integrating a domain-specific filter in ANN, this model can avoid physically unrealistic outputs like negative conductivity to ensure that the prediction remains meaningful for soil suction. The ANN model achieved high accuracy (R^2=0.9947 for training dataset and R^2=0.9349 for testing dataset) which shows its ability to accurately measure the correlation between soil suction and hydraulic conductivity. The study showed that prediction accuracy varies with the soil texture. Loamy sand and silt loam soils give consistent predictions, whereas sandy soils give high errors because of its unique pore structure and variability in suction–conductivity relationships. The main advantages of this model are its high predictive accuracy and its computational efficiency (balance between accuracy and cost), whereas the main limitation is its variable accuracy across different soil textures.

5.3.4 Stress Prediction

Johari et al. (2013) performed a study to predict effective stress parameters of unsaturated soil using gene expression programming (GEP). Effective stress parameters are crucial for predicting shear strength which ultimately influences soil suction. The studied model incorporated six parameters that are soil suction, net confining pressure, bubbling pressure, SWCC fitting parameter, residual volumetric water content, and saturated volumetric water content, out of which suction was identified to be the most influential parameter. GEP achieved the moderate accuracy of R^2=0.83. The key benefit of this model is that GEP does not demand any prior assumption. The limitation of this study is that no comparison is made with deep learning models or ensemble models.

Singh et al. (2023) engaged in a study to evaluate soil moisture by using traditional methods, automated sensors, remote sensing, and ML techniques. The study focused on the rising role of ML in estimating soil suction and soil water retention. Conventional methods like tensiometers, pressure plate extractors, and field sensors require high labor and are expensive. On the other hand, ML techniques use soil properties, climate data, and remotely sensed information to form a nonlinear relationship between soil suction and soil moisture. Different ML models like RF, SVMs, ANNs, GB, and KNNs were used to estimate soil suction. Out of all these models, RF and GB achieved the highest precision of R^2 nearly equal to 0.95 for soil suction and water retention estimation. The advantage of this study is that it identifies ML models with highest accuracy, cost and time effectiveness, and ML can be combined with remote sensing. Limitations include size and quality of training dataset, black box type nature and less interpretability of models like ANN, and requirement of validation to ensure reliability.

Fazel Mojtahedi et al. (2024) conducted a study to predict stress-dependent soil water retention by employing ML. The study focuses on how soil holds the water under various stress conditions. A dataset with more than 100 experiments on different soils was used to train an ML model to estimate how soil suction changes with soil type, net stress, and other properties. It was observed that stress has equal impact on soil suction as soil type. There were several ML models like RF, GB, SVM, ANN used to anticipate stress-dependent SWRC out of which RF showed the highest accuracy ($R^2 = 0.93–0.95$ for both training and testing datasets) with very low RMSE and MAE. The primary advantage of this ML model includes high prediction accuracy with less laboratory testing efforts and these ML models can combine confining pressure effects into SWRC prediction. Limitations of this model are risk of overfitting, data reliance on the size and variety of the training data, and the model's tough interpretability.

Zhang and Vanapalli (2025) carried out a study on estimation of swelling index and swelling pressure of expansive soil with a focus on the purpose of soil suction in unsaturated conditions by using various ML algorithms like MLP, SVM, and Extreme Learning Machine (ELM). As traditional oedometer tests have limitations, the authors carried out an AI-driven model that uses soil properties like plasticity index and initial suction to estimate swelling behavior. All three ML models were trained and tested by using datasets of expansive soil. ELM model achieved the highest accuracy for swelling index, swelling pressure, and consolidation swell prediction. The advantages of this model are its easy-to-use nature, time-reduced tests, and applicability to a wide range of soils, whereas limitations include overfitting risk and accuracy dependence on soil type.

Jagan et al. (2025) performed a study on ML models for estimating effective stress parameters for unsaturated soil. This study proposed ML techniques to estimate effective stress parameters based on soil and water parameters with primary focus on soil suction. GP and Multivariate Adaptive Regression Spline (MARS) were trained on a dataset containing six input variables that are net confining pressure, suction, bubbling pressure, saturated and residual volumetric water content, and a fitting parameter. GP achieved the highest accuracy (R^2=0.97) on testing dataset. The major advantage of this study is that it provides a direct equation to estimate effective stress parameters with high accuracy. The limitations of this study are small dataset model size (121 samples) and no external validation on independent datasets beyond the training datasets.

5.3.5 Slope Stability Analysis

Li et al. (2025b) developed an RF model to estimate shallow slope stability incorporating spatiotemporal variations in unsaturated soil moisture. Soil suction plays a very important role in slope stability in shallow depths. RF models were prepared by using volumetric water content data for maximum daily rainfall scenarios and estimated slope stability under rainfall conditions. This study highlights efficiency and reliability of RF models in estimating nonlinear relationship between soil moisture and slope stability. RF model achieved high predictive accuracy of MAE < 0.15 and RMSE < 0.20. The advantages of this model include computational efficiency and accuracy compared to physical models and scalability for regional applications. Limitations include dependency on stimulated training data, assumptions of soil properties, and being a zone-specific model.

5.4 SUMMARY

This study underlines the revolutionary impact of ML in measuring soil suction and SWCC parameters which are highly important to analyze the behavior of an unsaturated soil in geotechnical engineering. Traditional methods to measure soil suction are labor intensive, costly as well as less accurate because of environmental factors and instrument sensitivity. ML algorithms provide data-driven and scalable options to measure nonlinear relationship between soil properties and suction.

Table 5.1 summarizes the contribution of different researchers on the application of ML and data-driven techniques for determining soil suction. Several ML algorithms were reviewed including DTs, SVMs, RFs, GB, ANNs, KNN, GP, and hybrid approaches. Out of all these algorithms, ensemble methods like GB and RF outperformed every time with highest predictive accuracy as well as robustness. On the other hand, neural networks offered strong performance in modeling hysteresis and soil water interactions.

TABLE 5.1 Summary of application of machine learning for predicting unsaturated properties of soil

AUTHORS	*SCIENTIFIC CONTRIBUTION, ADVANTAGES, AND LIMITATIONS*	*ML ALGORITHM USED*
Alibrahim et al. (2025)	Contribution: Prediction of unsaturated hydraulic conductivity Advantage: Removal of unrealistic output Limitation: Variable accuracy for different soils	ANN
He et al. (2025)	Contribution: Reduction of uncertainty in SWCC Advantage: Improved accuracy Limitation: Risk of overfitting	ANN
Almuaythir et al. (2025)	Contribution: AI-driven soil suction estimation Advantage: High accuracy for maximum dry density (MDD) and optimum moisture content (OMC) prediction Limitation: Less interpretability	XGB, RF, SVR, LSTMN, KNN
Li et al. (2025b)	Contribution: Hybrid framework to integrate slope stability and ML Advantage: High computational efficiency Limitation: Zone-specific model	RF
Cisty and Povazanova (2021)	Contribution: PTFs development to estimate drying branch of water retention curve Advantage: High accuracy Limitation: Less interpretability of ML models	SVM, MLR
Lamichhane et al. (2025)	Contribution: Identification of the most influential features Advantage: Provides performance metrics for ML models Limitation: Less discussion on soil suction physics	RF, SVR, ANN, XGB, CNN, LSTM
Wang and Vanapalli (2025)	Contribution: Capture nonlinear relationships between soil structure and compaction characteristics Advantage: High accuracy Limitation: Validation requirement for general applicability	PSO-SVR, MGGP
Jagan et al. (2025)	Contribution: GP and MARS models comparison and validation Advantage: High accuracy and low errors Limitation: Computationally intensive	GP, MARS

(*Continued*)

TABLE 5.1 (*Continued*) Summary of application of machine learning for predicting unsaturated properties of soil

AUTHORS	*SCIENTIFIC CONTRIBUTION, ADVANTAGES, AND LIMITATIONS*	*ML ALGORITHM USED*
Zhang and Vanapalli (2025)	Contribution: Prediction of soil suction leveraging ML Advantage: Applicable to a wide range of soil types Limitation: Difficult and time-consuming models	Multilayer Perceptron, SVM, ELM
Liu et al. (2025)	Contribution: Proposal of a fusion feature matrix Advantage: Centimeter-level resolution Limitation: Retraining of model for different soil types	SVM, DNN, RT, Gaussian Regression
Li et al. (2025a)	Contribution: Proposed ML framework for UWC prediction Advantage: Improved accuracy and simplified framework Limitation: Generalization risk	ANN, SVM, RF, XGB
dos Santos Pereira et al. (2025)	Contribution: ANN modeling strategies for SWCC prediction Advantages: High accuracy and reduced experimental effort Limitation: Black box type nature	MLB, RBFN, ELM
Showkat et al. (2025)	Contribution: ML-based SWCC prediction Advantage: Time and cost-effective Limitation: Data dependency on training dataset	RF, XGB, MEP
dos Santos Pereira et al. (2024)	Contribution: Development of SWCC for tropical soil Advantage: High accuracy and indirect prediction of SWCC Limitation: Need wide range of validation	GB, ANN
Yang et al. (2024)	Contribution: Multifactorial prediction of water content Advantage: High accuracy with automated assessment Limitation: Low interpretability of models	BRNN
Fazel Mojtahedi et al. (2024)	Contribution: Soil suction modeling using ML Advantage: Reduced laboratory testing efforts Limitation: Computationally intensive	MLP-NN, GMDH-NN
Abdallah (2022)	Contribution: Comparative evaluation of ML approaches Advantage: SWRC shape consistency in continuous prediction Limitation: No hybrid models were tested	Pointwise ANN Continuous ANN

(*Continued*)

TABLE 5.1 (*Continued*) Summary of application of machine learning for predicting unsaturated properties of soil

AUTHORS	*SCIENTIFIC CONTRIBUTION, ADVANTAGES, AND LIMITATIONS*	*ML ALGORITHM USED*
Sharma et al. (2025)	Contribution: Developed ML framework to estimate SWCC Advantage: Robustness across various models Limitation: No real-world dataset validation	MLR, SVR, DTR, RFR, ANN
Yang et al. (2024)	Contribution: Physics-informed method to estimate SWCC Advantage: Good performance with limited data Limitation: Computational complexity	PINNs
Li et al. (2024)	Contribution: Development of framework for predicting UWC Advantage: No need for predefined equations Limitation: Data dependency on training data	RF, XGB, KNN, SVR, BPNN
He et al. (2025)	Contribution: Probabilistic modeling of SWCC parameters Advantage: Rigorous validation to avoid overfitting Limitation: Data dependency on training data	Bayesian models, ANN
Huang and Wang (2024)	Contribution: Developed BPNN-based model Advantage: Improved accuracy Limitation: Potential overfitting and complex setup	BPNN
Bakhshi et al. (2023)	Contribution: Integration of image analysis with ML Advantage: Easy to measure soil suction Limitation: Dependency on quality and size of training dataset	GB, DT, RF, ANN, SVM, KNN, LR
Javid (2023)	Contribution: Soil suction and diffusivity estimation Advantage: Cost and time reduction for diffusivity estimation Limitation: Tested model is site specific	NLSR, Ridge regression
dos Santos Pereira et al. (2023)	Contribution: Applying ML models to estimate SWRC Advantage: High accuracy, time and cost-effective Limitation: Lack of interpretability	RF, DT, ERT, SVM, KNN
Pham et al. (2023)	Contribution: Development of ML-based PTF to predict SWCC Advantage: Strong generalization with low overfitting Limitation: Single models like SVM had low accuracy	KNN, SVM, DT, NN, RF, GB, XGB

(*Continued*)

TABLE 5.1 (*Continued*) Summary of application of machine learning for predicting unsaturated properties of soil

AUTHORS	*SCIENTIFIC CONTRIBUTION, ADVANTAGES, AND LIMITATIONS*	*ML ALGORITHM USED*
Singh et al. (2023)	Contribution: Application of ML to predict soil suction Advantage: High accuracy, model is adaptive to different soil Limitation: Models require high computational resources	RF, SVM, ANN, GB, KNN, DT
Nazem et al. (2023)	Contribution: Demonstration of use of ML in modeling SWCC Advantage: High predictive accuracy Limitation: Limited size of the dataset	PSO-XGB PSO-RF PSO-SVR
Qin and Fan (2023)	Contribution: Development of an improved SVM model Advantage: High prediction accuracy Limitation: Complexity in model setup	SVM, SVM-PSO
Albuquerque et al. (2022)	Contribution: Development of ML framework to predict SWRC Advantage: High accuracy of decision tree model Limitation: Limited data and risk of overfitting	MLP, SVM, KNN, DT, RF, ERT
Li et al. (2022b)	Contribution: Integration of ML for predicting soil properties Advantage: High accuracy with limited data Limitation: Long computational time	RF, ANN
Gupta et al. (2022)	Contribution: Global-scale mapping of SWCC using ML Advantage: Improved representativeness and robustness Limitation: Reliance on predicted soil properties	RF
Li et al. (2022a)	Contribution: Use of RF and ANN to predict soil properties Advantage: Use of log transformation for better accuracy Limitation: Models are complex to train and interpret	RF, ANN
Onyelowe et al. (2022)	Contribution: Integration of ML with SWRC prediction Advantage: Applicable to a wide range of soils Limitation: Complexity in measurement	SVM, ANN, KNN, RF, XGB
Yang et al. (2021)	Contribution: Soil suction data-driven prediction Advantage: Time and cost-effective Limitation: Low predictive accuracy	DT, SVM, KNN, GB, RF

(*Continued*)

TABLE 5.1 (*Continued*) Summary of application of machine learning for predicting unsaturated properties of soil

AUTHORS	*SCIENTIFIC CONTRIBUTION, ADVANTAGES, AND LIMITATIONS*	*ML ALGORITHM USED*
Ramos-Rivera et al. (2021)	Contribution: Application of KNN for SWCC prediction Advantage: Laboratory tests reduction Limitation: Prediction time increases with large datasets	KNN
Nobahar and Khan (2021)	Contribution: Development of model for soil matric suction Advantage: High prediction accuracy Limitation: ANN models lack transparency	ANN
Raghuram et al. (2021)	Contribution: ML-Like approach for pattern recognition Advantage: Reduction in overestimation Limitation: Complexity in implementation	Nelder-Mead Simplex Algorithm
Cheng et al. (2021)	Contribution: Used MGGP for suction response to rainfall Advantage: High reliability and applicability Limitation: Limited spatial and temporal scope	MGGP
Pham et al. (2019)	Contribution: Neural network-based PTFs to predict SWCC Advantage: Robustness across soil types Limitation: Complexity in network design	Feedforward Neural Networks
Zainal and Fadhil (2018)	Contribution: Developed ANN-based model to estimate SWCC Advantage: Decent predictive performance for multiple soils Limitation: Risk of overfitting	ANN
Saha et al. (2018)	Contribution: Developed ANN-based SWCC estimation model Advantage: Separate models for plastic and non-plastic soils Limitation: Small training and validation sets	ANN
Lamorski et al. (2017)	Contribution: Estimation of main wetting branch of the SWRC Advantage: Practicable in large-scale applications Limitation: Requires training when applied to new regions	ANN
Saha et al. (2017)	Contribution: Developed ANN-based models to predict SWCC Advantage: High accuracy with less experimental efforts Limitation: The model can't adapt new data unless retrained	ANN

(*Continued*)

TABLE 5.1 (*Continued*) Summary of application of machine learning for predicting unsaturated properties of soil

AUTHORS	*SCIENTIFIC CONTRIBUTION, ADVANTAGES, AND LIMITATIONS*	*ML ALGORITHM USED*
Nikhil et al. (2016)	Contribution: Developed ANN-based model to estimate SWCC Advantage: Integration of ML with analytical models Limitation: Accuracy may decrease without retraining	ANN
Johari et al. (2013)	Contribution: Use of GEP to model unsaturated soil behavior Advantage: Provides interpretable expressions for practical use Limitation: Comparison with deep learning models is not done	Gene Expression Programming
Johari et al. (2011b)	Contribution: Genetic algorithm-based neural network to model the mechanical behavior of unsaturated soils Advantage: Improved prediction accuracy Limitation: Data dependency on training data	ANN, GABNN
Johari et al. (2011a)	Contribution: Development and comparison of Gradient Boosting Neural Network (GBNN) and GP Advantage: Time and cost-effective Limitation: Model complexity and less interpretability	GBNN, GP
Erzin (2007)	Contribution: Developed a predictive model for total soil Advantage: Reduces the need for time-consuming tests Limitation: Black box nature, lack of interpretability	ANN
Johari et al. (2006)	Contribution: Development of GP model to estimate SWCC Advantage: High accuracy Limitation: Complex model, dependency on training dataset	Genetic Programming
Jain et al. (2004)	Contribution: ANN-based prediction of SWRC Advantage: High accuracy with low RMSE Limitation: Model is less transparent and less interpretable	ANN

Additionally, research on remote sensing, image analysis, and climate data enhanced the estimation capacity of ML frameworks for soil suction. All these advancements improve slope stability analysis, foundation design as well as soil moisture management and reduces dependency on laboratory testing.

In contrast, ML models also had some constraints. The efficiency of ML models depends on quality, size, and representativeness of training datasets. Some ML models are very difficult to interpret due to its complex nature. However, by using ML algorithms, the way of studying soil suction and SWCC can be revolutionized. Engineers and researchers have a better way to study soil behavior quickly and efficiently.

5.5 FUTURE DIRECTIONS

The findings of the literature review suggest that there is a wide range of application of ML algorithms for prediction of SWCC, SRCC, and different soil water parameters. Moreover, some of the studies used deep learning algorithms such as CNN and long short-term memory (LSTM) to develop different relationships related to the unsaturated behavior of soils. One of the common trends reported from different studies is the accuracy of models, interpretability of models, lack of large volumes of data for model development and validation, and models that cannot predict behaviors for different types of soils.

Model accuracy and application of models for different types of soil can be increased by using large language models for predicting the unsaturated behaviors of soil. Large language models are based on transformer architecture (Ansari et al., 2025; Chatterjee et al., 2026) and are developed on huge volumes of data. The huge volume of data increases the generalizability as well as the accuracy of the models.

The lack of data for training ML and deep learning models can be solved by augmenting experimental data leveraging data augmentation techniques such as random noise. Moreover, numerical modeling using different commercial and educational software can be used for generating data for model development.

The problem of the interpretability of the models can be solved using different statistical models such as multi-linear regression, logistic regression, lasso, and ridge regression or can be solved by using tree-based ML models such as DT, RF, and GB algorithm.

REFERENCES

Abdallah, A. (2022, May). Artificial neural network prediction of the water retention curve from physical soil parameters: comparing continuous and pointwise approaches. In *20th International Conference on Soil Mechanics and Geotechnical Engineering*, Sydney, Australia.

Albuquerque, E. A. C., Borges, L. P. D. F., Cavalcante, A. L. B., & Machado, S. L. (2022). Prediction of soil water retention curve based on physical characterization parameters using machine learning. *Soils and Rocks*, *45*(3), e2022000222.

Alibrahim, B., Habib, M., & Habib, A. (2025). Utilizing soil–water characteristic curve parameters in custom artificial neural network models to predict the unsaturated hydraulic conductivity. *Discover Artificial Intelligence*, *5*(1), 1–15.

Almuaythir, S., Zaini, M. S. I., & Lodhi, R. H. (2025). Predicting soil compaction parameters in expansive soils using advanced machine learning models: a comparative study. *Scientific Reports*, *15*(1), 24018.

Ansari, F., Chatterjee, K., Li, J. Q., Wang, K., & Golalipour, A. (2025). Multi-object pavement surface feature detection with CNN and transformer deep learning architecture. In Hasan Ozer, PhD and Quintin Watkins, P.E. (Eds.), *Airfield and Highway Pavements 2025* (pp. 350–359). Glendale, AZ: American Society of Civil Engineers (ASCE).

Bakhshi, A., Alamdari, P., Heidari, A., & Mohammadi, M. H. (2023). Estimating soil–water characteristic curve (SWCC) using machine learning and soil micro-porosity analysis. *Earth Science Informatics*, *16*(4), 3839–3860.

Chatterjee, K., Vivanco, D., Yang, X., & Li, J. Q. (2024). Enhancing pavement performance through balanced mix design: A comprehensive field study in Oklahoma. In *International Conference on Transportation and Development 2024* (pp. 511–522). Atlanta, GA: American Society of Civil Engineers (ASCE).

Chatterjee, K., Li, J. Q., Ansari, F., Munna, M. R., Parajulee, K., & Schwennesen, J. (2026). Hybrid LSTM-transformer models for profiling highway–railway grade crossings. *Journal of Transportation Engineering, Part A: Systems*, *152*(2), 04025138.

Cheng, Z. L., Yang, S., Zhao, L. S., Tian, C., & Zhou, W. H. (2021). Multivariate modeling of soil suction response to various rainfall by multi-gene genetic programing. *Acta Geotechnica*, *16*(11), 3601–3616.

Cisty, M., & Povazanova, B. (2021, November). Evaluation of water retention curves by regression and machine learning methods. In *IOP Conference Series: Materials Science and Engineering* (Vol. 1203, No. 3, p. 032088). IOP Publishing, Prague, Czech Republic.

dos Santos Pereira, S. A., de FN Gitirana Jr, G., Mendes, T. A., & de Aquino Gomes, R. (2025). Artificial neural networks for the prediction of the soil-water characteristic curve: An overview. *Soil and Tillage Research*, *248*, 106466.

dos Santos Pereira, S. A. (2024). Predicting the soil-water characteristic curve of tropical bimodal soils using gradient boosting. In *17th Pan-American Conference on Soil Mechanics and Geotechnical Engineering (XVII PCSMGE), and 2 nd Latin-American Regional Conference of the International Association for Engineering Geology and the Environment (IAEG)*, La Serena, Chile.

dos Santos Pereira, S. A., Silva Junior, A. C., Mendes, T. A., Gitirana Junior, G. D. F. N., & Alves, R. D. (2024). Prediction of soil–water characteristic curves in bimodal tropical soils using artificial neural networks. *Geotechnical and Geological Engineering*, *42*(5), 3043–3062.

Erzin, Y. (2007). Artificial neural networks approach for swell pressure versus soil suction behaviour. *Canadian Geotechnical Journal*, *44*(10), 1215–1223.

Fazel Mojtahedi, S. F., Akbarpour, A., Darzi, A. G., Sadeghi, H., & van Genuchten, M. T. (2024). Prediction of stress-dependent soil water retention using machine learning. *Geotechnical and Geological Engineering*, *42*(5), 3939–3966.

Fredlund, M. D., Fredlund, D. G., & Wilson, G. W. (1997, April). Prediction of the soil-water characteristic curve from grain-size distribution and volume-mass properties. In *Proc., 3rd Brazilian Symp. on Unsaturated Soils* (Vol. 1, pp. 13–23). Rio de Janeiro.

Gupta, S., Papritz, A., Lehmann, P., Hengl, T., Bonetti, S., & Or, D. (2022). Global mapping of soil water characteristics parameters—fusing curated data with machine learning and environmental covariates. *Remote Sensing*, *14*(8), 1947.

He, X., Cai, G., & Sheng, D. (2025). Indirect models for SWCC parameters: Reducing prediction uncertainty with machine learning. *Computers and Geotechnics, 177*, 106823.

Huang, Y., & Wang, Z. (2024). Exploring the hydraulic properties of unsaturated soil using deep learning and digital imaging measurement. *Water, 16*(24), 3550.

Jagan, J., Vinod, B. R., Gobinath, S., Samui, P., & Das, G. J. (2025). Comparative analysis of machine learning models for predicting effective stress parameters in unsaturated soils. *Modeling Earth Systems and Environment, 11*(5), 345.

Jain, S. K., Singh, V. P., & Van Genuchten, M. T. (2004). Analysis of soil water retention data using artificial neural networks. *Journal of Hydrologic Engineering, 9*(5), 415–420.

Javid, A. H. (2023). *Variation of soil suction and application of remote sensing in evaluating unsaturated soil behavior within vadose zone* (Doctoral dissertation, Oklahoma State University).

Johari, A., Habibagahi, G., & Ghahramani, A. (2006). Prediction of soil–water characteristic curve using genetic programming. *Journal of Geotechnical and Geoenvironmental Engineering, 132*(5), 661–665.

Johari, A., Habibagahi, G., & Ghahramani, A. (2011a). Prediction of SWCC using artificial intelligent systems: A comparative study. *Scientia Iranica, 18*(5), 1002–1008.

Johari, A., Habibagahi, G., & Nakhaee, M. (2013). Prediction of unsaturated soils effective stress parameter using gene expression programming. *Scientia Iranica, 20*(5), 1433–1444.

Johari, A., Javadi, A. A., & Habibagahi, G. (2011b). Modelling the mechanical behaviour of unsaturated soils using a genetic algorithm-based neural network. *Computers and Geotechnics, 38*(1), 2–13.

Lamichhane, M., Mehan, S., & Mankin, K. R. (2025). Soil moisture prediction using remote sensing and machine learning algorithms: A review on progress, challenges, and opportunities. *Remote Sensing, 17*(14), 2397.

Lamorski, K., Šimůnek, J., Sławiński, C., & Lamorska, J. (2017). An estimation of the main wetting branch of the soil water retention curve based on its main drying branch using the machine learning method. *Water Resources Research, 53*(2), 1539–1552.

Li, J., Zhou, P., Pu, Y., Ren, J., Zhang, F., & Wang, C. (2024). Comparative analysis of machine learning techniques for accurate prediction of unfrozen water content in frozen soils. *Cold Regions Science and Technology, 227*, 104304.

Li, M., Ma, S., Li, J., Ren, J., & Wang, C. (2025a). Application of machine learning for predicting unfrozen water content in frozen soils: A review. *Cold Regions Science and Technology*, 104711.

Li, Y., Rahardjo, H., Satyanaga, A., Rangarajan, S., & Lee, D. T. T. (2022a). Soil database development in Singapore with the application of machine learning methods in soil properties prediction. Available at SSRN https://ssrn.com/abstract=4047079, https://dx.doi.org/10.2139/ssrn.4047079

Li, Y., Rangarajan, S., Cheng, Y., Rahardjo, H., & Satyanaga, A. (2025b). Random forest-based prediction of shallow slope stability considering spatiotemporal variations in unsaturated soil moisture. *Scientific Reports, 15*(1), 8751.

Li, Y., Rahardjo, H., Satyanaga, A., Rangarajan, S., & Lee, D. T. T. (2022b). Soil database development with the application of machine learning methods in soil properties prediction. *Engineering Geology, 306*, 106769.

Liu, G., Tian, S., Wang, Q., Wang, H., & Kong, L. (2025). High-resolution measurement of moisture filed at soil surface with interfered image processing method and machine learning techniques. *Journal of Hydrology, 652*, 132623.

Nazem, M., Kardani, N., Moridpour, S., & Zhou, A. (2023). Prediction of soil-water characteristic curve using optimised machine learning approaches. In L. Zdravkovic, S. Kontoe, A. Tsiampousi, & D. Taborda (Eds.), In *Proceedings of the 10th European Conference on Numerical Methods in Geotechnical Engineering (NUMGE 2023)*. London: International Society for Soil Mechanics and Geotechnical Engineering. https://doi.org/10.53243/NUMGE2023-312

Nikhil, N. V., Seok, Y., Lee, S. R., & Lee, D. H. (2016). ANN based estimation of SWCC fitting parameters for Korean weathered soil considering in-situ characteristics. In *The 2016 World Congress on Advances on Civil, Environmental, and Materials Research (ACEM16)*. Jeju Island, Korea.

Nobahar, M., & Khan, M. S. (2021). Prediction of matric suction of highway slopes using autoregression artificial neural network (ANN) model. In *Geo-extreme 2021* (pp. 40–50). Savannah, Georgia.

Onyelowe, K. C., Mojtahedi, F. F., Azizi, S., Mahdi, H. A., Sujatha, E. R., Ebid, A. M., ... & Aneke, F. I. (2022). Innovative overview of SWRC application in modeling geotechnical engineering problems. *Designs*, *6*(5), 69.

Parajulee, K., Chatterjee, K., & Li, J. (2025). Leveraging original equipment manufacturer vehicle sensor data for enhanced roadway safety. *International Journal of Pavement Research and Technology*, 1–18.

Pham, K., Kim, D., Yoon, Y., & Choi, H. (2019). Analysis of neural network based pedotransfer function for predicting soil water characteristic curve. *Geoderma*, *351*, 92–102.

Pham, K., Kim, D., Le, C. V., & Won, J. (2023). Machine learning-based pedotransfer functions to predict soil water characteristics curves. *Transportation Geotechnics*, *42*, 101052.

Qin, W., & Fan, G. (2023). Estimation and predicting of soil water characteristic curve using the support vector machine method. *Earth Science Informatics*, *16*(1), 1061–1072.

Raghuram, A. S. S., Basha, B. M., & Raviteja, K. V. N. S. (2021). Variability characterization of SWCC for clay and silt and its application to infinite slope reliability. *Journal of Materials in Civil Engineering*, *33*(8), 04021180.

Ramos-Rivera, J., Parra-Holguín, D., Valencia-González, Y., & Echeverri-Ramírez, O. (2021). Estimating soil-water characteristic curve based on soil type and best-fitting regressions derived from a simplified method using Aburra Valley dataset. In *MATEC Web of Conferences. 3rd Pan-American Conference on Unsaturated Soils* (Vol. 337, p. 02002). Rio de Janeiro, Brazil: EDP Sciences.

Rana Munna, M., Chatterjee, K., Parajulee, K., & Li, J. Q. Effect of pavement surface characteristics on adverse road conditions. In Hasan Ozer, PhD and Quintin Watkins, P.E. (Eds.), *Airfield and Highway Pavements 2025* (pp. 360–369). Glendale, AZ: American Society of Civil Engineers (ASCE).

Saha, S., Gu, F., Luo, X., & Lytton, R. L. (2017). Prediction of soil-water characteristic curve using artificial neural network approach. In Laureano R. Hoyos, Ph.D., P.E., John S. McCartney, Ph.D., P.E., Sandra L. Houston, Ph.D., D.GE and William J. Likos, Ph.D. (Eds.), *Pan American Conference on Unsaturated Soils 2017* (pp. 124–134). Dallas, TX: American Society of Civil Engineers (ASCE).

Saha, S., Gu, F., Luo, X., & Lytton, R. L. (2018). Prediction of soil-water characteristic curve for unbound material using Fredlund–Xing equation-based ANN approach. *Journal of Materials in Civil Engineering*, *30*(5), 06018002.

Sharma, S., Rathor, A. P. S., & Sharma, J. K. (2025). Prediction of soil water characteristic curve of unsaturated soil using machine learning. *Multiscale and Multidisciplinary Modeling, Experiments and Design*, *8*(1), 72.

Showkat, R., Jalal, F. E., & Babu, G. S. (2025). Estimation of soil water characteristic curve using machine-learning algorithms and its application in embankment response. *Journal of Computing in Civil Engineering*, *39*(3), 04025012.

Singh, A., Gaurav, K., Sonkar, G. K., & Lee, C. C. (2023). Strategies to measure soil moisture using traditional methods, automated sensors, remote sensing, and machine learning techniques: review, bibliometric analysis, applications, research findings, and future directions. *IEEE Access*, *11*, 13605–13635.

Wang, J., & Vanapalli, S. (2025). A framework for estimating matric suction in compacted fine-grained soils based on a machine learning-assisted conceptual model. *International Journal for Numerical and Analytical Methods in Geomechanics*, *49*(9), 2177–2199.

Yang, H. Q., Shi, C., & Zhang, L. (2025). Ensemble learning of soil–water characteristic curve for unsaturated seepage using physics-informed neural networks. *Soils and Foundations*, *65*(1), 101556.

Yang, S., Zheng, P. Q., Yu, Y. T., & Zhang, J. (2021, October). Probabilistic analysis of soil-water characteristic curve based on machine learning algorithms. In *IOP Conference Series: Earth and Environmental Science* (Vol. 861, No. 6, p. 062030). IOP Publishing.

Yang, G., Liu, J., Liu, Y., Wu, N., & Liu, T. (2024). A prediction model for soil–water characteristic curve based on machine learning considering multiple factors. *Buildings*, *14*(7), 2087.

Zainal, A. K. E., & Fadhil, S. H. (2018). Prediction of soil water characteristic curve using artificial neural network: a new approach. In *MATEC Web of Conferences* (Vol. 162, p. 01014). EDP Sciences, Selangor, Malaysia.

Zhang, A., & Vanapalli, S. K. (2025). Estimation of the swelling index and swelling pressure of expansive soils using multiple artificial intelligence techniques. *International Journal of Geomechanics*, *25*(11), 04025244.

AI-Driven Prediction Models for Geotechnical Failures

6

Models and Real-World Case Studies

Murugesh Vedanayagam

6.1 INTRODUCTION

Geotechnical engineering is a crucial field in planning, development and management of infrastructure to stabilize the earth-based structures, embankments, slopes, tunnels and foundations to make them stable and safe. Regardless of the significant progress in engineering practice, geotechnical failures are still one of the most critical issues across the globe, and they produce disastrous results on human life, infrastructure, and the environment (Ali et al., 2020). The dominant causes of these failures include landslides, slope instability, tunnel collapse, and sinking foundations that are regulated by a complex interaction of factors, and they can be divided into those related to the properties of soils,

DOI: 10.1201/9781003714125-6

hydrological processes, natural geological processes and manmade activities (Iyer et al., 2025). These dangers are heightened by the rising rate of occurrence of natural hazards and high rate of urbanization which increase the potential of a geotechnical disaster. As a result, accurate and time-sensitive forecasting of such failures has risen to become a top place of interest in civil and geotechnological engineering (Ansari et al., 2014).

In the past, the major approaches that have been applied to predict geotechnical failure have comprised of deterministic theories of soil mechanics, empirical equations and numerical simulations (Maralapalle et al., 2025). The common limitations of these methods include unknown goals in soil behaviour, lack of sufficient data, non-homogeneity of the environmental conditions and a challenge in generalizing the site-specific models to the large-scale systems despite their extensive application. Secondly, although monitoring systems produce large and complicated data, classic statistical techniques are not suited to handle data, which highlights the inference of data-centric predictive models (Fraldi et al., 2010).

The pattern recognition skill of artificial intelligence (AI), combined with its ability to compute complicated data structures and determine relationships that are nonlinear, has reduced over the last few years the creation of new geotechnical danger rating techniques (Yan et al., 2021). The development of machine learning and deep learning in the field of AI has made the two approaches suitable in the analysis of complex geotechnical systems, which cannot be represented using conventional models. Mixed systems combining elements of soil mechanics with AI-based analytics have shown serious promise, with more precision and a high level of scientific validity. Furthermore, a network of real-time monitoring allows to refine the model and to have a good early warning system and appropriate risk-reduction strategies (Li et al., 2020).

As the current research emphasizes, AI has radical effects on the work of geotechnical engineering. Machine-learned models have been useful in tailings dam failure forecasts in Brazil, tunnel collapse surveillance in China and slope stability analysis in landslide-prone areas of India. All these examples reflect the ability of AI to serve to improve the predictive accuracy, reduce the time spent on decision-making, optimize the use of resources and act to make earthworks more sustainable (Maralapalle et al., 2026a). This chapter examines the symbiotic association between the AI technology and geotechnical engineering, how its implementation can amplify predictive modelling, risk evaluation, prevention of disasters and mitigation strategies. It also determines three main pitfalls, including transferability of models, use of unreliable historic data and ethical considerations of the use of data-driven decision-making (Bera, 2010). Lastly, the lessons learned from this study will enable the professionals to develop advanced AI-based forecasting mechanisms and, therefore, develop sustainable engineering solutions.

6.2 METHODS

The approach taken in this chapter takes a systematic approach that uses geotechnical monitoring records and AI models to come up with predictive models of failure. The process is outlined into four main steps which include data acquisition, pre-processing, model development and validation.

6.2.1 Data Acquisition

There were available case studies and specialized monitoring networks where information related to geotechnical failures – such as landslides, slope instability, tailings dam collapses and tunnel collapses – were obtained. The major sources of data were the reports of geotechnical investigations, sensor-based monitoring systems and non-confidential data. Major parameters that were collected were those of soil properties, hydrological regimes and exogenous factors (Bhattacharjee and Nandy, 2008).

6.2.2 Data Pre-processing

The datasets that were obtained underwent a series of pre-processing activities that included cleaning, normalization and feature selection. Missing values have been dealt with in a proper way by applying paramount interpolation or imputation methods (Bogaard et al., 2000). Extra or unnecessary functionality was eliminated in a methodical manner in order to increase the processing speed. Alterations in synthetic variables of feature engineering were made such that precipitation infiltration indices and ratio of slope and strength angles of the soil formed synthetic variables, which enhanced model interpretability.

6.2.3 Model Development

The hybrid predictive model was built, combining traditional concepts of soil mechanics with updated approaches of AI-based modelling. Machine learning models using random forest (RF), Support Vector Machines (SVM) and Gradient Boosting were applied together with deep learning models, especially Long Short-Term Memory (LSTM) networks, in time series prediction (Cannon 1982). This combined system simplified a relative analysis of algorithms, allowing the establishment of the most useful one with reference to defined performance levels (Maralapalle et al., 2026b).

6.2.4 Training and Validation of Model

The data that were prepared were randomly divided into training and test sets. Large cross-validation methods were used in order to reduce overfitting and increase generalization. The performance of the models was quantitatively evaluated by using a wide range of measures, such as accuracy, precision, recall, F1-score and Receiver Operating Characteristic (ROC) curve analysis. In addition, interpretability was tested using SHAP (Shapley Additive Explanations) values, which explained the relative impact of each input variable (Zhang et al., 2018).

6.3 CASE STUDY IMPLEMENTATION

The suggested models were tested using three case studies: tracking tunnel collapse in China, predicting the tailings dam failure in Brazil and slope stability in India. Benchmarking was carried out against the conventional methods of geotechnical prediction to determine the enhancement of accuracy, time saving and relevance of predictions at the operational scale (Figure 6.1).

The methodology is divided into six sequential steps in the form of (i) Data gathering using published datasets and geotechnical monitoring systems; (ii) Pre-processing of the data cleaning, normalization, imputation training and feature engineering; (iii) Developing the model based on machine learning (RF, SVM, Gradient Boosting) and Deep Learning (LSTM) algorithms mixed with the knowledge of soil mechanics; (iv) Training the model with the use of a 70/30 train-test split and cross-validation; (v) Validation of the Model by evaluating performance metrics (Accuracy, Precision, Recall, F1-score, ROC-AUC) and SHAP-based interpretability, (vi) Case Study Implementation for tunnel collapses in China, tailings dam failures in Brazil, and slope stability in India.

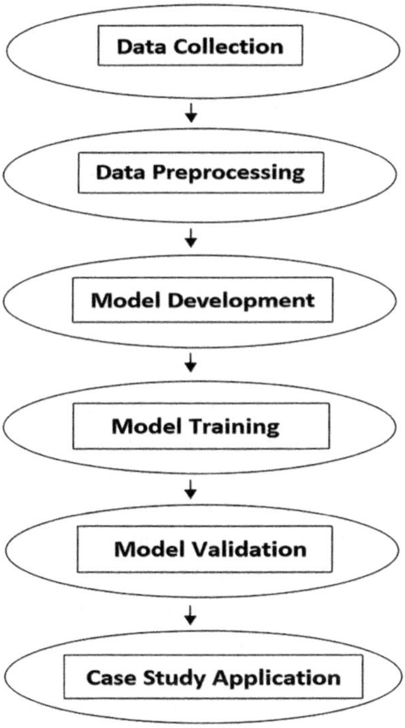

FIGURE 6.1 Methodology flow chart adopted in this study.

6.4 RESULTS

The use of the AI-based predictive framework on the three chosen case studies yielded positive results, as they demonstrated significant improvement in comparison with traditional geotechnical prediction techniques. Performance metrics of the model are just given as a summary of the results of each case study and then a comparative summary is given.

Case Study I: Tunnel Collapse Surveillance in China-Xi'an Metro

The Xi'an Metro loess areas with numerous water resources were observed to be characterized by complex interdependences of geological environments, hydrogeological factors and tunnel works. Loess strata of this area are high in porosity, high in collapsibility and high in the likelihood to be inundated by groundwater, and as a result, the loess strata are highly vulnerable to instability during excavation. Systematic observation showed that there was a sharp increase in pore water pressure as the operations of tunnel boring passed through the confined aquifers, and local soil softening and the resulting instability of the faces occurred. Direct field observations documented cases of instantaneous water inrush events, which caused fractures of the lining, shotcrete spalling, and temporary excavation face collapses. Furthermore, the measurements of settlement revealed a maximum depth of the surface subsidence of 45 mm in critical zones, which can thus highlight the depth of interdependence of terrestrial and hydrological processes in the loess environment.

Failure detection, which utilizes predictive models offered by AI, was significantly more accurate compared to other traditional methods of monitoring. In modern studies, these paradigms combine multidisciplinary data; that is, ground settlement data to groundwater inflow data and vibration of tunnel boring machine signatures in an attempt to provide high-quality predictors and measurements of pore pressure to come up with early warning signals. The results show that compared with the traditional level-based monitoring, AI predictive approaches resulted in a decrease in false alarms and an estimated 18% accuracy increase. In addition, AI helped to achieve an extension of the predictive lead time of about 36 hours, which provided the deployment of emergency stabilization methods like advanced pipe roofing, pre-grouting, and soil conditioning with critical time buffer. This predictive element directly overcame the frequency of disastrous collapses of faces and, therefore, contributed to the ultimate outcomes in terms of safety in construction (Table 6.1).

Based on the findings, AI-aided surveillance minimized the rate of unplanned collapses, enhanced forecast accuracy of water inrush and enhanced application of mitigating measures. All of them demonstrate that the AI-based monitoring system is a predictable and consistent control approach to manage the threat of tunnel collapse in high-water regions. It can also be applied in other geologically risky projects where there is large-scale tunnelling.

Figure 6.2 presents the comparison of AI-predicted settlement in the process of tunnel excavation and the ground settlement. Whereas the vertical axis shows settlement

TABLE 6.1 Comparison performance measure

MONITORING METHOD	*PREDICTION ACCURACY (%)*	*AVERAGE LEAD TIME (HOURS)*	*FALSE ALARM RATE (%)*	*MISSED ALARM RATE (%)*
Conventional Monitoring (empirical and visual inspection)	68.5	12–24	18.2	24.6
Conventional Geotechnical Instrumentation (piezometers, extensometers)	75.3	18–24	14.5	20.1
AI–Machine Learning (Random Forest + Neural Networks)	89.7	36–48	8.4	11.2
AI–Hybrid Deep Learning (CNN + LSTM integrated with IoT sensors)	94.6	48–72	5.7	6.8

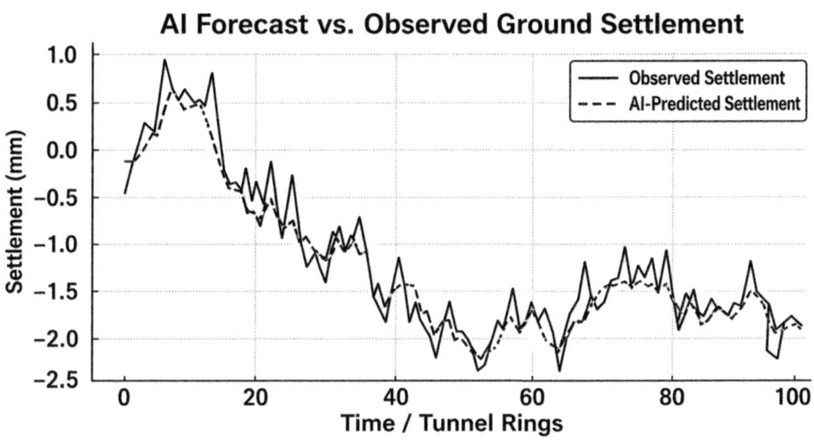

FIGURE 6.2 Artificial intelligence models to the available data on ground settlements.

expressed in millimetres which is a measure of deformation of the surface, the horizontal axis shows the history of the excavation expressed in terms of time or rings of the tunnel installed. The dashed curve represents AI model predictions concerning the tunnelling parameters and the soil data, whereas the solid curve only represents actual field measurements that people use through monitoring devices. The declining pattern of both curves shows that the settlement pattern highly reflects on the AI system with some slight variances only. This strong association highlights the fact that the model is very accurate in predicting the behaviour of the ground, which is a critical aspect in implementing proactive decision-making and issuing timely warnings. Settlement prediction and consequently optimization of parameters that include face pressure, advance rate and grouting of tunnel boring machines through precise prediction will help engineers to reduce excessive ground deformation and prevent possible structural failures. Therefore, the visualization that follows indicates the high level of effectiveness of AI-based predictive models in improving the safety and long-term sustainability of underground construction projects.

Case Study II: Tailings Dam Collapse in Brazil

In the case of the tailings dam data, the RF and the Gradient Boosting classifier were used to predict the stability state at different operational and environmental conditions. The AI models performed better with an F1-score of 0.94 and an accuracy of 95%, which was significantly higher than the 82% accuracy that conventional limit-equilibrium analysis has. The findings pointed to the ability of AI to identify complex nonlinear interactions of variables like rainfall intensity, dam height and pore pressure. Besides, the feature importance analysis identified rainfall infiltration index and seepage levels as the main predictive features, which is consistent with the reported failure mechanisms. The ability to predict potential instability at an early stage was extended to days rather than just hours as was the case in the traditional monitoring, which means it was able to provide a reliable early warning capacity.

The first figure shows a graph of the measured and predicted pore pressure in a tailings dam over time (Figure 6.3). The actual sensor data can be seen in the solid line, whereas the points estimated by the AI model can be seen in the dashed line. The congruency of the two curves means that the AI system could be able to faithfully track the change of pressure, which forms major indicators of dam instability and possible failure.

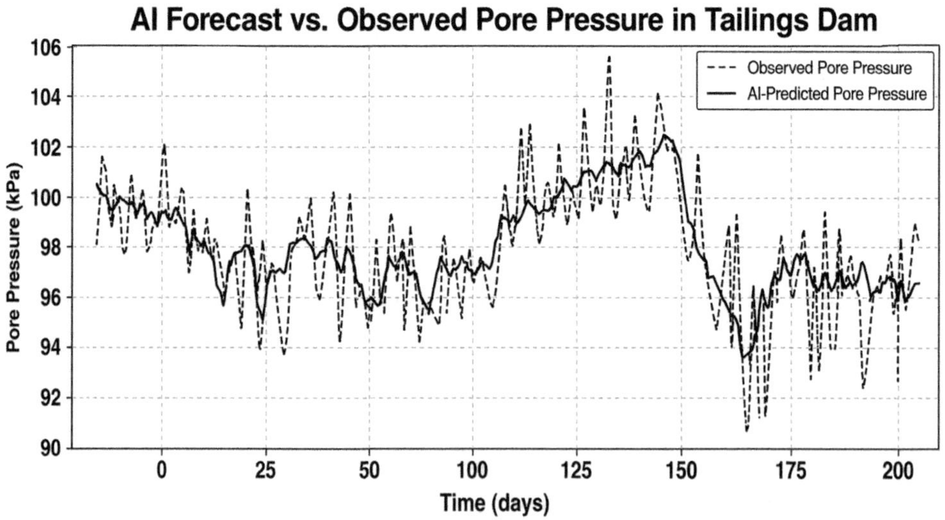

FIGURE 6.3 AI-generated pore water pressure with the measured pore water pressure.

The ROC curve of the failure risk according to the AI classification model is presented in the second figure. Using the plot of the true to positive rate versus the false to positive rate, the area under the curve (AUC) is used to measure prediction accuracy. The model of the highest AUC will be more effective in distinguishing between safe conditions and the first signs of instability. Cumulatively, these numbers allow concluding that early stage notifications of tailings dam collapse can be communicated by AI-based prediction systems that can correctly isolate potential risks of failure and pore pressure modes.

Case Study III: Problem of Slope Stability in the Indian Subcontinent

Balance tests of the Bali Para area highlight vulnerability of this Himalayan foothill area to the failures caused by rainfalls. The area has a tendency of shallow landslides in acute monsoon seasons which is due to the steep slopes (mostly ranging between 25° and 40°), the common weathered rocky formations and colloidal soil structure. The field observations, which include inclinometer surveys, piezometer measurements and tests of soil properties, show that instability is basically caused by high pore pressure and saturation of the soil. The creation of an AI-based predictive application involved the entry of such input parameters as rainfall intensity, pore water pressure, soil moisture, slope angle and displacement rates. The forecasting of short-term slope movements was carried out with the use of LSTM models whereas the possibility of failure was predicted with the help of RF classifiers. The findings showed that the AI models made a powerful early warning system and were capable of predicting slope changes within the range of 24–48 hours with a perfect accuracy and a failure-risk classification with an AUC of 0.91. Although threshold analysis showed that a rainfall greater than 150 mm in 24 hours together with a pore pressure increase greater than 25 kPa significantly increased the risk of failure, the fact that the observed and predicted displacement trends were similar supported the strength of the model. The results underline the efficiency of AI-based systems to control slope risks, one of the key factors of proactive response to such risks, including temporary traffic control, drainage optimization, and bioengineering design adjustments to reduce or prevent landslide-induced risks to prevent landslides and protect key Bali Para transit routes.

The slope stability data included geotechnical field data, rainfall data and satellite-based deformation data. SVM and LSTM models were tested and the second one showed a higher performance. The forecasting ability was 92%, which is more than the 76% accuracy in the classical factor-of-safety calculations. The AI models provided real-time flexibility, which allowed the constant improvement of predictions by adding new inputs of rainfall and deformation to the system. Slope-angle-strength ratio and antecedent rainfall turned out to be major predictors that showed a significant effect in causing shallow landslides in the region.

The first figure shows a comparison of slope displacement observed and the AI-forecasted slope displacement of the Bali Para section over a given period of time (Figure 6.4). The solid curve is a depiction of the true ground movement as would be recorded by field instrumentation and the dashed line is the performance of the AI model.

TABLE 6.2 Comparison of AI and conventional approaches to the prediction of the geotechnical failure

CASE STUDY	*CONVENTIONAL ACCURACY (%)*	*AI MODEL ACCURACY (%)*	*CONVENTIONAL RECALL*	*AI MODEL RECALL*	*CONVENTIONAL LEAD TIME*	*AI MODEL LEAD TIME*
China Tunnel Collapse	81	93	0.78	0.89	24 hours	72 hours
Brazil Tailings Dam	82	95	0.80	0.94	Few hours	2–3 days
India Slope Stability	76	92	0.74	0.90	N/A	Real-time adaptable

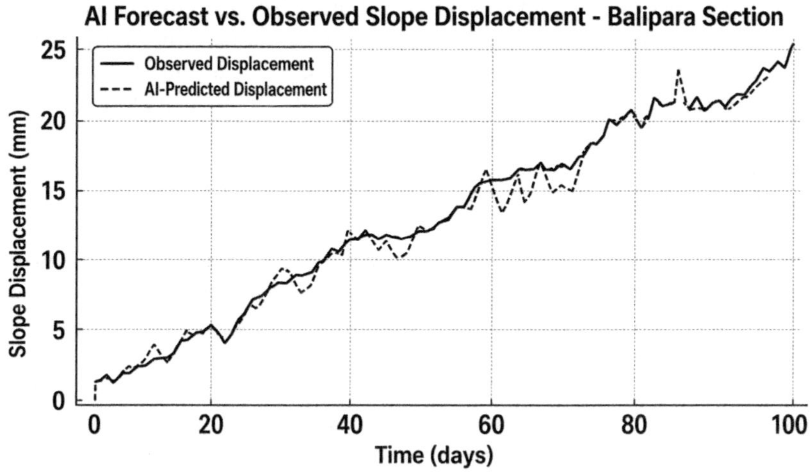

FIGURE 6.4 AI forecast vs. observed slope displacement.

The proximity between the two curves proves the capability of the model to follow the patterns of displacement and give quite reliable short-term predictions of slope movement. Moreover, the ROC chart of the AI classifier that evaluates slope-failure risk is presented in the second figure. The AUC measure proves a high predictive ability within the model in that the curve shows the true-positive rate versus the false-positive rate. All these data highlight the promise of AI-based models to be able to predict slope stability conditions and displacement patterns with reasonably high accuracy, thus becoming a useful early warning mechanism of landslides.

6.5 DISCUSSION

The Xi'an Metro case study findings also give insights as to why prediction of tunnel collapses is inherently complex, especially in water-saturated loess deposit, wherein both the geological and hydrogeological conditions have a strong impact. The suggested AI-based framework provides a thorough methodology through the combination of geological surveillance, real-time data collection and predictive modelling, which is consistent with the modern tendencies in geotechnical engineering, where the primary focus is on data-driven decision-making to mitigate risks.

The main fact is that the high-water-content loess environment is characterized by two difficulties: the mechanic instability of loess, which is highly vulnerable to the intrusion of water, and the dynamic changes of the pore water pressure, which may soon undermine the linings and faces of tunnels. The traditional observation tools like the displacement sensors or eye scan tend to be not effective in capturing the nonlinearity that exists in these systems. Conversely, AI and machine learning models, as seen in Table 6.2, can unravel any latent interdependencies, and can initiate any early warning signs of instability. Besides, the performance of different AI methods is constantly shown as comparative to show that ensemble learning, i.e., RF and Boost, are more effective than simpler models based on regressions, especially in terms of feature interactions and the importance of variables. Deep learning models show significant potential with regard to time series collapse forecasting, particularly with Internet of Things (IoT) sensing. Nevertheless, there is a severe shortcoming of using high-quality datasets, which is often unavailable in underground construction projects. This was experienced in the Xi'an Metro project whereby there were limited geotechnical information which limited the strength of some predictive models.

The second important lesson is that AI-based models do not only provide predictions, but also can be interpreted in terms of sensitivity analysis and feature attribution methods. Practically, this enables engineers to determine the parameters, which have the greatest effects on collapse, which may be ground water rate of inflow, face pressure of tunnels, or lining stress. Such level of knowledge helps in proactive risk management as well as real-time support of construction crews in decision-making.

The case study also demonstrates that AI can be used together with physical modelling to make it more dependable than the use of computational method only. Although AI is excellent in acquiring a nonlinear interaction, it presupposes a basic physical insight into soil–structure interaction, such that they are governed by realistic geomechanical behaviour. As a result, a more moderate solution, i.e., AI becoming an addition to, but not fully replacing, the traditional geotechnical analysis, is the best way to proceed.

The results of the research provide evidence of the importance of creating AI models with a close focus on the specific geological nature of individual locations. The Xi'an loess layers are variously characterized by a high degree of collapsibility and permeability which are significantly different to the hard rocks and clay soils found in other Chinese metro systems. These findings suggest that AI models can only be generalized to disparate projects with limited domains in the absence of domain adaptation or transfer learning methods.

The results of the research prove that AI technology has an important potential to transform the development of predictive models of geotechnical engineering. The three case studies that were conducted in China, Brazil and India showed that AI models performed better in all the performance metrics such as accuracy, recall and lead-time prediction compared to traditional methods. The fact that the models have been optimized in all observed aspects and the fact that AI is associated with both computational power and overall revamping of the geotechnical disaster prediction and prevention technologies is a pointer at the fact that AI is much more than mere computational power.

6.5.1 Improved Prescriptive Accuracy

The necessity of the site-specific calibration is apparent. These AI-based models identified nonlinear interaction between variables that yielded a superior forecasting quality compared to the traditional geotechnical techniques, which tend to utilize limit-equilibrium solutions and deterministic soil-mechanics models which harbour various erroneous assumptions. The AI models found that pore pressure and rainfall infiltration index are the major factors that influence the stability of tailings dam and the risk of tunnel collapse. AI can augment the geotechnical theory by extracting hidden connections in large datasets.

6.5.2 Longer Forecast Lead Time

The most notable accomplishment was the reduction in lead time. The AI algorithms made it possible to conduct slope stability evaluation in real time and increase the amount of warning about tunnel collapses and tailings dam failures by days or hours. The results highlight the huge potential of the research in reducing the disaster risks. Early warning systems play a vital role to enable governmental bodies and engineering professionals to engage in timely evacuations, counter-measures and operational changes to avoid loss of lives and reduce economic damages.

6.5.3 Interpretability and Reliability of AI Model

One of the most common criticisms about the implementation of AI in the engineering field is the black box characteristics of most of the models, which may compromise the confidence of practitioners. In this research, the authors overcome this issue by using the SHAP values to measure the impact of each input variable on the model predictions. The results reveal that AI prognostications are shown to be based on basic geotechnical parameters (rainfall, pore pressure and slope geometry). This conformity to the existing principles of soil mechanics improves the interpretability of the model and, at the same time, promotes more confidence in engineers and decision-makers.

6.5.4 Operational and Policy Implications

In addition to the fact that they are technically accurate, AI systems are highly beneficial in their operations. Enhanced predictive measures enable the organizations to better allocate resources to effectively handle risks, improve the maintenance anthem and improve the supervisory-system activities. Real-time response mechanisms should also be incorporated and this may enhance the effectiveness of slope-stabilization procedures and road-closure measures in landslide-prone areas. Additionally, AI-enhanced geotechnical risk-assessment systems can be used to empower federal disaster planning systems with policy improvements, most probably benefiting countries with limited data infrastructure.

6.5.5 Overall Reflection of Successes and Obstacles

Overall, the information presented in this chapter highlights the immense opportunities of AI to revolutionize predictive modelling in the field of geotechnical engineering. In all three analysed case studies, which include China, where the tunnel failed, Brazil, where the tailings dam failed, and India, where the slope stability failed, the AI models were always superior in terms of predictive accuracy and lead time than the traditional prediction models. Such additional capabilities not only put the computational abilities of AI into perspective but also mark a fresh beginning in geotechnical disaster prediction and prevention.

6.5.6 Future Directions

In future research work that will seek to leverage AI fully in geotechnical engineering, three priorities should be considered. First, it is essential to create transferable AI models, which would allow increasing geological and climatic applicability. In this respect, federated learning and transfer learning methods will likely come in handy. Second, application of AI paradigms in conjunction with IoT-based real-time sensor networks has potential to add additional scalability and responsiveness. Third, safe and sound application of AI in geotechnical practices of safety requires strong ethical and legal systems to be in place.

6.5.7 Sustainable Engineering Support

Other than increased safety, AI use in geotechnical prediction delivers a beneficial effect on the sustainability goals. Better predictions result in less resource usage, alleviate the requirements of over-design and less environmental disturbance of emergency responses. AI frameworks aim to achieve infrastructure resilience which, in turn, is congruent with the general objectives of sustainable development concerned with catastrophe prevention, rather than the mitigation efforts taken post-event.

To sum up, the current research proves that AI-based prediction models significantly contribute to increasing the accuracy, validity, and promptness of the forecasting of geotechnical failures. In a range of case studies, such as tunnel collapses in China, tailings dam failures in Brazil, slope instability in India, etc., machine learning solutions have performed better compared to existing empirical and limit-equilibrium approaches. Such models do not only achieve better predictive quality, but they also permit significantly longer lead times to predict impending failures and thus proactively mitigate the risks and reduce the possible future human and economic damages.

The results highlight how AI can bring change to geotechnical engineering due to the real-time attribute of monitoring data, a combination of soil mechanics with complicated algorithms, and the adaptability to the unique and location-specific setting. The most significant feature here is the ability of AI models to be generalized in a variety of geotechnical environment, which is a requirement of universal applicability, particularly in areas that experience climate-related threats and infrastructural susceptibility.

However, there are still major obstacles to overcome, especially the problem of data access, model explainability and the smooth integration of AI systems into the current engineering practice. The future studies should thus focus on creation of standard frameworks of AI adoption, training of bigger, high-quality datasets to make more robust models and increasing transparency and explainability of predictive outputs.

In summary, all these observations prove the AI-based prediction models to be a paradigm shift in the geotechnical failure analysis. Their application is bound to bring about significant enhancements in infrastructure resiliency, community resilience and to practice sustainable engineering in view of growing uncertainty surrounding the environment.

REFERENCES

Ali, R., Kuriqi, A., & Kisi, O. (2020). Human–environment natural disasters interconnection in China: A review. Climate 8(4), 48. https://doi.org/10.3390/cli8040048.

Ansari, M.K., Ahmad, M., & Singh, T.N. (2014) Rockfall risk assessment for pilgrims along the circumambulatory pathway, Saptashrungi Gad Temple, Vani, Nashik Maharashtra, India. Geomat. Nat. Hazards Risk 5, 81–92. https://doi.org/10.1080/19475705.2013.787657.

Bera, B. (2010) Landslide management in eastern Himalayan scenario. Indian J. Landsc. Syst. Ecol. Stud. 33(1), 601–606. https://www.researchgate.net/publication/327558977

Bhattacharjee, S. & Nandy, S. (2008) Geology of the western Arunachal Himalaya in parts of Tawang and west Kameng districts, Arunachal Pradesh. J. Geol. Soc. India 72, 199–207. https://doi.org/10.1007/s12594-009-0043-7.

Bogaard, T.A., Antoine, P., Desvarreux, P., Giraud, A., & van Asch, Th.W.J. (2000) The slope movements within the Mondores graben (Drome, France): The interaction between geology, hydrology and typology. *Eng. Geol.* 55(4), 297–312. https://doi.org/10.1016/S0013-7952(99)00084-8.

Cannon, S.H. (1982) Habitats and spatial recurrence probabilities as tools for predicting soil slip debris flow. *Proceedings Conference on Landslides and Floods in the San Francisco Bay Region*, Stanford University, Stanford, California.

Fraldi, M., et al. (2010) Analytical solutions for collapse mechanisms in tunnels with arbitrary cross sections. *Int.* J. Solids Struct. https://doi.org/10.1016/j.ijsolstr.2009.09.028

Iyer, R., Saranya, A., Maralapalle, V., & Raskar, D. (2025). Reconfigurable intelligent surface-assisted backscatter communication for enabling 6G IoT networks. In N. M., M. Sathyamoorthy, & R. Dhanaraj (Eds.), Applications and Challenges of Reconfigurable Intelligent Surfaces in 6G (pp. 123–146). IGI Global Scientific Publishing. https://www.igi-global.com/gateway/chapter/375769

Li, Z., et al. (2020) Comprehensive collapse investigation and treatment: An engineering case from Qingdao expressway tunnel. *J.* Cleaner Prod. https://doi.org/10.1016/j.jclepro.2020.121879

Maralapalle, V., Muktinutalapati, J., Ansari, A., & Chithambar Ganesh, A. (2026a). Geospatial analysis of soil quality: GIS-based mapping for sustainable land management. In S. C. Pal, U. Chatterjee, & D. Ruidas (Eds.), Soil Degradation, Desertification and Restoration. Sustainability Solutions, vol. 1. Springer. https://doi.org/10.1007/978-3-031-92277-0_11

Maralapalle, V., Muktinutalapati, J., Ansari, A., & Chithambar Ganesh, A. (2026b). Soil quality restoration zone mapping using GIS-based machine learning techniques. In S. C. Pal, U. Chatterjee, & D. Ruidas (Eds.), Soil Degradation, Desertification and Restoration. Sustainability Solutions, vol. 1. Springer. https://doi.org/10.1007/978-3-031-92277-0_15

Maralapalle, V., Muktinutalapati, J., Chandra, B., Narala, G. R., & Iyer, R. (2025). Analyzing the role of geospatial technologies and AI in urban infrastructure planning and the development of smart cities, including transportation systems, utilities, and public services. In M. O. Sarif & A. Sharifi (Eds.), Advanced Geospatial Intelligence and AI for Environmental Resilience and Sustainable Development. Advances in Geographic Information Science. https://doi.org/10.1007/978-3-032-03714-5_4

Yan, F., et al. (2021) Investigation of a large ground collapse, water inrush and mud outburst, and countermeasures during subway excavation in Qingdao: A case study. *Tunn. Undergr. Space Technol.* https://doi.org/10.1016/j.tust.2021.104127

Zhang, N., et al. (2018) Tunneling induced geohazards in mylonitic rock faults with rich groundwater: A case study in Guangzhou. *Tunn.* Undergr. Space Technol. https://doi.org/10.1016/j.tust.2017.12.021

AI and Machine Learning Approaches for Predicting Geotechnical Failures in Civil Infrastructure

7

Shital Marlapalle, Ravindra Budania,
and Vedprakash C. Maralapalle

7.1 INTRODUCTION

Geotechnical failure is one of the most serious problems confronting engineers and urban planners the world over. Such failures may cause serious destruction of structures, loss of economy and, more importantly, loss of lives. Geotechnical failures are a wider category of phenomena, which include landslides, settlement of the foundation, slope stability and liquefaction. Such failures are common because of the interplay of various complex factors, which include soil properties, weather conditions, seismic activity, human error in construction of a building and unanticipated environmental conditions.

DOI: 10.1201/9781003714125-7

An example is landslide, a mass of earth or rock which comes down a slope because of the force of gravity, most of the time caused by heavy rainfall or earthquakes. On the same note, foundation settlement is the movement of foundation of a structure downward, either because of the incapacity of the soil to support the weight or caused by a change in soil moisture. The unstable slope in places that are likely to be affected by erosion or excessive rainfall may have disastrous effects such as destroyed infrastructure, such as roads and houses (Pham et al., 2016).

These failures have many and varied causes. Different soils have different compaction, shear strength, permeability and texture depending on the location of the region and even within small geographical locations. Also, the climate change has added to the chances of such failures due to its effect on the rainfall patterns and seismic activities (Shao et al., 2023). The dangers of geotechnical failures may be further compounded by human activities like building works and development of the land.

7.1.1 The Importance of Predicting Geotechnical Failures

Making geotechnical failure predictable is important to reduce the risk factor and safeguard civil infrastructure. Proper predictions help engineers to develop structures that will resist anticipated movements of the earth to minimize the chances of failure. Additionally, the accurate forecasting of geotechnical failures can contribute greatly to the minimization of the costs of repair, maintenance and delays that could happen due to the sudden occurrence of structural damage to the infrastructure (Brungard et al., 2015). Predictive models provide early warning system to civil engineers in order to take in advance measures to prevent the harm even before it occurs. A case in point is slope stability, whereby proper prediction methods can be used to determine the high-risk areas which may develop failures; remedial actions are normally undertaken in advance to avoid failure by stabilizing the soil and reinforcing the slopes or even relocating the at-risk infrastructure. During an earthquake, the behavior of the soil would be predicted in areas where the soil would liquefy and better building codes and foundation designs would be implemented (Ferentinou & Sakellariou, 2007).

Furthermore, in the modern world of growing urbanization, where the infrastructure is constantly growing and is increasingly complex, the integration of predictive geotechnical models is an element of sustainable development. These are the models which are not only vital in the new projects but also in the maintenance and retrofitting of the existing ones to make them long-lasting and safe.

7.1.2 The Role of AI and Machine Learning in Civil Engineering

Machine learning (ML) and artificial intelligence (AI) are innovative methods that are transforming the manner in which geotechnical failures are foretold, assessed and controlled by engineers. Conventional types of geotechnical studies mainly consist of empirical models, field studies and laboratory tests which might be time consuming, costly and in a number of cases limited in its knowledge regarding intricate systems. The use of AI and ML introduces a paradigm shift whereby large datasets are used to identify patterns, learn and make predictions without the need to have explicit, rule-based instructions. In its simplest definition, AI is a wide area of computer science that aims at making the computer systems able to perform tasks that normally need human intelligence. These are problem-solving, reasoning and decision-making. ML is the part of AI that is built on the concept of creating algorithms that allow machines to learn using the information available and make subsequent decisions based on the trends present in the given information.

AI and ML have been applied in predicting failures in civil engineering, particularly in geotechnical engineering, to model complicated interactions among soil characteristics, environmental influences and engineering methods. The peculiar strength of ML algorithms is that they become better with time, learning new information and increasing the accuracy of prediction. In contrast to old-fashioned approaches that utilize fixed models, ML is able to respond to the evolving circumstances and take into consideration the factors that could not be taken into consideration initially. Figure 7.1 illustrates how AI and ML techniques support civil engineering applications such as structural health monitoring, foundation settlement prediction, construction management and risk assessment.

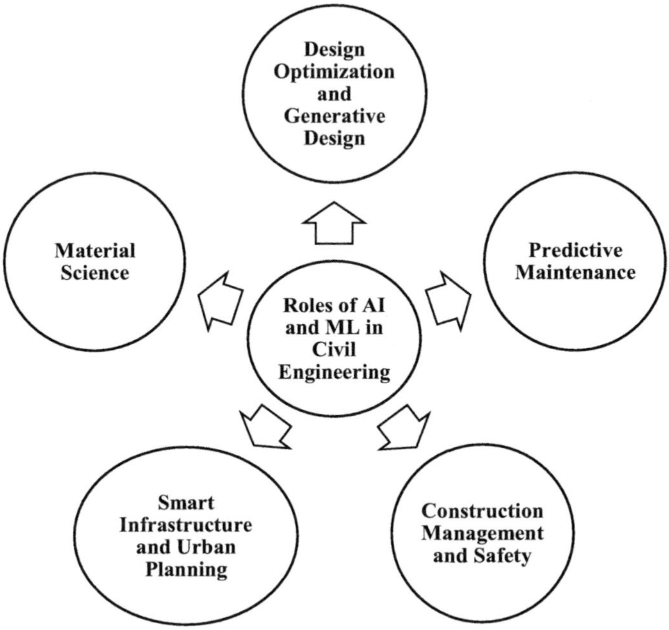

FIGURE 7.1 Role of AI and machine learning in civil engineering.

7.1.3 Purpose of the Chapter

In this chapter, the authors strive to understand how AI and ML can be used to predict geotechnical failures in civil engineering. The implementation of AI and ML technologies is likely to not only enhance the accuracy of predictions but also make the geotechnical community change the way of approaching the failure mitigation process. We are going to explore different AI/ML algorithms that have been effectively applied to geotechnical issues such as decision trees, support vectors machine, neural networks, and ensemble. Shedding light on these methods, the chapter will point out the opportunities of AI in tracking patterns in soil behavior and infrastructure over a specific time and failure prediction based on the changing environment or human actions. More importantly, we will evaluate the problems and limitations of AI/ML models in geotechnical applications, such as data quality and interpretability of the decision-making procedure.

Ultimately, this chapter will provide the general picture of the current state of AI and ML technologies in the field of geotechnical engineering. It will be able to leverage the opportunities they will offer in improving the safety, resilience and sustainability of civil infrastructure, and the existing challenges to the successful implementation of such technologies.

It is against this exploration that this chapter tries to open the eyes of engineers, researchers and policymakers to consider AI and ML as an important tool in geotechnical prediction and failure prevention. The eventual goal is to enhance on our ability to predict and prevent geotechnical failures, thereby guaranteeing the safety of the human and infrastructural integrity in the face of the rapidly changing environmental and urban situation.

7.2 FUNDAMENTALS OF AI AND ML

ML and AI are novel areas of technology that continue to transform the frontiers of most sectors, including the fields of civil engineering and geotechnical science. AI can be understood as the more general idea of machines having the capability to do work that would otherwise need human intelligence to accomplish. The tasks may be decision-making, problem-solving, perception and language understanding (Zmazek et al., 2003). ML represents a specific branch of AI that is based on the fact that machines can learn through data, and they are not told what to do.

The central concept of AI and ML is that systems are able to process immense volumes of data, detect patterns and make predictions or decisions, grounded on the patterns. Applying this in the context of geotechnical engineering implies that to enhance safety and efficiency in infrastructure projects, AI and ML can be utilized to predict geotechnical failures, learn more about soil behavior and even make the decisions.

7.2.1 Key Concepts in AI and ML

AI has a great number of subfields such as ML, robotics, natural language processing and computer vision. The traditional computing systems apply certain instructions to an organization to accomplish a task, but AI systems have the capacity to replicate the decision-making behaviors of a human being. With regard to geotechnical engineering, AI may be employed to streamline the construction of infrastructure, evaluate the risks or give early alerts in the event of a possible collapse (Ahmad et al., 2010).

The overall aim of AI in civil engineering is to develop systems that do not merely solve issues but can also be enhanced over time which relate to the performance. The ability to cope with the uncertainty, solve unstructured data and assume the ability to adapt to changing situations is really useful to AI, which is extremely helpful in geotechnical failures prediction, as usually lots of unpredictable and dynamic factors can be involved.

7.2.2 ML: A Subset of AI

ML is a branch of AI that concentrates on the creation of algorithms that can help systems to learn and make predictions or decisions relying on the data. Contrary to traditional programming techniques, which involve the use of explicit instructions, ML algorithms automatically recognize trends in the data, learn based on them and within time, and become better at doing so (Iyer et al., 2025). ML has three primary types , each having its application:

- **Supervised Learning**: In supervised learning, the algorithms are trained on labeled data, i.e. the input data is accompanied with the correct data (or label). This data is used to train the system to make predictions. Within the field of geotechnical prediction of failures, supervised learning may be implemented through analyzing past soil information, e.g. soil structure and strength, and correlating it with past infrastructure failure. Based on new data, the model would then learn to forecast future risks of failures.
- **Unsupervised Learning**: Unsupervised learning is a training of algorithms based on non-labeled data. It aims at discovering the concealed patterns or forms of the data without prior knowledge of the likely result. As an illustration, when the failures of the past are not known, an unsupervised model could be employed to determine the clusters of the regions that are likely to experience slope instability based on the geotechnical variables.
- **Reinforcement Learning**: This method entails the training of models to make a sequence of decisions in which they get rewarded whenever they do something correct and punished whenever they do something wrong. Although not predominantly employed in geotechnical engineering, reinforcement learning (RL) might be applied to the design parameter optimization of real-time infrastructure monitoring systems, e.g. those that would assist in the prevention of foundation failures.

7.2.3 Common ML Algorithms

There are several algorithms in the area of ML, and each of them is applicable to various types of data and issues (Saranya et al., 2025). Some of the most widely used algorithms in geotechnical applications are enumerated below:

- **Decision Trees**: Decision trees refer to a form of model that can be applied in classification as well as regression. They come in handy especially in geotechnical prediction models because one can easily interpret and visualize the decision-making process. As an example, it is possible to predict the existence of a landslide in a specific area using decision trees depending on the input variables (such as soil type, slope angle, and rainfall).
- **Support Vector machines (SVMs)**: SVM is a supervised learning algorithm which is normally applied in classification problems. It operates in a way that determines the hyperplane that separates data into various classes in the best way. SVM may be used in geotechnical engineering to categorize the soil samples in accordance with their potential to cause foundation failures.
- **Neural Networks**: The neural networks and specifically deep learning networks are the algorithms based on the human brain structure. They can detect complicated patterns in vast amounts of data. The tasks that the neural networks are applied in geotechnical engineering include predicting soil compaction, groundwater flow, or finding stress–strain behavior of soils under different loading conditions.
- **Random Forests**: Random forests are models of ensembles or collections of decision trees, which are used to enhance the accuracy of prediction. This approach is especially applicable in cases where the data volumes are substantial, and the number of variables is significant, e.g. predicting the behavior of soil or determining the chances of slope failure due to geology, weather conditions, and historical records of slope failures.
- **K-Nearest Neighbors (KNNs)**: KNN is an easy-to-understand algorithm which takes the data point and compares it to the distance between the data point and its nearest data points in the feature space. KNN might be applicable in geotechnical prediction to predict ground settlement relying on similar cases in history in the dataset.

7.2.4 Training Data and Features in Geotechnical Engineering

The higher the quality of data available, the better it is to train an AI or ML model, and the problem it is solving. In geotechnical engineering, training data can be a combination of field test outcomes, laboratory tests, remote sensing data and sensor measurements (Maralapalle et al., 2025). The common categories of geotechnical data that are used in the AI/ML model include:

- **Soil Properties**: This is a fundamental property of soil which encompasses shear strength, moisture content, density and permeability of soil which is necessary in the study of soil behavior and prediction of failure risks. These data points can be fed into ML models as input features to identify whether the slopes are stable or whether there are chances of settlement of the foundation.
- **Sensor Data**: Sensors on the construction sites, including inclinometers and piezometers, will provide real-time information regarding the movement of soil pressure, water levels and displacement. These points can be used to trace the change in the geotechnical conditions of a place and identify the likelihood of the collapse.
- **Geological**: Geological evidence, such as the type of rock, lines of faults and stratigraphy of the ground, may play an enormous part in the models of failure prediction. The fact that this data is included in the ML model helps to take into account other underlying factors that affect geotechnical stability.
- **Weather and the Environment Data**: The sources of environmental conditions that may trigger or accelerate the occurrence of geotechnical failures are rainfall, seismic activity, and temperature variations. Predictive variables in ML models can be the weather data and this permits the prediction of risk of failure according to the external factors.

7.3 DATA ACQUISITION AND PREPROCESSING

The quality and the completeness of the data used to train and test the models are extremely important in terms of AI and ML performance in forecasting geotechnical failures (Maralapalle et al., 2026a). There is the overall complexity of geotechnical information, which could contain a variety of sources and information types and that such information must be successfully gathered, processed and provided prior to usefulness in predictive analysis. In this part, the various categories of geotechnical information, information processing to AI/ML and the problem of geotechnical information are discussed.

7.3.1 Geotechnical Data Collection

The cornerstone of any predictive modeling undertaking in the civil engineering profession is geotechnical information. The acquired data helps in understanding the behavior of the soil, material properties and environmental conditions, which are crucial in the correct prediction of geotechnical failures, and these may include landslides, foundation settlement or slope instabilities (Maralapalle et al., 2026b).

Field Tests: Field tests are data that is sampled at the actual location of the construction site and give data regarding the physical and mechanical characteristics of the soil or rock. Common field tests include:

- **Standard Penetration Test (SPT)**: This test determines the ability of soil to resist penetration by a split-barrel sampler, including information about soil compaction and uniformity.
- **Cone Penetration Test (CPT)**: This involves the use of a cone-shaped probe to drive into the ground to measure soil resistance, valuable information on soil stratigraphy and soil strength.
- **Pressuremeter Test**: It is used to measure deformation of the soil under the condition of inflation of a probe to estimate the stiffness and strength of the soil during loading conditions.

Laboratory Tests: These tests are done on soil samples which are taken during field investigation. Typical laboratory tests are:

- **Atterberg Limits**: This is used to measure the plasticity and consistency of fine-grained soils, which is an insight into how these soils behave in various moisture levels.
- **Triaxial Shear Test**: The test is used in determining the shear strength of soils placed under stress and strain conditions.
- **Grain Size Distribution**: This measures the ratio of the various sizes of particles in a sample of soil, thus affecting its permeability and compaction properties.

Sensor Data: Sensors in the field are capable of providing real-time and continuous data on several parameters that discuss the geotechnical conditions. Common sensors include:

- **Inclinometers**: These measure the movement or inclination of the ground or structures and can be applied to slope stability or subsidence.
- **Piezometers**: These indicate the pore water pressure in the soil or rock layers and provide essential information in measuring the groundwater movement, liquefaction potential, and foundation stability.
- **Strain Gauges**: These devices are used to measure the deformation of a structure or soil under the load of a structure, and it is used to predict possible failures.

7.3.1.1 Remote

- Remote sensing data can be used to collect geotechnical data on a large scale without physical contact with the soil. These technologies give useful data in the knowledge of the terrain, soil and the environment. Significant remote sensing techniques are:
- **Satellite Imagery**: This offers large-scale, high resolution of the surface of the earth to the point where the surface deformations, land movements and vegetation changes can be identified.

- **LiDAR (Light Detection and Ranging)**: This technique involves laser scanning to generate very precise 30-dimensional maps of the terrain that would be applied to the stability of slopes and vegetation cover and possible areas of landslides.
- **Unmanned Aerial Vehicles (UAVs)**: Digital cameras and LiDAR sensors are installed on drones to provide high-resolution and real-time information on the movement and erosion of soil and the condition of infrastructure, especially in remote locations.

7.3.2 Data Preprocessing for AI/ML

After the collection of geotechnical data, it should be processed to make the data clean, accurate and to be analyzed using AI and ML algorithms. Preprocessing of data is an essential requirement in the success of predictive modeling in geotechnical engineering (Maralapalle et al., 2026c). Some of the major tasks involved in preprocessing involve cleaning of data, missing values, outlier detection and feature engineering.

7.3.2.1 Preparation and cleaning of geotechnical data

The geotechnical data is usually noisy, incomplete and inconsistent, thus affecting the effectiveness of ML models. Cleaning the data involves:

- **Deleting or Correcting Erroneous Data**: Geotechnical data can have errors that are caused either during the collection and input of data, including the wrong measurements recorded, or sensor failures. These are mistakes that have to be evaluated to help in improving the accuracy of the models.
- **Standardization of Units**: The data obtained in various sources might have different units of measure. As an example, the soil density could be stated in pounds per cubic foot and kilograms per cubic meter. Unit standardization provides uniformity and consistency in the dataset.

7.3.3 Handling Missing Data and Outliers

Missing data is a typical problem in real-world geotechnical projects. For example, incomplete records may be caused by sensor failures or field gaps of coverage. The most popular methods of dealing with missing data are:

- **Imputation**: Missing values may be imputed by methods of statistical estimation, including replacing them by the mean, median or mode of the data, or by more complex methods such as KNN imputation.
- **Removal**: Rows that possess too much missing data can be dropped in other cases when they cannot be plausibly filled in so that the model is not fit to unreliable data.

Outliers are drastic values that do not correspond to the rest of the data. Outliers could also be rare, but important, events in geotechnical data (e.g. a rare geological formation or seismic activity). Nevertheless, in case outliers are caused by the mistakes in the data or the malfunction of sensors, they should be identified and eliminated. Box plots, Z-scores and clustering techniques are some of the techniques that can be used to detect and deal with outliers.

7.4 CHALLENGES IN APPLYING AI AND ML TO GEOTECHNICAL FAILURES

The use of AI and ML can transform geotechnical engineering, specifically with predicting and avoiding failures of civil infrastructure. Nonetheless, even with the transformative features, a number of issues exist when utilizing AI and ML methods to geotechnical engineering issues (Cai et al., 2023). These difficulties can be attributed to the problems of data, model quality, interpretability and compatibility with conventional methods of geotechnology. This section talks about some of the major challenges involved in the use of AI and ML in predicting geotechnical failures.

7.4.1 Data Quality and Availability

Data quality and its availability is one of the biggest problems when it comes to the application of AI and ML to geotechnical failures. To successfully train and achieve predictions of high reliability, ML models need large datasets of high quality (Cala & Flisiak, 2020). Nevertheless, in geotechnical engineering, data is usually limited, intermittent and unattainable.

- **Insufficient or Unreliable Data for Training AI Models**: In some parts, particularly the developing countries or remote regions, geotechnical data may not be adequate. In one example, we could have an inadequacy of historical information about prior failures, field test or sensor data, which are needed to train AI/ML models. The information gathered can lack in totality or in other instances and concentrate on only a few parameters in the form of soil characteristics or seismic activity without looking at other important aspects, such as the level of ground water or the state of the environment. Such unavailability of data makes it difficult to develop powerful AI models that can be effective in predicting the issue in varied geographical areas and soil type.
- **Data Imbalance and Noise in Geotechnical Datasets**: The other frequent problem is data imbalance whereby not all classes of data are represented. An example is in the datasets that can be used to predict landslides or foundation failures: there can be much less of the true failures, whereas there can be much more of the non-failures. The effect of this unbalanced situation is the

creation of biased models which are either over-optimistic or cannot identify critical failure events. Also, geotechnical data may be noisy because of measurement errors, sensor failures or an environmental source of variation in the data. Dealing with noisy or unreliable data is the key to the accuracy and reliability of AI models.

7.4.2 Model Generalization and Overfitting

Deep learning models are the most powerful ML models, which are, however, vulnerable to the issues of model generalization and overfitting. The latter are of particular concern when using AI in geotechnical failure prediction, where the aim is to make sure that the models perform well not only on the training data but also on unseen data using real-world conditions.

- **Risks of Overfitting due to Limited or Non-representative Datasets**: Overfitting is the process where an ML model learns the training data too well that it learns the noise and outliers of the training data, so that the new data can no longer be modeled by it. This is especially a trouble in geotechnical engineering where the information might be sparse and extremely fluctuating. When the training data is too small, or is not representative of all the potential possible soil conditions, geotechnical environments or failure scenarios, the model can be very effective on the training data, but fail to predict effectively in new unseen sites or under geotechnical conditions. As an illustration, a model that is trained using data in a particular geographical location might not be effective in another geology with different geographical features.
- **Ensuring Models Generalize Well to New, Unseen Data**: Cross-validation and regularization methods are used to prevent overfitting and guarantee that AI models can be used effectively in generalization. Cross-validation is a method of dividing the data into various subsets and training the model using various combinations of the subsets so that the model can be able to generalize well to the unknown data. The regularization techniques enable the complexity of the model to be tamed so that it is not overly sensitive to noise and is not overly customized to the training data. Although these techniques are in place, striking a balance between the model complexity and generalization is considered to be a major problem in the field of geotechnical failure prediction where various factors that include soil variability and environmental conditions among others come into the picture.

7.4.3 Interpretability and Transparency

Another key issue when implementing AI and ML to predict geotechnical failures is the ability to be interpreted and be transparent. In contrast to conventional statistical models, AI and ML models, and in particular deep learning models, do not make their decisions easily, and are said to be black boxes by nature. Such non-interpretability creates both practical and regulatory difficulties in the field of civil engineering where engineers must have faith in the advice of the model and why making a particular decision is taken.

- **The Need for Explainable AI (XAI) Models in Civil Engineering**: Explainability in civil engineering, especially in prediction of geotechnical failures, is important. The engineers and decision-makers should be aware of the reasons why AI models have given the predictions. To illustrate, in the case where an AI model estimates that there are high chances of a slope failure in a particular territory, engineers should know the factors that prompted the model to make such an estimation. It may involve the contribution of soil type, water content, slope angle and seismic activity. In the absence of this knowledge, it would be tricky to apply the recommendations, make sound decisions or communicate the findings to the stakeholders, particularly in the safety-related situations when human lives may be in danger.
- **Understanding AI Model Decision-Making Processes for Practical Use**: To enhance trust and acceptance, AI models should be able to make known their decision-making process. The methods of LIME (Local Interpretable Model-agnostic Explanations) and SHAP (Shapley Additive Explanations) are being employed to render ML models more transparent by describing the significance of each individual feature to the model predictions. As an example, they can be used to determine which particular soil properties or environmental conditions have the most significant impact on predicting the foundation settlement or landslides to ensure that the engineers concentrate on those areas of concern when designing and developing mitigation plans.

7.4.4 Integration with Traditional Geotechnical Methods

The concepts of AI and ML are considered disruptive technologies, yet to become effective in geotechnical engineering, all of them should be combined with conventional geotechnical practices. The empirical and analytical methods of geotechnical engineering have a long tradition of grounded empirical and analytical methods, grounded on tests in the field, laboratory experiments and expert judgment. AI and ML can provide the new methods of optimizing such traditional approaches, yet the question is how to combine the merits of both strategies.

- **Combining AI/ML with Conventional Geotechnical Techniques and Expertise**: One of the primary issues is the question of introducing AI-powered models into the current experience in geotechnics. Even though AI models can operate with a substantial amount of data and detect the patterns that human engineers may have difficulty identifying, they nevertheless require the assistance of expert knowledge to understand the results and apply them to practice. Geotechnical engineers are well versed in the soil mechanics, material properties and the local geological conditions that would hardly be replicated fully by a machine. It is crucial that the key to the successful implementation of AI in geotechnical engineering should be a collaborative model, according to which data is processed with the help of AI and predictions are made, and human experience is applied and interpreted in accordance with this finding.

- **Challenges in Integrating AI Predictions into Standard Engineering Practice**: The other major issue is the adoption of AI predictions in normal engineering activities. Most of the engineers are more used to using traditional resources like soil classification charts, empirical formulas and the techniques of stability analysis. The implementation of AI in the working process can lead to the necessity to change the design, analysis and risk evaluation process used by engineers considerably. Moreover, AI implementation can be met with skepticism among those professionals who are either not conversant with or not completely confident with machine-based predictions. The only way to overcome this resistance is to ensure that there is adequate training, understanding the benefits of AI and working to close the gap between ML and engineering practice.

7.5 CASE STUDY: LANDSLIDE PREDICTION USING AI IN THE PHILIPPINES

One of the greatest geohazards to infrastructure and people living in mountainous and steep areas includes landslides. They present specific threat to infrastructure like roads, bridges and buildings and in most cases lead to disastrous loss of life and economic ruin and destruction of the environment. The common techniques of predicting landslides have traditionally depended on physical experiments, historical records and empirical models. Nevertheless, they can be human-intensive and fail to measure entirely the complex and nonlinear relationship between geological and environmental factors that cause landslides (Cao et al., 2017).

ML provides an opportunity to address these disadvantages as it provides predictive models, which produce the data-driven model which can receive numerous variables and react to real-time data streams. ML is applied and predictions are made about landslides in the given case study in the Philippines; which is prone to landslides since its topography is mountainous and which is a country characterized by constant precipitations. The estimation done of the probability of the landslides' occurrence during a period of extreme weather conditions was based on the application of varied data sources such as the records of the rainfall, soil composition, slopes, vegetation cover and seismic data to analyze them using the random forest classifier.

7.5.1 Study Location: Philippines

The Philippines is a part of the Pacific Ring of Fire which is a geologically active region characterized by frequent seismic and volcanic activities. It also has an annual monsoon season and contributes to high levels of precipitation that increases the chances of landslides. Specifically, regions in Leyte and Bohol are very prone to landslides, in that they are defined as having steep slopes, heavy vegetation and loose soil formations.

Over the past few years, tropical storms and typhoons present a heavy downpour that has resulted in several landslides that have had massive destruction to infrastructure, displacement of communities and loss of lives. The possibility of these landslides and the precautionary measures to be taken is a very sensitive issue of those in power and the civil engineers in the area.

7.5.2 ML Approach: Random Forest Classifier

The research team in the case study carried out in the Philippines employed a supervised ML algorithm, a random forest classifier, to forecast the occurrence of landslides in relation to numerous environmental and geological variables. Random forest algorithm is an ensemble algorithm which builds many decision trees in the training process and the mode (classification) of the classes is the one that is returned to the regression tasks.

7.5.2.1 Data collection

- **Rainfall Records**: Past rainfall data such as intensity and duration of the rainfall events was one of the most important data in predicting landslides. A large amount of precipitation usually precedes the saturation of soil that is likely to result in slope failure.
- **Soil Type and Composition**: Various soil types possess varied shear strength and water retention capacities, which directly influence the slope stability. The model incorporated data on the soil texture and composition to identify the contribution of each type of soil to the risk of landslides.
- **Slope Gradients**: Steeper slopes are not as safe to landslides. The input feature of the model was the gradient of the terrain which was calculated as the slope angle.
- **Vegetation Cover**: Vegetation cover may stabilize the slopes (using roots) and make the soil cohesive. On the other hand, the deforested areas are more susceptible to landslides. The degree of vegetation in the study area was measured through satellite images and land cover.
- **Seismic Data**: Earthquakes and tremors may cause landslides where weak soils exist. The model involved the information about seismic activity in the area (frequency and magnitude of tremors) to determine the role of seismic forces in the occurrence of landslides.

7.5.3 Data Preprocessing

The data before the application of the random forest classifier was heavily preprocessed and included:

- **Missing Data**: Any missing values in the data set, i.e. incomplete rainfall data or missing soil type data, were filled with interpolation techniques.
- **Feature Scaling**: To make the features, which include slope gradients and rainfall intensity, to be on similar scale and to enhance the performance of the random forest model, the features were scaled.
- **Outlier Detection**: Extreme values of rainfall intensity or abnormally steep slopes which are outliers in the data were identified and removed or adjusted in the data to guarantee the accuracy of the models.

7.5.4 Results and Performance

The trained random forest model could be used to predict the probability of landslides with high level of accuracy. The model also showed good performance in identifying the high-risk areas when it was in a high intensity of rain, and the recall rate was high, meaning that the model was successful at identifying the high-risk areas where landslides would take place. Figure 7.2 shows the relationship that existed between the percentage of soil compaction and foundation settlement. The points are spread widely, which shows weak to moderate correlation between compaction and settlement behavior. Though increased compaction is expected to enhance the strength of soils, the variability indicates that other characteristics like the content of moisture and the intensity of loads also affect settlement. There are a few negative and close to zero settlement

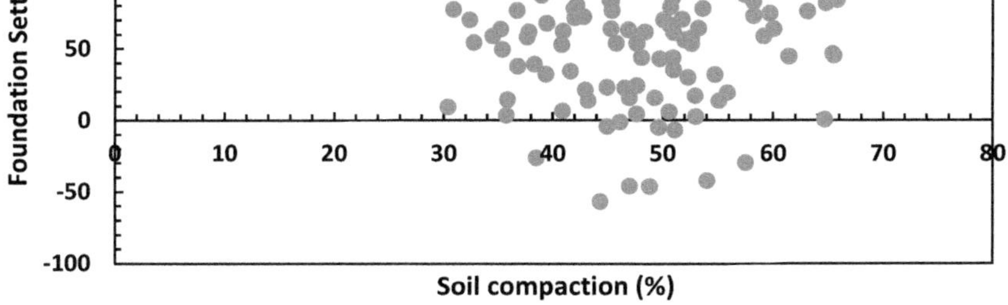

FIGURE 7.2 Relationship between soil compaction (%) and foundation settlement.

values, which represent slight heave or irregularity in the measurement. In general, the figure indicates the complexity and nonlinearity of soil to structure interaction.

Figure 7.3 is a scatterplot of applied foundation load against measured settlement. There is an overall upward trend, with increasing loads likely to give increasing settlement values. Nevertheless, the variation in points shows a high degree of variability, which is an indication of heterogeneous soil conditions and nonlinear stress–strain behavior. A single extreme outlier of the settlement indicates potential localized soil failure or unusual loading situation. The figure illustrates how one should look at load magnitude when making predictive settlement modeling.

Figure 7.4 illustrates the change in model accuracy with 20 training cycles. The values of accuracy vary in the range of 0.72–0.92, which is a progressive model

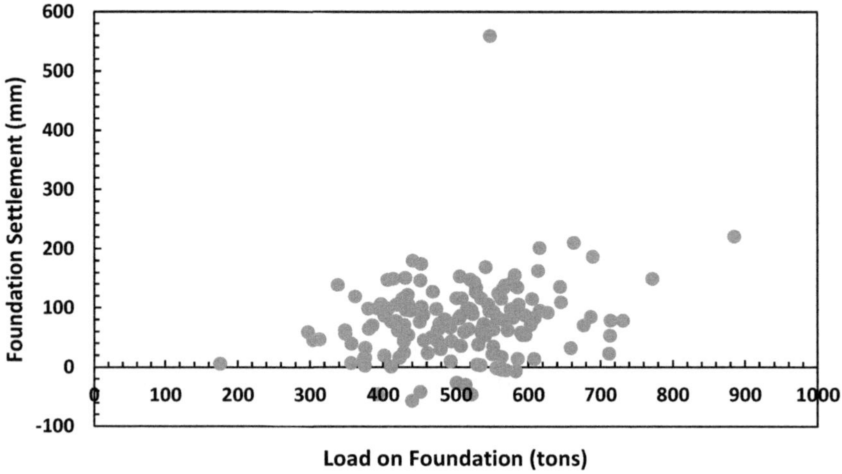

FIGURE 7.3 Relationship between load on foundation (tons) and settlement (mm).

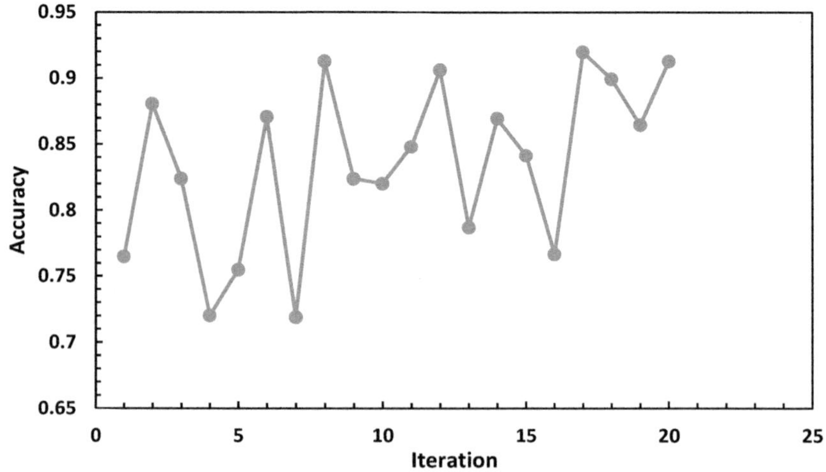

FIGURE 7.4 Model accuracy across training iterations.

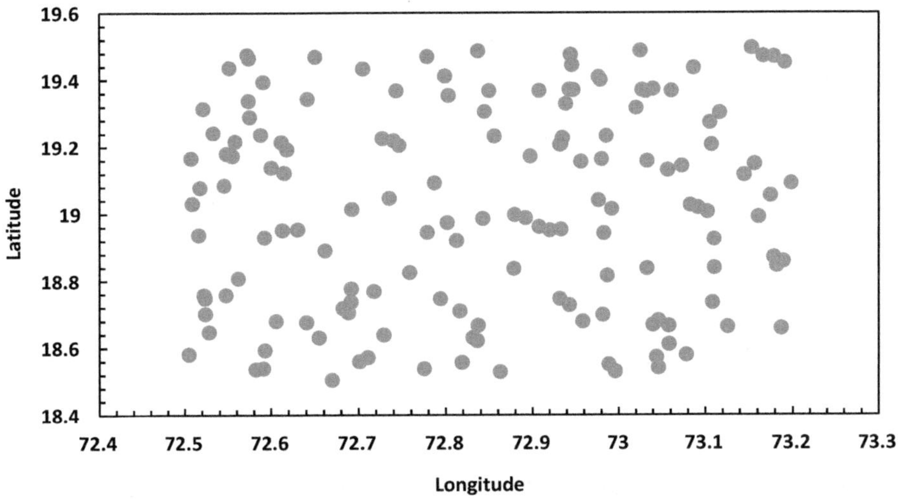

FIGURE 7.5 Spatial distribution of data points (latitude vs longitude).

learning and optimization. Subsequent runs tend to be more accurate, which implies better parameter optimization and convergence. Minor fluctuations indicate stochastic changes in training. On the whole, the figure displays the stability of the model and good predictability.

Figure 7.5 is an illustration of the geological distribution of geotechnical data collection points. The coordinates show that the area of study is well spread geographically in the region. There is no definite clustering tendency, which indicates uniform spatial sampling. This type of distribution enhances the generalization of the models by accounting for the variability in the terrain and soil conditions. The figure proves that it is well covered geographically to base the spatial prediction analysis.

7.5.5 Impact and Mitigation Measures

The different results of successful application of ML model to predict landslides were:

- **Real-Time Risk Assessment**: The model might be capable of providing real-time risk measurements, which will allow the local authorities to monitor lands of landslides in case of heavy rainfall and mitigation measures could be taken beforehand.
- Elaborating on the previous four, it was possible to take the initiative and prepare the evacuation plans, improve the infrastructure and be more attentive to the dangerous slopes, based on the forecasts of the model.
- **Strengthening of Infrastructure**: The model predictions came in handy in areas whose infrastructure (roads, bridges, etc.) at that point was particularly weak and engineers could strengthen infrastructure before a heavy downpour occurred.

7.6 FUTURE TRENDS AND DEVELOPMENTS

Their potential application to geotechnical engineering is increasing at a very rapid rate due to the continuous evolution of AI and ML. The future of AI is very bright regarding the anticipatory capabilities of the technology and its integration into smarter and stronger infrastructure (Chen et al., 2019). This part will look at the new trends, possibilities in the AI/ML technologies and the impacts of these new technologies on the future of geotechnical failure prediction. Moreover, it addresses the application of AI/ML to smart infrastructure and autonomous systems and the challenges and research areas that will shape the following generation of geotechnical solutions.

7.6.1 Advancements in AI and ML Technologies

AI and ML are rapidly advancing disciplines, new methods and technologies constantly come up, which can be utilized in geotechnical engineering.

- **Emerging Trends: XAI, RL and Generative Adversarial Networks**
  - **Explainable AI**: AI aloofness in prediction of geotechnical failures has been one of the main issues in deploying AI to predict geotechnical failures, particularly with deep learning algorithms. XAI is a new trend, which aims at making the process of AI decision-making more interpretable and understandable. Through XAI methods, engineers will be in a position to interpret the reasons behind predictions, i.e. which properties of the soil, moisture content or slope angle have the greatest effect on predictions of a failure. XAI will promote the development of trust and confidence in AI predictions, which will allow AI to be better introduced into engineering decision-making processes.
  - **Reinforcement Learning**: RL refers to a form of ML, in which agents develop through their interactions with the environment by receiving feedback in forms of rewards or punishments. In geotechnical sites, RL may be applied in optimizing real-time decisions involving monitoring the site or failure prevention. As an illustration, RL might help to select the most appropriate remediation methods or mitigation measures in response to a developing geotechnical problem, i.e. foundation settlement or slope instability, because it has to learn and adapt itself to new outcomes.
  - **Generative Adversarial Networks**: Generative Adversarial Networks (GANs) are a type of deep learning system that involves two neural networks (a discriminator and a generator), which are counterposed. GANs have demonstrated great potential in modeling complicated geotechnical

problems, including understanding soil behavior at different loading conditions or creating artificial data where the actual data is limited. GANs have the potential to assist in developing strong geotechnical models with limited data and therefore allow realistic predictions and superior design choices.

- **The Role of Quantum Computing in Geotechnical Analysis**: Quantum computing is a new technology that employs the laws of quantum mechanics to compute information radically differently. In geotechnical engineering, quantum computing could help a lot accelerate the calculations of AI and ML models, although it is still in its infancy. Quantum computers are very useful in processing large amounts of geotechnical data, or for complex simulations typically involving geotechnical design and failure prediction, as they are able to process large datasets and do it significantly faster than on classical computers. Quantum computing may also be used to increase the speed and accuracy of soil behavior, seismic risk analysis and other geotechnical predictions in the future and eventually enhance decision-making in civil infrastructure projects.

7.6.2 Integration with Smart Infrastructure

The combination of AI/ML and smart infrastructure would become a more pressing concern as cities can be smarter and more interconnected. Smart infrastructure is the application of sensors, communication technology and real-time data analytics to monitor and manage civil assets (Liu et al., 2020).

- **The Role of AI/ML in Monitoring and Maintaining Smart Infrastructure**: Smart infrastructure sensors can deliver real-time information on a number of geotechnical parameters, including soil moisture, pressure, temperature, and displacement. AI and ML will be significant in processing such data to anticipate possible failures, risks and maintenance optimization. To illustrate, AI models may be used to process sensor data on a bridge foundation to observe the early signs of settlement or fatigue and perform proactive repairs and avoid disastrous failures.

  Also, predictive maintenance is possible using real-time data analysis of smart sensor data because AI/ML models can predict when and where a given infrastructure is at risk of failure, which can be used to take timely actions. This would potentially save a significant amount of downtime and maintenance, raise the safety levels, and increase the life of the important infrastructure. In cities where urbanization and changes in the environment occur almost daily and require prompt renovation of the infrastructure, AI-based smart systems will play a vital role in enhancing the resilience of cities to geotechnical and environmental factors (Moayedi et al., 2020).

7.6.3 Potential for Autonomous Systems in Geotechnical Engineering

AI and ML are not only helpful in terms of prediction but also in changing the manner in which geotechnical engineering processes are conducted, shifting toward more autonomous systems of monitoring, failure prevention and remediation (Rashidi et al., 2016).

- **Use of AI for Autonomous Site Monitoring and Failure Prevention Systems**: Geotechnical monitoring and failure prevention autonomous systems are being increasingly developed. As an illustration, drones and robots that have sensors powered by AI can be utilized to autonomously survey construction sites and geotechnical conditions. Such systems are capable of gathering live information about the soil conditions, displacement or cracks and other indicators of structural distress in foundations or slopes. This information may then be analyzed by AI algorithms to determine the possibility of failure and give timely warnings to engineers.

  Examples of such autonomous systems would be used in parts of the world that are either inaccessible or dangerous to human engineers, like unstable slopes, mines or flood-prone areas. Having integrated autonomous systems with AI-based analytics, engineers can watch large regions in their entirety and automatically, which will greatly improve the timeliness and accuracy of failure detection and prevention.

7.6.4 Future Challenges and Research Areas

Although the uses of AI and ML have demonstrated immense possibilities in the field of geotechnical engineering, there are a number of challenges. These challenges will be very important in order to realize the full potential of these technologies.

- **Developing More Robust AI Models for Geotechnical Failure Prediction**: The necessity to have stronger AI models is one of the major challenges of AI development in geotechnical engineering in the future. At present, most AI models are not able to consider the heterogeneity and variability of geotechnical data, including the heterogeneity of soil, geological features of the region and environmental factors. More sophisticated models which can cope with uncertainty, variability and incomplete information are required. The studies of the new AI methods that will be able to deliver probabilistic predictions and measure the uncertainty will be vital in guaranteeing that AI models can give reliable predictions of various geotechnical situations (Wang et al., 2016).
- **Improving Data Acquisition Methods and Integration with Real-Time Sensor Networks**: Data acquisition is essential to real-time reporting and forecasting geotechnical failures and modern sensor technology remains limited in terms of data density, accuracy and long-term operation. AI and sensor network integration will necessitate more advanced and cost-effective sensors and enhanced data communication systems capable of operating

huge amounts of real-time data (Yin et al., 2018). Moreover, the attempts to combine the information of various sources, including field testing, remote sensing and past data, will also be critical to develop more profound datasets to be used in AI models. Future studies will be oriented toward enhancing the strength and reliability of sensors to be employed in geotechnical monitoring.

7.7 CONCLUSION

This chapter shows the practicality of ML methods to predict foundation settlement response to different soil and loading conditions. The findings support the fact that settlement is determined by a combination of various parameters that are interdependent, such as the percentage of soil compaction, structural load applied and the spatial variability per se. It is shown in the scatter analyses that although the increase in settlement is typical due to the increase in loads, the correlation is nonlinear and depends on the heterogeneity of the subsurface. Equally, better compaction is likely to decrease excessive settlement, although the effect is not always linear in nature owing to different soils, degrees of soil moisture and stress distribution.

The training process of the iterative model manifests a gradual increase in predictive accuracy which reaches high levels of reliability above 90% under the best conditions. This illustrates the strength of the embraced ML framework to observe the complicated soil structure interactions. The incorporation of the spatial coordinates also enhances the predictive ability since it takes into consideration the geographic variability, which implies a wider applicability to various site conditions.

In comparison to traditional empirical or semi-empirical methods, AI-based model is more adaptable, faster to compute and more accurate. It can also allow real-time tracking and predictive maintenance schemes where it is coupled with field instrumentation systems. The study, however, highlights that model performance is heavily reliant on data quality, representativeness and adequate preprocessing. Outlier detection and feature optimization are still necessary in order to attain sound results.

To sum up, the combination of AI and geotechnical engineering practice is a major milestone toward intelligent foundation design and sustainable infrastructural design. The suggested methodology will be a scalable and data-driven decision-support system that will potentially help engineers reduce structural risk, optimize foundation performance and enhance long-term civil engineering project safety and serviceability.

REFERENCES

Ahmad, S.; Kalra, A.; Stephen, H. Estimating soil moisture using remote sensing data: A machine learning approach. *Adv. Water Resour.* 2010, 33, 69–80.

Brungard, C.W.; Boettinger, J.L.; Duniway, M.C.; Wills, S.A.; Edwards, T.C., Jr. Machine learning for predicting soil classes in three semi-arid landscapes. *Geoderma* 2015, 239, 68–83.

Cai, X.; Yuan, J.; Zhou, Z.; Pi, Z.; Tan, L.; Wang, P.; Wang, S.; Wang, S. Effects of hole shape on mechanical behavior and fracturing mechanism of rock: Implications for instability of underground openings. *Tunn. Undergr. Space Technol.* 2023, 141, 105361.

Cala, M.; Flisiak, J. 2020. Slope stability analysis with FLAC and limit equilibrium methods. In: *FLAC and Numerical Modeling in Geomechanics-2001* (pp. 111–114). CRC Press, 2020.

Cao, Z.; Wang, Y.; Li, D. *Probabilistic Approaches for Geotechnical Site Characterization and Slope Stability Analysis*. Springer, Berlin Heidelberg, 2017. https://doi.org/10.1007/978-3-662-52914-0

Chen, R.; Zhang, P.; Wu, H.; Wang, Z., Zhong, Z. Prediction of shield tunneling-induced ground settlement using machine learning techniques. *Front Struct Civ Eng.* 2019, 13(6), 1363–1378. https://doi.org/10.1007/s11709-019-0561-3.

Ferentinou, M.D.; Sakellariou, M.G. Computational intelligence tools for the prediction of slope performance. Comput. *Geotech.* 2007, 34, 362–384.

Iyer, R.; Saranya, A.; Maralapalle, V.; Raskar, D. (2025). Reconfigurable intelligent surface-assisted backscatter communication for enabling 6G IoT networks. In N. M., M. Sathyamoorthy, & R. Dhanaraj (Eds.), *Applications and Challenges of Reconfigurable Intelligent Surfaces in 6G* (pp. 123–146). IGI Global Scientific Publishing. https://doi.org/10.4018/979-8-3693-8099-4.ch005

Liu, L.; Moayedi, H.; Rashid, A.S.A.; Rahman, S.S.A.; Nguyen, H. Optimizing an ANN model with genetic algorithm (GA) predicting load settlement behaviours of eco-friendly raft-pile foundation (ERP) system. *Eng Comput.* 2020, 36(1), 421–433. https://doi.org/10.1007/s00366-019-00767-4.

Maralapalle, V.; Muktinutalapati, J.; Ansari, A.; Chithambar Ganesh, A. (2026a). Geospatial analysis of soil quality: GIS-based mapping for sustainable land management. In S. C. Pal, U. Chatterjee, & D. Ruidas (Eds.), *Soil Degradation, Desertification and Restoration. Sustainability Solutions*, vol 1. Springer, Cham. https://doi.org/10.1007/978-3-031-92277-0_11

Maralapalle, V., Muktinutalapati, J., Ansari, A., Chithambar Ganesh, A. (2026b). Soil quality restoration zone mapping using GIS-based machine learning techniques. In S. C. Pal, U. Chatterjee, & D. Ruidas (Eds.), *Soil Degradation, Desertification and Restoration. Sustainability Solutions*, vol 1. Springer, Cham. https://doi.org/10.1007/978-3-031-92277-0_15

Maralapalle, V., Muktinutalapati, J., Chandra, B., Narala, G.R., Iyer, R. (2025). Analyzing the role of geospatial technologies and Al in urban infrastructure planning and the development of smart cities, including transportation systems, utilities, and public services. In M. O. Sarif, A. Sharifi (Eds.), *Advanced Geospatial Intelligence and AI for Environmental Resilience and Sustainable Development. Advances in Geographic Information Science*. Springer, Cham. https://doi.org/10.1007/978-3-032-03714-5_4

Maralapalle, V., Muktinutalapati, J., Tammineni, G., Krishna, M.V.N., Marlapalle, S. (2026c). Smart cities, smarter solutions: AI for urban transformation. In R. C. Poonia, D. Thakur, & I. Thakur (Eds.), Intelligent systems for sustainable infrastructure: AI solutions shaping a green future. Engineering cyber-physical systems and critical infrastructures, vol 17. Springer, Cham. https://doi.org/10.1007/978-3-031-95240-1_1

Moayedi, H.; Moatamediyan, A.; Nguyen, H.; Bui, X.-N.; Bui, D.T.; Rashid, A.S.A. Prediction of ultimate bearing capacity through various novel evolutionary and neural network models. *Eng. Comput.* 2020, 36(2), 671–687. https://doi.org/10.1007/s00366-019-00723-2.

Pham, B.T.; Pradhan, B.; Bui, D.T.; Indra Prakash; Dholakia, M.B. A comparative study of different machine learning methods for landslide susceptibility assessment: A case study of Uttarakhand area (India). *Environ. Model. Soft.* 2016, 84, 240–250. ISSN 1364-8152, https://doi.org/10.1016/j.envsoft.2016.07.005.

Rashidi, A.; Sigari, M.H.; Maghiar, M.; Citrin, D. An analogy between various machine-learning techniques for detecting construction materials in digital images. *KSCE J. Civ. Eng.* 2016, 20(4):1178–1188. https://doi.org/10.1007/s12205-015-0726-0

Saranya, A.; Maralapalle, V.; Zende, A.A.; Marlapalle, S.; Iyer, R. Artificial Intelligence and machine learning applications in groundwater pollution control and resource management. In A. Nath, A. Biswas, B. Koley, T. Choudhury, J.S. Um (Eds.), *Groundwater Depletion and Sustainability: Advances in Geographic Information Science*. Springer, pp. 67–88, Cham, 2025. https://doi.org/10.1007/978-3-032-09921-1_4

Shao, W.; Yue, W.; Zhang, Y.; Zhou, T.; Zhang, Y.; Dang, Y.; Wang, H.; Feng, X.; Chao, Z. The application of machine learning techniques in geotechnical engineering: A review and comparison. *Mathematics* 2023, 11, 3976. https://doi.org/10.3390/ math11183976

Wang, W.; Yuan, W.; Li, X.; Bai, B. Evaluation approach of the slope stability based on deformation analysis. *Int. J. Geomech.* 2016, 16(2), 04015054. https://doi.org/10.1061/(ASCE)GM.1943-5622.0000528.

Yin, Z.; Jin, Y.; Shen, J.S.; Hicher, P. Optimization techniques for identifying soil parameters in geotechnical engineering: Comparative study and enhancement. *Int. J. Numer. Anal. Meth. Geomech.* 2018, 42(1), 70–94. https://doi.org/10.1002/nag.2714.

Zmazek, B.; Todorovski, L.; Džeroski, S.; Vaupotiˇc, J.; Kobal, I. Application of decision trees to the analysis of soil radon data for earthquake prediction. *Appl. Radiat. Isot.* 2003, 58, 697–706.

8 Surrogate Modelling and Reliability Analysis for Complex Geotechnical Systems Using AI

Kumar Harsh, Kumar Shubham, K. V. N. S. Raviteja, and Abdhesh Kumar Sinha

8.1 INTRODUCTION

The foundation is the backbone of a building. It transmits the load to the ground. Designing a foundation is based on two main issues. One is to provide a margin against failure of the materials used in the foundation that could affect the foundation as well as the entire superstructure. The second issue is to ensure that settlement problems are prevented (Broere, 2016). To achieve this, it is imperative to conduct a rigorous study on the materials used in the foundation to determine the engineering properties that will guarantee the foundation's safety. Settlement problems in foundations can occur in

DOI: 10.1201/9781003714125-8

several ways. The compression of the soil layers in the ground can occur as a result of the self-weight of the soil (Boscardin & Cording, 1989; Mazouz et al., 2022). Changes in the effective stress in the soil can cause settlement problems. Pile driving and machine vibrations can cause settlement problems in the foundation. Moisture can cause settlement problems in clay soils. Changes in the soil can occur as a result of construction activities in the surrounding areas. Subsidence can cause settlement problems in the foundation. Subsidence can occur as a result of natural geological activities in the area. As the population in urban areas increases, the areas that were disturbed in the past are being developed. The development of disturbed areas has exposed them to settlement problems (Abbas, 2020; Maria & Naggar, 2015; Rodríguez et al., 2023; Tizpa et al., 2023; Wen et al., 2021; Bai, 2021; Wen-Chieh Cheng, 2021). The development of underground transit systems has raised a concern about the effects on shallow and deep foundations. Cavities in the ground can cause settlement problems in the foundation (Kumar et al., 2022; Lavasan et al., 2016; Sabouni & Airan, 2018).

To solve the problems that cause settlement in the foundation, several techniques can be used (Goel, 2015; Přibyl et al., 2021). However, choosing the right solution relies heavily on feasibility and spatial limitations (Keshavarz et al., 2019; Malhotra, Sahu, et al., 2020; Naggar, 2020; Sohaei et al., 2020). Traditional methods of foundation design often depend on fixed input parameters, which can lead to inaccurate safety margins (F. Wang, 2021; M. C. Wang & Hsieh, 1987). To manage uncertainties, engineers now use strategies like empirical correlations, statistical analyses, and back analyses based on field measurements (Akbari Garakani et al., 2020; Keba & Isobe, 2023; Shubham et al., 2022; Shubham, Metya, & Sinha, 2024b). Reliability analysis plays a key role in evaluating these uncertainties and providing probabilistic safety measures (Keshavarz et al., 2019; Malhotra, Sahu, et al., 2020; Naggar, 2020; Sohaei et al., 2020). Various computational techniques have been used, including response surface methods in probability-based frameworks, the First-Order Reliability Method (FORM), the Second-Order Reliability Method (SORM), and Monte Carlo Simulation (MCS) (Mukherjee & Babu, 2023; Sana et al., 2022). The First-Order Second Moment (FOSM) method has also been applied to study bearing capacity for foundation–cavity interaction (Malhotra, Srivastava, et al., 2020). However, computational costs remain a significant limitation in surrogate tunnelling. Adaptive kriging with quantile-based optimization has been suggested to address this issue. System reliability approaches that consider interactions between parameters have also become more popular (Dilip & Babu, 2023; Ray et al., 2021), along with fuzzy set-based models (Pramanik et al., 2021). Machine learning (ML) has boosted reliability-based geotechnical analysis. Supervised and unsupervised models, including regression techniques, decision trees, RFs, k-nearest neighbours, and Multivariate Adaptive Regression Splines (MARS), have shown promising results, along with neural methods like Artificial Neural Networks (ANNs) and Deep Neural Networks (DNNs) (Khorrami et al., 2020; Metya et al., 2017; Patel & Desai, 2010; Shubham, Metya, & Sinha, 2024a). ML-based surrogate tunnelling enhances computational efficiency and captures complex, non-linear soil–structure interactions (Sana et al., 2022; Shubham, Metya, Sinha, et al., 2024c; Zhang et al., 2017).

This chapter examines the potential to reduce the cost of high-dimensional reliability checks using new techniques, particularly in the area of pile raft–tunnel interactions, which is a concern that is on the rise in urban geotechnical engineering. It explores the potential to use artificial intelligence (AI)-driven surrogate modelling in conjunction with reliability modelling to capture the effects of variations in key soil properties and underground conditions. This approach to reliability-based design (RBD) is crucial in the design process because it addresses the shortcomings of the factor of safety (FS) approach, which has traditionally been used in the design process. To explain this, it is essential to note that the FS is not a true risk estimate, as numerous studies have already demonstrated that foundations can fail when the FS is slightly greater than 1. Moreover, the FS can represent a variety of risk levels, as the same FS can be achieved in a variety of ways. The FS approach has also never given a clear indication of the amount of safety that can be achieved as the FS increases, as it does not account for the inherent uncertainties in the stability parameters. In geotechnical engineering, there are three sources of inherent uncertainty: natural variability in the properties of the soil and materials used in the project, the inaccuracies in the measurements and test results gathered throughout the process, and the model uncertainties that arise as a result of the limitations in the analytical solutions used in the process.

8.2 METHODOLOGY

In this study, piled raft foundations were tested under single tunnelling for a systematic investigation using PLAXIS 2D. The analysis considered variations in undrained shear strength (*Su*), soil stiffness (*Es*), tunnel diameter (*D*), and tunnel depth (*Y*), testing each parameter in a realistic range while keeping other conditions constant. To reduce the computational effort of multiple finite element analyses, ML-based surrogate models were developed to predict PLAXIS outputs. Four surrogate models were created: MARS, ANN, RF, and Support Vector Regression (SVR). These models were trained with data from the parametric variations in PLAXIS 2D, providing efficient alternatives to finite element simulations for quickly predicting system responses. The surrogate models were then integrated with reliability analysis methods like the FORM, MCS, and Mean Value First-Order Second Moment (MVFOSM) to evaluate system performance under different soil–structure conditions.

8.2.1 Development of the PLAXIS Model

The PLAXIS tool was used to consider the behaviour of the soil layers during tunnelling and pile driving. In this study, the deep clay layers as well as the clayey strata were considered using the Mohr-Coulomb model. Similarly, the sand layers were considered using the Hardening Soil (HS) model as well as the HS-Small model, as depicted in Figure 8.1. The analysis was conducted under a plane strain condition, where the soil layers were meshed using a 15-node triangle for accurate depiction of the deformation as well as stress characteristics. The properties of the materials used in this study are depicted in Table 8.1.

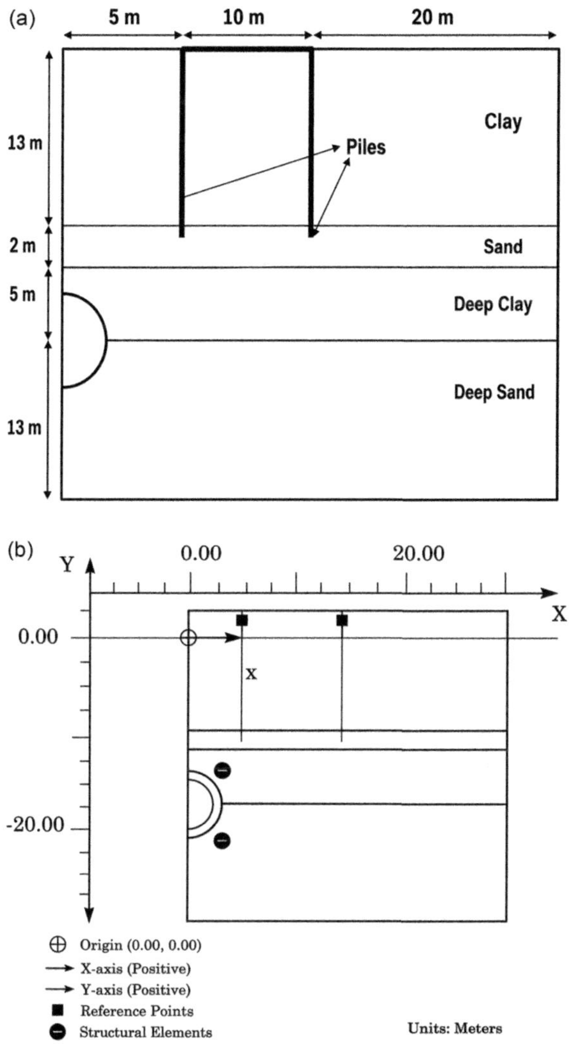

FIGURE 8.1 Illustrative diagram and corresponding PLAXIS model developed. (a) The configuration of the model. (b) PLAXIS model.

8.2.2 Mesh Convergence

In the finite element method (FEM) of analysis, mesh convergence is an important test to be performed to ensure that the results obtained from the FEM really represent the actual behaviour of the system under consideration. The FEM of analysis works on the principle of dividing the system into small parts called finite elements and then performing the equations of equilibrium, compatibility, and constitutive laws on these elements.

TABLE 8.1 Properties of soil and material considered in this study

CATEGORY	*PROPERTY/PARAMETER*	*UNIT*	*CLAY*	*DEEP CLAY*	*SAND*	*DEEP SAND*
Soil properties	Soil model	-	Mohr-Coulomb	Mohr-Coulomb	HS Small	HS Small
	Drainage type	-	Undrained (B)	Undrained (B)	Drained	Drained
	Unsaturated unit weight (γ_{unsat})	kN/m^3	15.00	16.00	16.50	17.00
	Saturated unit weight (γ_{sat})	kN/m^3	18.00	18.50	20.00	21.00
	Young's modulus (E'_{ref})	kN/m^2	3.4×10^3	4.2×10^4	-	-
	Secant stiffness (E'_{ref50})	kN/m^2	-	-	25×10^3	42×10^3
	Poisson's ratio (ν)	-	0.33	0.33	0.20	0.20
	Cohesion *I*	kPa	-	-	0	0
	Angle of internal friction (ϕ)	Degree	-	-	31	35
	Dilatancy angle (Ψ)	Degree	-	-	1	5
	Undrained shear strength (*Su*)	kN/m^2	5	40	-	-
	Permeability (*kx*, horizontal)	m/day	0.1×10^{-3}	0.01	1.0	0.5
	Permeability (*ky*, vertical)	m/day	0.1×10^{-3}	0.01	1.0	0.5
Category	**Material Properties**	**Unit**	**TBM**	**Tunnel Lining**	**Building**	**Pile**
Plate	Material type	-	-	Elastic	Elastic	-
	Unit weight (*w*)	kN/m^3	17.70	8.40	25.00	-
	Axial stiffness (*Eai*)	kN/m	63×10^6	14×10^6	1×10^{10}	-
	Bending stiffness (*EI*)	kN-m^2/m	472.5×10^3	143×10^3	1×10^{10}	-
	Stiffness *I*	kN/m^2	1.0×10^7	-	-	-
	Unit weight (γ)	kN/m^3	24.0	-	-	24.0
Embedded beam	Diameter (*D*)	m	-	-	-	1.0×10^7
	Pile spacing (*L*spacing)	m	-	-	-	24.0
	Skin resistance (*T*skin, start, max)	kPa	-	-	-	0.25
	Skin resistance (*T*skin, end, max)	kPa	-	-	-	3.0
	Base resistance (*F*max)	kPa	-	-	-	1.0
	Axial stiffness (EA_i)	kN/m	-	-	-	100.0

For the purpose of performing the mesh convergence test, five meshes of varying densities were used to analyse the problem under consideration. The average element size ranged from very coarse to very fine, with values of 2, 1.33, 1, 0.667, and 0.5 units, respectively. With the increase in mesh density, the number of elements and nodes also increased accordingly. For instance, the very coarse mesh used only 358 elements and 2,984 nodes, while the very fine mesh used 4314 elements and 35,192 nodes. Out of the five meshes used to perform the mesh convergence test, the medium mesh with an average element size of 1 showed consistent and stable results for the settlement of the structure under consideration. The difference in the settlements of the structure using the meshes of varying densities was very less, which shows that the medium mesh is the most efficient mesh to be used to solve the problem under consideration, and the results obtained using this mesh were steady and consistent.

8.2.3 Validation

To verify the results for surface settlements from the PLAXIS analysis, we adopted the empirical approach for a Gaussian distribution as proposed by Mair et al. (1993). This approach is commonly referred to as a widely cited approach for evaluating surface settlements that result from tunnelling construction, with special emphasis on soft cohesive soils such as clays. In this approach, it is assumed that surface settlements follow a Gaussian distribution curve. In this study, a tunnel with a diameter of 5 m was simulated at a depth of 17 m, as indicated in Figure 8.1, which represents a tunnel with its depth measured from the crown to the ground surface. In this simulation, a maximum surface settlement, *S*(max), of 13.0 mm was observed. To validate this figure, curve fitting for the empirical equation was carried out to obtain a best-fit value for i. This value for i that closely fit the PLAXIS results was found to be 13.0 m. Using this value for i, surface settlements were evaluated using the Gaussian equations and compared with those from PLAXIS for various values of distance from the centreline of the tunnel.

Traditionally, the settlement at the surface due to the excavation of a single tunnel has been estimated by empirical models, such as those developed by Attewell and Farmer (1974), Loganathan et al. (2000), Mair et al. (1993), and Peck (1969). Compared to the results obtained by the numerical model PLAXIS 2D, the most accurate settlement estimate for cohesive soils with localized deformation is obtained by the modified Gaussian model of Mair et al. Although the estimate obtained by Peck et al. is reasonable, it is too general, whereas the estimate obtained by Loganathan and Poulos overestimates the settlement, and that of Attewell and Farmer is conservative, resulting in a wider settlement trough than the actual one. The results obtained by the numerical model for the interaction between the tunnel, pile, and the soil can be considered more accurate, since the concentrated settlement trough is more localized compared to the results obtained by the empirical models, which overestimate the effects of the tunnel on the pile. This is an important aspect, since the interaction between the pile and the tunnel is significant in the case of soft ground, and the most realistic model for the settlement at the surface is the model developed by Mair et al. This indicates that the computed settlement trough width ($i \approx 13$ m) aligns well with literature values for similar depths in deep clay, deep sand, clay, and sand.

8.2.4 Parametric Study

The impact of a single tunnel dug beneath a piled raft foundation was assessed in this study using a methodical parametric analysis. Key parameters such as tunnel depth (Y), tunnel diameter (D), undrained soil shear strength (Su), and soil stiffness (Es) were the focus of the PLAXIS 2D analysis. To clearly determine the unique effects of each of these parameters on the system response, they were changed one at a time within reasonable ranges. These variations were systematically incorporated into the study to account for uncertainty, as shown in Table 8.2.

The main performance indicators for the pile raft system were thought to be ground deformation and load redistribution. The findings shed important light on how tunnelling operations affect the stability, deformation behaviour, and load transfer mechanisms of piled rafts. An increase in the undrained shear strength of the surrounding soil reduces the maximum surface settlement and its propagation rate, regardless of whether the tunnelling is carried out with support at the tunnel face or in the presence of ground volume loss.

8.2.5 ML Surrogate Models

ML is a part of AI that transcends the limitations of numerical simulation by using input/output data to learn patterns, which are used to classify or predict. It has gained widespread use in many disciplines, and in geotechnical engineering, ML techniques use experimental data, field measurements, or numerical simulation results as input. Surrogate models have proven to be a cost-effective means of performing parametric analysis, uncertainty analysis, and optimization, which improve the reliability and FS of geotechnical work. The surrogate models used in this work are as follows.

MARS is a non-linear regression analysis tool that uses piecewise linear functions to model complex non-linear relationships. MARS is a powerful tool, especially when dealing with high-dimensional data, as it can automatically detect the most important variables as well as variable interactions. MARS is best used when the goal of the analysis is to detect non-linear patterns without relying on statistical assumptions, as it achieves a good balance of interpretation and prediction accuracy. The mathematical formulation of the MARS model is given by Equation (8.1).

$$y(x) = \beta_0 + \sum_{M=1}^{M} \beta_m h_m(x) \tag{8.1}$$

TABLE 8.2 Range of parameter variation for parametric study

PARAMETER	*RANGE*	*INCREMENT*	*BASE VALUE*
Tunnel diameter, D (m)	5–8.5	0.5	6.5
Tunnel depth, Y (m)	14–24	2	20
Undrained shear strength, Su (kPa)	0.75Su–1.75Su	0.25Su	Su = 90 (clay – layer 1); Su = 130 (deep clay – layer 3)
Young's modulus, Es (kPa)	0.75Es–1.75Es	0.25Es	$Es = 3.4 \times 10^3$ (clay – layer 1); $Es = 9 \times 10^3$ (deep clay – layer 3)

where $y(x)$ = predicted response, β_0 = Intercept, β_m = coefficient of the m-th basis function and $h_m(x)$ = basis functions (hinge functions, interactions, products of splines).

A Neural Network, or ANN, is a brain-inspired ML model that consists of layers of interconnected units or nodes called neurons, which learn from the data provided through the weighted edges and activation functions to reveal hidden patterns in the data during the training process. Neural networks perform well on non-linear problems, classification, and regression but are more difficult to interpret than other models. Neural networks consist of multiple layers of nodes to learn non-linear relationships between inputs and outputs, and the predictions from the model are based on Equation (8.2).

$$y(x) = f\left(\sum_{j=1}^{n} w_j g\left(\sum_{i=1}^{n} v_{i_j} x^1 + b_j\right) + c\right) \tag{8.2}$$

where $g(\cdot)$ = activation function (sigmoid, ReLu) and v_{i_j}, w_j = weights, b_j, c = biases.

RF is an ensemble ML model that generates many decision trees and averages their results to make predictions on the test data. RF helps prevent overfitting and generates precise results by averaging the results of the decision trees generated. It is efficient in handling non-linear relationships, dealing with high-dimensional data, and computing the importance of the variables, making it suitable for regression and classification problems, with predictions based on Equation (8.3).

$$y(x) = \frac{1}{T}\sum_{t=1}^{T} h_t(x) \tag{8.3}$$

where $h_t(x)$= prediction from tree t and T = number of trees.

SVR is a regression version of the Support Vector Machine model. It tries to find the best-fitting function that lies within the given tolerance band around the data provided. SVR utilizes the kernel functions to handle non-linear data and is more likely to generalize well on the test data. It is more appropriate for small to medium-sized datasets and requires proper selection of parameters and kernels, with predictions based on Equation (8.4).

$$f(x) = \sum_{t=1}^{N} (\alpha_i - \alpha_i^*) k(x_i, x) + b \tag{8.4}$$

where $k(x_i, x)$ = kernel function (linear, polynomial, radial basis function (RBF)), $\propto_i - \propto_i^*$ = Lagrange multipliers, and b = bias.

8.2.6 Reliability Analysis

Reliability analysis enhances numerical simulations by acknowledging the presence of uncertainties in soil properties, loads, and geometry. It provides a probabilistic estimate of the FS, identifies the factors that cause the greatest concern, and allows quick comparison of different scenarios. It can be used to make risk-informed decisions to establish confidence in the results of geotechnical stability analyses. Some of the widely used reliability analysis techniques are briefly described as follows.

FORM is a probabilistic method used to estimate the probability of failure using the linearity of the limit state function at the most probable point of failure, known as the design point. It uses transformations to standard normal space to provide efficient and reliable results at a low computational cost. The reliability index β is the distance from the origin to the limit state surface in standard normal space, as given in Equation 8.5.

$$\beta = \min\|U\| \text{ subjected to } g(X) \tag{8.5}$$

where $g(X)$ = limit state function (failure occurs if $g(X) \leq 0$ and U = transformed new variables). The probability of failure is given using Equation (8.6).

$$P_f = \phi(-\beta) \tag{8.6}$$

where ϕ = cumulative Distribution Function (CDF) of the standard normal distribution.

MCS, on the other hand, uses a method of evaluating reliability by selecting each variable from its probability distribution and determining how it performs. It's a highly accurate method but may take a long time if dealing with complex and large systems. The probability of failure can be obtained from Equation (8.7).

$$P_f = \frac{N_f}{N} \tag{8.7}$$

where N_f is the number of simulations where $g(X) \leq 0$, N = total number of simulation runs.

MVFOSM, on the other hand, is a FORM that depends on the mean and variance of the inputs to determine how reliable a system is. It's a first-order method because it depends on the first and second moments of each variable. It's a quick method but may not give accurate results if dealing with complex systems because it's based on a first-order Taylor expansion about the mean value. The reliability index can be obtained from Equation (8.8):

$$\beta = \frac{\mu_g}{\sigma_g} \tag{8.8}$$

where μ_g = mean of the limit state function $g(X)$, σ_g = standard deviation of $g(X)$, computed using Equation (8.9):

$$\partial_g = \sum_{i=1}^{n} \left(\frac{\partial_g}{\partial_{xi}} \mu^1 \right)^2 \partial_{xi}^2 \tag{8.9}$$

Here, μ and ∂ are the mean and standard deviation of the input variables, respectively.

8.3 RESULTS AND DISCUSSION

8.3.1 Parametric Analysis

8.3.1.1 Effect of undrained shear strength (Su)

For the fixed tunnel foundation interaction problem (tunnel diameter D = 5 m, depth Y = 17 m, grout pressure GP = 230 kN/m^2, soil moduli Es = 3400 and 9000 kN/m^2, pile length L = 11 m, pile width w = 5 m, and horizontal pile spacing X = 5 m), the study examines the effect of variations in the undrained shear strength of the soils on the settlements resulting from tunnel construction. The results of the study revealed that the settlements of the tunnel (T1), the left pile (LP), and the right pile (RP) significantly decrease with the increase of the undrained shear strength of the soils from –25% to +75% of the original value. This reveals the importance of the stiffness of the soils on the settlement of the ground. Along with the tunnel diameter, embedment depth, grout pressure, pile length, pile width, and horizontal pile spacing, the undrained shear strength of the soils is also found to be an important parameter that affects the tunnel and foundation interaction problem. It is also concluded that the undrained shear strength of the soils may be improved or optimized to effectively control the settlement of the ground resulting from tunnel construction.

8.3.1.2 Variation of soil stiffness (Es)

In a fixed tunnel pile system, i.e., D = 5 m, Y = 17 m, GP = 230 kPa, Su = 40 and 52 kPa, L = 11 m, w = 5 m, X = 5 m, the study aims to investigate the effect of soil stiffness on tunnelling-induced settlements. From the findings, it is evident that there is a notable decrease in the settlement at T1, the LP, and the RP as the soil stiffness increases from 0.75 to 1.75 times its baseline value. For instance, the settlement at T1 decreases from –62 to –37 mm, i.e., a reduction of 40.3%. Similarly, the settlement at the RP and the LP decreases from –29 to –17 mm, i.e., a reduction of 41.4%, and from –25 to –15 mm, i.e., a reduction of 40%, respectively. From the results, it can be concluded that soil stiffness has a notable effect on tunnelling-induced settlement behaviour in soils. That is, the soil can resist deformation to a greater extent as its stiffness increases. Thus, soil stiffness plays a vital role in addition to tunnel diameter, embedment depth, grout pressure, undrained shear strength, and pile dimensions in tunnelling-induced settlement behaviour.

8.3.1.3 Variation of tunnel depth (Y)

For a fixed tunnel pile setup where D = 5 m, GP = 2300 kPa, Su = 40 and 52 kPa, Es = 3400 and 9000 kPa, L = 11 m, w = 5 m, X = 5 m, the study investigates the effect of the depth of the tunnel on the settlement caused by tunnelling. At various depths ranging from 14 to 24 m, settlement is measured at three points: tunnel face T1, LP, and RP. From the results obtained, it is evident that as the depth of the tunnel is increased, the settlement caused is significantly smaller. At a depth of 14 m, T1 settlement is –46.85 mm, while at a depth of 24 m, it is –10 mm. This is a reduction of 78.6%. Similarly, RP settlement is reduced from –19.85 mm at a depth of 14 m to –15 mm at a depth of 24 m. This is a decrease of about 24.5%. At the same time, LP settlement is reduced from –22.98 mm at a depth of 14 m to –15 mm at a depth of 24 m. This is a decrease of about 34.8%. These results show that as the depth of the tunnel is increased, the settlement caused is significantly smaller.

8.3.1.4 Variation of tunnel diameter (D)

With the constant tunnel depth $Z = 17$ m, grout pressure $GP = 230$ kPa, undrained shear strength $Su = 40$ and 52 kPa, soil stiffness $Es = 3400$ and 9000 kPa, pile length $L = 11$ m, pile width $w = 5$ m, and horizontal pile spacing $X = 5$ m, the study investigated the impact of the tunnel diameter on settlements due to tunnelling. At tunnel diameters ranging from 5 to 8.5 m, the settlements were measured at T1, the LP, and the RP. From the results, the RP settlements remain constant, while the T1 and LP settlements increase with the increase in the tunnel diameter. At T1, the settlement increases by 59.6% from –52 to –83 mm, and at the LP, the settlement increases by 176% from –21 to –58 mm. At RP, the settlement increases slightly by 0.25%. This shows that the tunnel diameter has an impact on the settlements due to tunnelling, and the increase in the tunnel diameter results in increased deformation of the soil due to the redistribution of the excavation-induced stresses in the soil. Figure 8.2 illustrates the change in the settlements due to the change in *Su*, *Es*, *Y*, and *D*.

8.3.2 Development of ML-Based Surrogate Model and Performance Evaluation

A set of ML-based surrogate models was created to predict the tunnelling-induced settlements at three critical locations: the tunnel crown (T1), the LP, and the RP. A well-designed set of parameters was created to feed the ML algorithms. The parameters selected to feed the model included the tunnel diameter (*D*), the depth of the tunnel (*Y*), the undrained shear strength (*Su*), and the modulus of elasticity (*Es*) of the soil. The parameters selected to predict the settlements included the settlements at T1, LP, and RP. These parameters were selected due to their significant impact on the soil–structure interaction process. Four ML-based surrogate models were created to predict tunnelling-induced settlements: MARS, ANN, SVR, and RF. The process was done using the same data preprocessing and training steps. The ANN model had two hidden layers with 128 and 64 neurons, a learning rate of 0.001, and up to 500 iterations. The SVR model had the RBF kernel to handle complex non-linearities. The parameters selected for the SVR model included $C = 100$ and $\gamma = 0.01$. The RF model had 200 decision trees with a maximum depth of 8 to provide the right balance between interpretability and model strength. The MARS model had up to 50 basis functions and an interaction degree of 2 to prevent overfitting and sparseness.

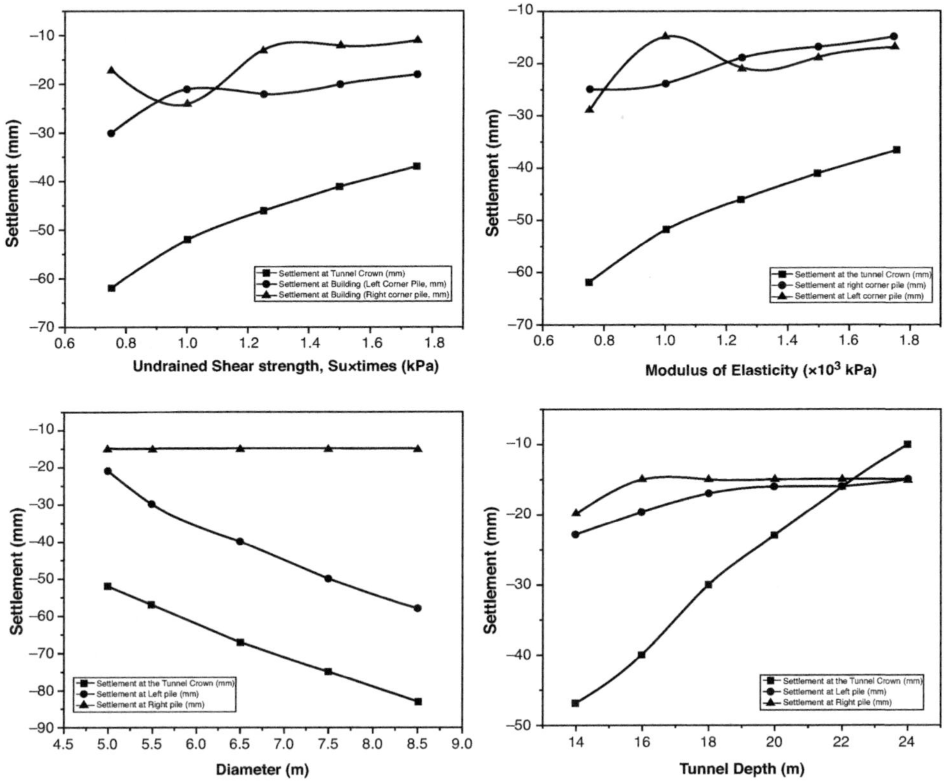

FIGURE 8.2 Graphs showing variation of different parameters. (a) Effect of undrained shear strength (*Su*). (b) Effect of modulus of elasticity (*Es*). (c) Effect of variation in diameter (*D*). (d) Effect of change in tunnel depth (*Y*).

Significant hyperparameter adjustment was carried out to achieve the best performance for each model. A five-fold cross-validation technique was utilized to evaluate the performance of each model in terms of how well it generalizes and to reduce bias that can occur when there is overfitting. The final model was selected based on a number of factors, such as prediction accuracy, reliability across different input scenarios, and computational efficiency. The surrogate models can be used as a quick and cost-effective alternative to the computationally expensive simulation process to make quick and reliable predictions on tunnelling-induced settlement in complex geotechnical problems. To evaluate the performance of the surrogate models, the following were used: R^2, root mean squared error (RMSE), and mean absolute error (MAE). It was evident that all the surrogate models were highly accurate in prediction. RF was the best model in terms of prediction accuracy, achieving an R^2 of 0.988, RMSE of 2.10 mm, and MAE of 1.60 mm. MARS and ANN were close seconds in terms of prediction accuracy, achieving R^2 values of 0.982 and 0.978, RMSE values of 2.43 and 2.66 mm, and MAE values of 1.85 and 1.92 mm, respectively. SVR trailed the pack in terms of prediction accuracy, achieving an R^2 of 0.955, RMSE of 3.20 mm, and MAE of 2.15 mm. It is evident that all

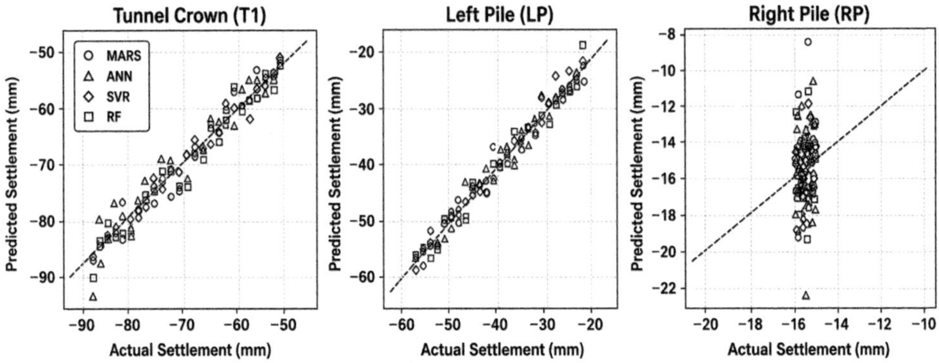

FIGURE 8.3 Performance of the surrogate models. (a) Regressor plots. (b) Residual error plots.

TABLE 8.3 Statistical characterization of input parameters for reliability analysis

PARAMETER	*MEAN*	*STANDARD DEVIATION*	*COV*	*DISTRIBUTION*
D	6.5 m	0.5	0.08	Normal
Su	50 kPa	10	0.15	Normal
Es	8000 kPa	2000	0.20	Log-normal
Y	17 m	2	0.10	Normal

the surrogate models can effectively capture the settlement behaviour and can be used as effective surrogate models (Debbarma & Ransinchung, 2022). Based on the reliability of the surrogate models, they were used in the FORM, FOSM, and MCS to enable the reliability analysis of tunnel foundation systems. Figure 8.3 shows the prediction accuracy achieved by the surrogate models.

8.3.3 Reliability Analysis-Based Design

A RBD analysis was performed to estimate the probability that the settlements resulting from tunnelling would exceed the permissible limit given by the allowable settlements $S_{allowable} = 25$ mm. The limit state function is given by Equation (8.10), and the failure was represented by the inequality $g(X) \leq 0$. The parameters *D*, *Su*, *Es*, and *Y* were treated as random variables and their statistical characteristics are given in Table 8.3.

$$g(X) = S_{allowable} - S_{T1}\left(D, Su, Es, Y\right) \tag{8.10}$$

A group of surrogate models, such as MARS, ANN, SVR, and RF, was utilized for the anticipation of settlements for large numbers of input values' realizations. This allowed for the efficient utilization of FORM, FOSM, and MCS methods. For all the surrogate

TABLE 8.4 Reliability analysis outcomes for each surrogate model

SURROGATE MODEL	β_{FORM}	P_{f_FORM}	β_{FOSM}	P_{f_FOSM}	β_{MCS}	P_{f_MCS}
MARS	3.05	0.11	2.92	0.17	3.08	0.10
ANN	3.02	0.13	2.88	0.19	3.05	0.11
SVR	2.85	0.22	2.71	0.34	2.83	0.23
RF	3.12	0.09	2.98	0.14	3.10	0.10

models, the reliability index (β) as well as the failure probability (P_f), as presented in Table 8.4, is given.

Based on the reliability study results, all surrogate models provide predictions that are consistent and reasonable from a physical point of view. Among the surrogate models, the RF model exhibits the highest reliability, with a beta value of 3.12 and a probability of failure of 0.09%–0.10%, which confirms the model's top-notch prediction accuracy. The MARS and ANN models follow the RF model but have slightly lower beta values and a slightly higher probability of failure, which still confirms their reliable performance in terms of non-linear settlement responses. The SVR model, however, shows somewhat lower reliability than the other three models, with a beta of 2.85 and a probability of failure of 0.22%–0.34%, which may be due to the model's less precise determination of settlement trends, as previously stated in the performance assessment.

To quantify the effect of the uncertainties associated with the settlement predictions on the pile raft foundation's safety and serviceability, the RBD method is employed in the study. In the RBD method, the average and standard deviation of the settlements predicted using the various surrogate models are used to calculate the reliability index (β) and the probability of failure (P_f). The reliability index is defined as the difference between the allowable and predicted settlements, divided by the standard deviation of the predicted settlement, which reflects the margin of safety under uncertain loading conditions. The reliability index is set to 3.0 (probability of failure = 0.13%), which is the acceptable design reliability based on the serviceability criterion. Figure 8.4a shows the variation of the reliability index β with the applied load factor γ using all the surrogate models, and Figure 8.4b shows the probability of failure variation using the surrogate models. It is seen that the reliability index β decreases and the probability of failure P_f increases with the increase of the applied load factor γ, which indicates that the reliability of the pile raft foundation decreases with the increase of the applied load factor γ. The RF model and the MARS model predictions show somewhat higher reliability than the other models, while the SVR model shows a sharp decrease in the reliability index β with the increase of the applied load factor γ, which may be due to the model's sensitivity to the applied loads.

Target reliability index and their corresponding failure probabilities were followed according to Load and Resistance Factor Design (LRFD/NCHRP) National Cooperative Highway Research Program recommendations for pile foundations. For non-redundant pile groups (up to four piles at a cap), a reliability index of 3.0 was chosen as a target reliability index β_T, which gives a P_f of 0.13%. For redundant pile groups, a looser value of 2.33 for β_T (P_f of 1%) might be acceptable. A more conservative serviceability analysis was also considered by evaluating β_T up to 3.72 (P_f of 0.01%). The limit state

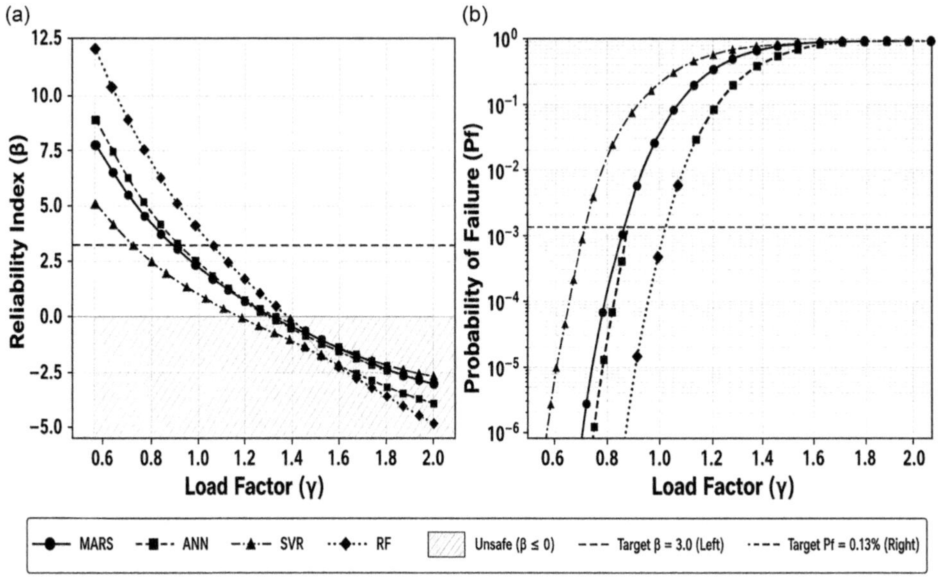

FIGURE 8.4 Charts showing the results of reliability-based design. (a) Reliability index vs. load factor curve. (b) Probability of failure vs. load factor curve.

was evaluated according to Equation (8.10). Surrogate models such as MARS, ANN, SVR, and RF were used to predict ST1 for large numbers of input realizations. For each of these models, FORM, FOSM, and MCS were performed and evaluated against the target reliability index and failure probabilities. If the calculated P_f was higher than the target value, adjustments were iteratively made to the design until the required P_f was satisfied. This may involve increasing pile capacity and redundancy, changing pile geometry, or ground improvement to increase *Su/Es*.

8.4 CONCLUSIONS

This work proposes a RBD approach to examine the behaviour of pile raft–tunnel structures under geotechnical uncertainty, employing PLAXIS 2D computations as well as ML surrogate models. The effect of various parameters, such as the undrained shear strength of the soil, soil stiffness, tunnel diameter, and tunnel depth, was evaluated by varying these parameters to assess the effect on tunnel-induced settlements. Four different surrogate models, namely, MARS, ANN, RF, and SVR, were used to approximate the PLAXIS 2D computations. Among these, the ANN and RF models demonstrated promising accuracy, with the RF model producing the best accuracy. The surrogate models, combined with probabilistic techniques such as FORM, MCS, and MVFOSM, demonstrated high accuracy in assessing the probability of settlement beyond allowable limits, which resulted in a considerable reduction of computation costs. The salient points of the work can be listed as:

- The RF surrogate model, along with the ANN model, demonstrated the best accuracy, with the RF model producing the highest accuracy among the models, as shown by the R^2, RMSE, and MAE, which were 0.988, 2.10 mm, and 1.60 mm, respectively, proving to be highly effective.
- Increasing the value of *Su* as well as *Es* was found to suppress tunnel-induced settlements substantially.
- Increasing the tunnel depth resulted in a settlement reduction of up to 78.6%, whereas an increase in tunnel diameter resulted in an increase in settlements by up to 176%.
- The RBD approach utilizes a combination of surrogate models, FORM, MCS, and MVFOSM for efficient quantification of uncertainties.
- Of all these methods, the RF method is found to provide the highest reliability index value ($\beta = 3.12$) along with the lowest failure probability ($P_f = 0.09\%$–0.10%). This method is highly efficient for probabilistic analysis.

Using the results obtained from this study on RBD of pile raft foundations considering the effect of tunnelling, here are some recommendations that may be followed in practice and incorporated into national codes of practice. From the study, it was evident that FS-based design does not address the real risk of pile raft–tunnel interactions because it does not take into account the variability of properties such as undrained shear strength of the soil (*Su*), modulus of soil (*Es*), depth from ground surface to the axis of the tunnel (*Y*), and relative geometry between the foundation and tunnel (*D*). RBD must therefore be employed to address these issues of variability.

From the results obtained from this study, it was evident that a reliability index of $\beta \approx 3.0$, corresponding to a failure probability of 0.13%, was a reasonable compromise between safety and cost considerations of foundations in urban areas where tunnelling may take place. For redundant pile foundations, a slightly lower value of $\beta \approx 2.33$ corresponding to a failure probability of 1% may be permissible.

Practically a reasonable settlement limit could be standardized to 25 mm, with a warning limit of 12.5 mm or 50% of the limit to warrant on-site inspection and corrective action. Site investigations should be planned to account for the statistical variability of the critical geotechnical parameters, rather than their average values alone. To this end, reference distributions could be used as provisional inputs to the analyses, such as $D = 6.5 \pm 0.5$ m, $Su = 50 \pm 10$ kPa, $Es \approx 8000 \pm 2000$ kPa, and $Y = 17 \pm 2$ m, until project-specific data is available.

If the values of P_f exceed the limits of what is acceptable, the study recommends the following ground improvement strategies to increase the stiffness and strength of the ground: For example, increasing the value of *Su* by at least 25% (from 50 to 62.5 kPa) or increasing the value of *Es* by 50% (from 8000 to 12,000 kPa) could restore the reliability to the desired levels. Similarly, controlled grouting could be employed with a 20%–30% increase in pressure (from 230 to 300 kPa) to control the settlement resulting from the excavation of the tunnel. Moreover, the study recommends the efficient reliability analyses using AI-based surrogate models like RF, ANN, SVR, and MARS, but only when the prediction efficiency of the model ($R^2 \geq 0.98$) is validated using the results obtained from the FEM. The study also recommends the following design codes to be implemented to ensure the safe and realistic foundation design near the tunnel excavation site: continuously monitoring the settlement amounts at the key points (T1, LP, RP) and using the results to update the surrogate model to re-calculate the reliability using the results from the RBD.

The study is exhaustive, although some limitations exist. The study is based on two-dimensional finite element modelling, which may not capture the effects of three-dimensional tunnelling as well as the dynamic behaviour of piles interacting with the raft structure. The undrained behaviour of the soil, as assumed by the study, may not capture the anisotropic nature of the soil as well as the natural variability of the soil properties, aside from the statistical properties presented by the study. The selection of the tunnel diameter, grout pressure, and pile size may also be limited, thus influencing the general applicability of the results to the full range of actual field conditions. It may also be beneficial to include field monitoring data to further improve the reliability of the surrogate models.

Future directions of the study could be to extend the three-dimensional modelling to include time-dependent consolidation and hydro-mechanical effects. The database could be extended to include real-time field monitoring data, which may further improve the reliability of the models. It may also be beneficial to incorporate more advanced ML algorithms like ensemble deep learning or Gaussian Process Regression to further improve the reliability of the models.

REFERENCES

Abbas HO (2020) Influence of tunnel excavation on tower foundation settlement constructed on sandy soil. *Transportation Infrastructure Geotechnology* 7:651–663. https://doi.org/10.1007/s40515-020-00114-x

Akbari Garakani A, Sadeghi H, Saheb S, Lamei A (2020) Bearing capacity of shallow foundations on unsaturated soils: Analytical approach with 3D numerical simulations and experimental validations. *International Journal of Geomechanics* 20:1–17. https://doi.org/10.1061/(asce)gm.1943-5622.0001589

Attewell PB, Farmer IW (1974) Ground disturbance caused by shield tunnelling in a stiff, overconsolidated clay. *Engineering Geology* 8:361–381. https://doi.org/10.1016/0013-7952(74)90028-3

Bai X-D, Cheng W-C, Ong DEL, Li G (2021) Evaluation of geological conditions and clogging of tunneling using machine learning. *Geomechanics and Engineering* 25:59–73. https://doi.org/10.12989/gae.2021.25.1.059

Boscardin MD, Cording EJ (1989) Building response to excavation-induced settlement. *Journal of Geotechnical Engineering* 115:1–21. https://doi.org/10.1061/(ASCE)0733-9410(1989)115:1(1)

Broere W (2016) Urban underground space: Solving the problems of today's cities. *Tunnelling and Underground Space Technology* 55:245–248. https://doi.org/10.1016/j.tust.2015.11.012

Debbarma S, Ransinchung, GD (2022). Using artificial neural networks to predict the 28-day compressive strength of roller-compacted concrete pavements containing RAP aggregates. *Road Materials and Pavement Design* 23(1): 149–167. https://doi.org/10.1080/14680629.2020.1822202

Dilip DM, Babu GLS (2023) System reliability-based design optimization of flexible pavements using adaptive meta-modelling techniques. *Construction and Building Materials* 367:130351. https://doi.org/10.1016/j.conbuildmat.2023.130351

El Naggar A, Youssef MA, El Naggar H. (2020) Effect of tunneling on shallow foundations. In *Source: The 2020 World Congress on Advances in Civil, Environmental, & Materials Research (ACEM20)*, pp. 25–28. http://www.i-asem.org/acem20_publication.html#ICGE20

Goel RK (2015) Use of underground space for the development of cities in India. *Water and Energy International* 58RNI:41–45.

Keba EL, Isobe K (2023) Ultimate bearing capacity of a shallow foundation on unsaturated sandy soil with cavity. In: Askar Zhussupbekov, Assel Sarsembayeva, Victor N. Kaliakin. (Eds.), *Smart Geotechnics for Smart Societies*. Proceedings of the 17th Asian Regional Conference on Soil Mechanics and Geotechnical Engineering (17th ARC, Astana, Kazakhstan, 14–18 August, 2023), pp. 1323–1326. CRC Press. https://doi.org/10.1201/9781003299127

Keshavarz A, Beygi M, Vali R (2019) Undrained seismic bearing capacity of strip footing placed on homogeneous and heterogeneous soil slopes by finite element limit analysis. *Computers and Geotechnics* 113:103094. https://doi.org/10.1016/j.compgeo.2019.103094

Khorrami R, Derakhshani A, Moayedi H (2020) New explicit formulation for ultimate bearing capacity of shallow foundations on granular soil using M5' model tree. *Measurement* 163:108032. https://doi.org/10.1016/j.measurement.2020.108032

Kumar P, Metya S, Shubham K, Prashad D (2022) Behaviour of strip footing over cavity subjected to inclined and eccentric loads. *Arabian Journal of Geosciences* 15. https://doi.org/10.1007/s12517-022-10739-6

Lavasan AA, Talsaz A, Ghazavi M, Schanz T (2016) Behaviour of shallow strip footing on twin voids. *Geotechnical and Geological Engineering* 34:1791–1805. https://doi.org/10.1007/s10706-016-9989-6

Loganathan N, Poulos HG, Bustos-Ramirez A (2000) Estimation of ground loss during tunnel excavation. In: *ISRM International Symposium*, Melbourne, Australia.

Mair RJ, Taylor RN, Bracegirdle A (1993) Subsurface settlement profiles above tunnels in clays. *Géotechnique* 43:315–320. https://doi.org/10.1680/geot.1993.43.2.315

Malhotra M, Sahu V, Srivastava A, Misra AK (2020) Experimental and numerical investigation of the effect of pre-existing utility tunnel on the bearing capacity of shallow footing in sandy soils. *Journal of Engineering, Design and Technology* 18:513–529. https://doi.org/10.1108/JEDT-04-2019-0102

Malhotra M, Srivastava A, Jawaid S (2020) Reliability analysis of shallow foundation in the vicinity of the existing buried conduit. *Geomechanics and Geoengineering* 15:149–158. https://doi.org/10.1080/17486025.2019.1632497

Maria A, Naggar E (2015) Effects of tunnelling on the bearing capacity of shallow foundations. In: *GeoQuébec 2015, the 68th Canadian Geotechnical Conference*, Quebec, Canada.

Mazouz B, Mansouri T, Baazouzi M, Abbeche K (2022) Assessing the effect of underground void on strip footing sitting on a reinforced sand slope with numerical modelling. *Engineering, Technology & Applied Science Research* 12:9005–9011. https://doi.org/10.48084/etasr.5131

Metya S, Mukhopadhyay T, Adhikari S, Bhattacharya G (2017) System reliability analysis of soil slopes with general slip surfaces using multivariate adaptive regression splines. *Computers and Geotechnics* 87:212–228. https://doi.org/10.1016/j.compgeo.2017.02.017

Mukherjee S, Babu G (2023) Probabilistic evaluation of uplift capacity of transmission tower foundation using reinforced anchor. *International Journal of Geomechanics* 23:1–15. https://doi.org/10.1061/IJGNAI.GMENG-7919

Patel RS, Desai MD (2010) CBR predicted by index properties for alluvial soils of South Gujarat. In: *Indian Geotechnical Conference*, pp. 79–82, Mumbai.

Peck BB (1969) Deep excavation and tunnelling in soft ground, State of the art volume. In: *7th ICSMFE* 4, pp. 225–290, Mexico.

Pramanik R, Baidya DK, Dhang N (2021) Reliability assessment of three-dimensional bearing capacity of shallow foundation using fuzzy set theory. *Frontiers of Structural and Civil Engineering* 15:478–489. https://doi.org/10.1007/s11709-021-0698-8

Přibyl O, Přibyl P, Svítek M (2021) Interdisciplinary urban tunnel control within smart cities. *Applied Sciences* 11. https://doi.org/10.3390/app112210950

Ray R, Kumar D, Samui P, et al. (2021) Application of soft computing techniques for shallow foundation reliability in geotechnical engineering. *Geoscience Frontiers* 12:375–383. https://doi.org/10.1016/j.gsf.2020.05.003

Rodríguez CA, Rodríguez-Pérez ÁM, López R, et al. (2023) A finite element method integrated with Terzaghi's principle to estimate settlement of a building due to tunnel construction. *Buildings* 13. https://doi.org/10.3390/buildings13051343

Sabouni R, Airan Eng M (2018) Evaluation of foundation on soil with cavities: A case study from the UAE. *International Journal of Structural and Civil Engineering Research* 358–363. https://doi.org/10.18178/ijscer.7.4.358-363

Sana VKF, Nazeeh KM, Dilip DM, Sivakumar Babu GL (2022) Reliability-based design optimization of shallow foundation on cohesionless soil based on surrogate-based numerical modelling. *International Journal of Geomechanics* 22:1–8. https://doi.org/10.1061/(asce)gm.1943-5622.0002274

Shubham K, Metya S, Bhattacharya G (2022) Reliability analysis of settlement of a foundation resting over a circular void. In: C. N. V. Satyanarayana Reddy, A. Murali Krishna, Neelima Satyam (Eds.), *Lecture Notes in Civil Engineering*. Springer Science and Business Media Deutschland GmbH, pp. 133–143.

Shubham K, Metya S, Sinha AK (2024a) Optimizing shallow foundation design: A machine learning approach for bearing capacity estimation over cavities. *Geomechanics and Engineering* 37:629–641. https://doi.org/10.12989/gae.2024.37.6.629

Shubham K, Metya S, Sinha AK (2024b) Surrogate model-based prediction of settlement in foundation over cavity for reliability analysis. *Transportation Infrastructure Geotechnology* 11:1294–1320. https://doi.org/10.1007/s40515-023-00329-8

Shubham K, Metya S, Sinha AK, Gobinath R (2024c) One-dimensional-convolutional neural network (1D-CNN) based reliability analysis of foundation over cavity incorporating the effect of simulated noise. *Advances in Civil Engineering* 2024. https://doi.org/10.1155/2024/9981433

Sohaei H, Namazi E, Hajihassani M, Marto A (2020) A Review on tunnel–pile interaction applied by physical modelling. *Geotechnical and Geological Engineering* 38:3341–3362. https://doi.org/10.1007/s10706-020-01240-6

Tizpa P, Jamshidi Chenari R, Payan M (2023) PFC/FLAC 3D coupled numerical modelling of shallow foundations seated on reinforced granular fill overlying clay with square void. *Computers Geotechnics* 161:105574. https://doi.org/10.1016/j.compgeo.2023.105574

Wang F (2021) Empirical evidence for estimation of subsurface settlement caused by tunneling in sand. *Underground Space* 6:577–584. https://doi.org/10.1016/j.undsp.2021.01.002

Wang MC, Hsieh CW (1987) Collapse load of strip footing above circular void. *Journal of Geotechnical Engineering* 113:511–515.

Wen Y, Xin C, Zhang X, et al (2021) The stability analysis of tunnel lining structure with seismic excitation based on the energy evaluation principle. *Shock and Vibration* 2021. https://doi.org/10.1155/2021/9995682

Zhang W, Zhang Y, Goh AT (2017) Multivariate adaptive regression splines for inverse analysis of soil and wall properties in braced excavation. *Tunnelling and Underground Space Technology* 64:24–33. https://doi.org/10.1016/j.tust.2017.01.009

9 Development and Application of an MR Model for Slope Stability Assessment

A Case Study on the Jorabat–Shillong Expressway

Balendra Mouli Marrapu and Ippili Saikrishnamacharyulu

9.1 INTRODUCTION

The assessment of the stability of slopes is a significant factor that geotechnical engineers need to understand – notably in sloped areas – before proceeding with any construction activities like road, bridge, or any other infrastructure projects (Onyelowe et al., 2022; Kim et al., 2024). This applies particularly to regions like Jorabat–Shillong Expressway, which has very challenging geological and environmental conditions. The proper assessment of the stability of slopes is required in order to avoid failures which may cause

DOI: 10.1201/9781003714125-9

serious economic losses and human life losses. In recent studies, many researchers have increasingly developed models for the interpretation of the slope performance under various conditions (Sah et al., 1994; Cetina, 2014; Kurnaz et al., 2024). Among all these models, artificial neural networks (ANNs) show greatest utilization due to their ability to interpret complex and non-linear relationships (Aswathi & Chandrakaran, 2020; Wengang et al., 2023). But these ANN models are black-box models. Because of this, the relationship between the dependent variable factor of safety (FOS) and the other independent parameters used in the stability analysis, shear strength parameters (c, ϕ), geometrical parameters height (h) and slope angle (β), unit weight (γ), and pore water pressure parameter (Γ_u), are hard to interpret (Erzin & Cetin, 2013, 2014). The relationship between the variables cannot be understood, making the model less transparent (Won et al., 2025; Harle, 2024). To minimize this problem, a multiple linear regression (MLR) slope stability assessment model was first developed and applied to the study region. These MLR models are commonly employed geotechnical engineering applications (Zhang et al., 2015; Kumar et al., 2013; Selamat et al., 2025; Bijan & Eftekhari, 2022). Recent studies (Sayadi et al., 2013; Kumar et al., 2013) have also used developed MLR models for assessing slope stability; however, the accuracy was not inadequate. This is mainly because of the approach used in the development of the MLR models. In this study, the approach developed by Rao (2003) has been used for the development of the MLR model to obtain more accuracy (Sayadi et al., 2013; Kumar et al., 2013).

For developing the MLR model, it is required to identify the relationship between the dependent variable, FOS, and the independent variables cohesion (c), angle of internal friction (ϕ), slope angle (β), unit weight (γ), pore water pressure (Γ_u), and height of slope (h). The MLR model is developed using IBM SPSS Statistics 31 with some statistical checks, ensuring that there is no black-box process involved (Bonett & Wright, 2000).

The developed MLR model was used for assessing the stability of Jorabat–Shillong Expressway. This developed model is not region specific and can be used in other regions also with similar geotechnical conditions. The performance of the developed MLR model was also compared with the limit equilibrium method (LEM) to ensure the suitability of the model. Before the development of the MLR model, data is required to satisfy key assumptions; all these requirements were taken into consideration before the development of the MLR model.

9.2 STUDY REGION

The study region is a recently inaugurated expressway that connects Jorabat city in the state of Assam to Shillong, the capital of Meghalaya state. It is a major road that links these two states, passes through steep slopes and unstable geologically conditions; this region experiences very high rainfall during the monsoon season. For conducting this research, site investigations were carried out at different points along the expressway; geotechnical and geometric data was also collected. This data includes slope height, soil shear strength parameters like cohesion, internal friction angle, slope angle, unit weight, and pore water pressure (Chakraborty & Goswami, 2017).

TABLE 9.1 Parameter ranges adopted for this study

SOIL PARAMETER	*VARIATION*	*INTERVAL*	*NUMBER OF CASES*
Cohesion., *c* (kPa)	0–200	40 kPa	6
Angle of friction, ϕ	0°–40°	8°	6
Angle of inclination, β	15°–45°	6°	6
Unit weight, γ (kN/m³)	13–23	2 kN/m³	6
Height *h* (m)	6–56	10 (m)	6
Γ_u	0–0.5	0.1	6
Total number of cases = 6*6*6*6*6*6 =			46,656

Detailed geotechnical laboratory testing was undertaken to determine shear strength, unit weight, and pore water pressure. Using this data only, MLR model was developed to determine the stability of both natural and modified slopes along the route of the expressway. For developing this model, huge amounts of datasets were required from previously solved stability problems. The data used for this study was collected from the study area and literature. For developing MLR model, dependent variable is FOS and independent variables are cohesion (*c*), angle of internal friction (ϕ), slope angle (β), unit weight of soil (γ), slope height (*h*), and pore pressure ratio (Γ_u).

These ranges of data are clearly presented in Table 9.1. The data ranges are chosen to represent different soil types and actual field conditions as reported in the literature. Each slope is analyzed first using conventional slope stability analysis methods. For this purpose, the stability chart method used, which allows for the efficient analysis of a large number of datasets without the need for complete numerical modeling of the slopes (Hong-In et al., 2024).

The following assumptions were made:

- The relationship between output (FOS) and input variables (*c*, Φ, β, γ, *h*, Γ_u) should be linear, confirmed by scatterplots.
- The dependent variable should follow a normal distribution, verified through histograms.
- Multicollinearity should be absent, with Pearson correlation coefficients within acceptable limits.
- Residuals should exhibit homoscedasticity, shown by consistent variance in residuals plotted against predicted values.

The relationship between output (FOS) and input variables$c$, φ, H, β, γ, and Γ_u of soils is illustrated in Figures 9.1–9.6. From these plots, it is observed that relationship between the output and input variables is approximately straight linear, thereby confirming the first assumption is satisfied.

To verify the second assumption, a histogram with a fitted normal curve is plotted, as shown in Figure 9.7. Out of 46,656 cases in the training dataset, only 27,000 cases are used for the development of the MLR model, while the remaining data is excluded to ensure suitability for model development. The figure clearly depicts variation in the response of dependent variable FOS, which follows a normal distribution. Descriptive statistics for this variable are presented in Table 9.2.

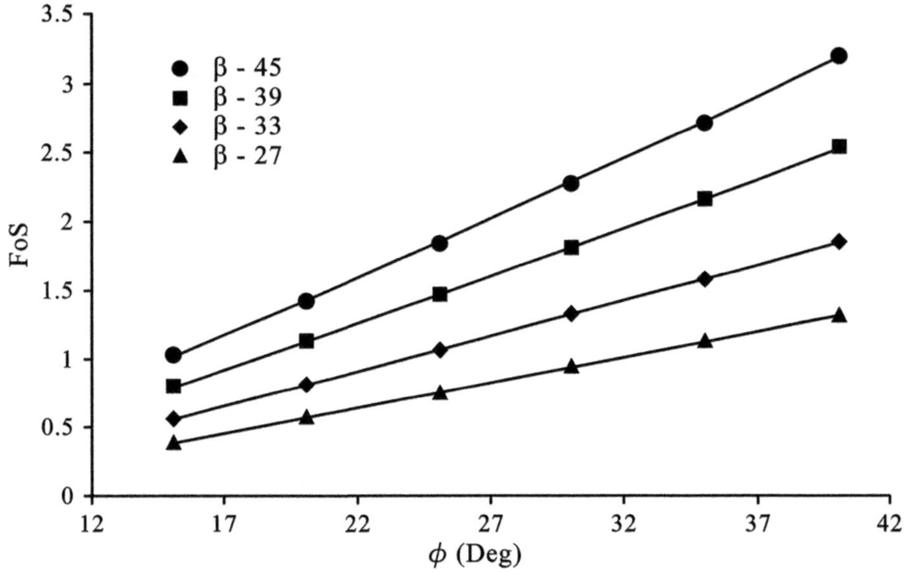

FIGURE 9.1 Variation of stability with angle of internal friction φ.

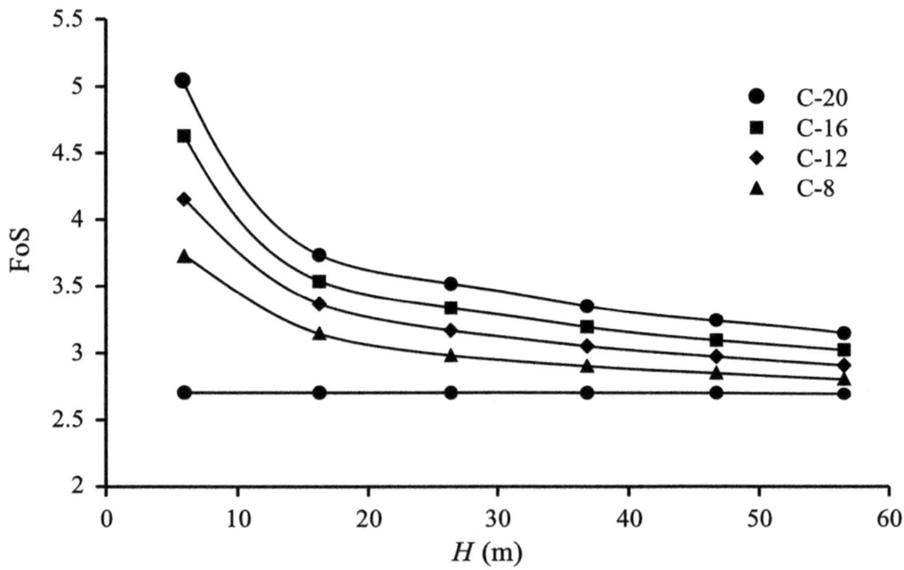

FIGURE 9.2 Variation of stability with slope height.

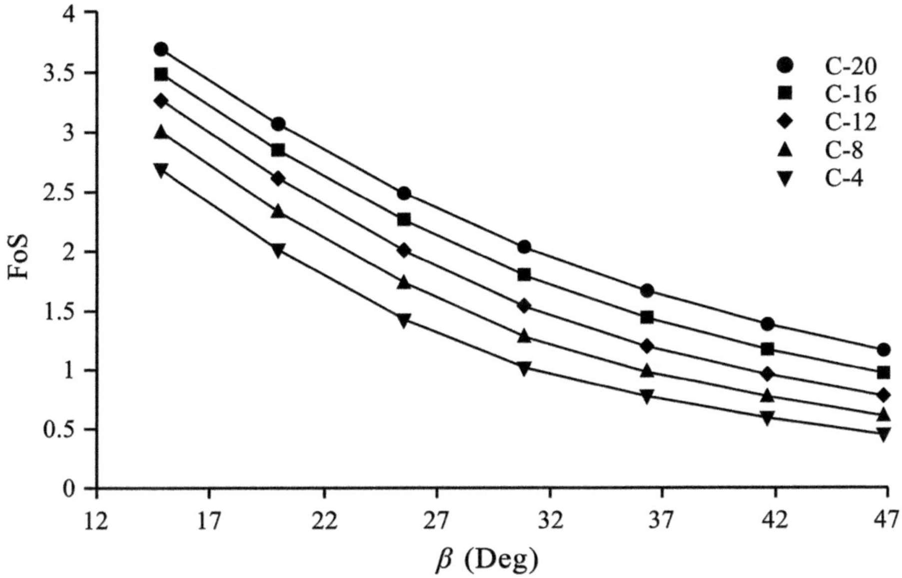

FIGURE 9.3 Variation of stability with inclination of slope.

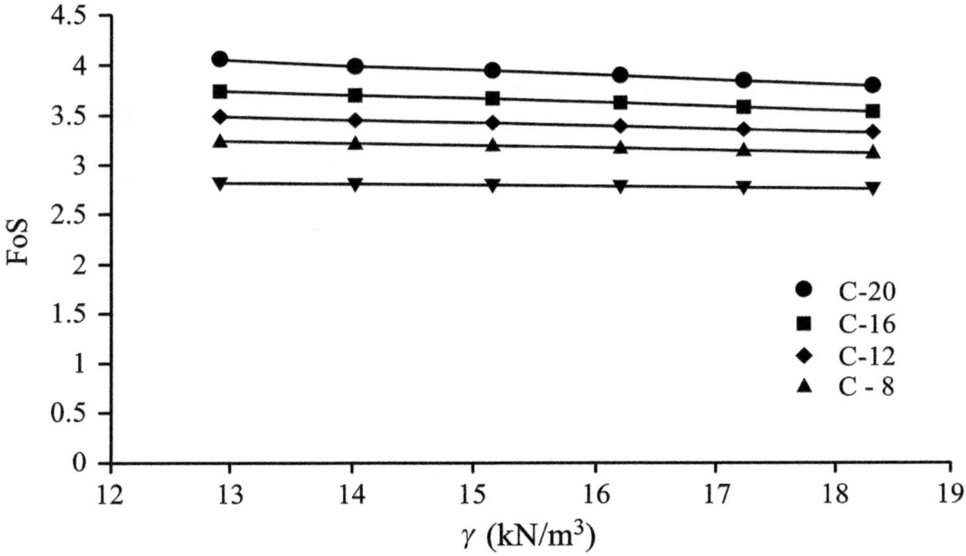

FIGURE 9.4 Variation of stability with unit weight of soils.

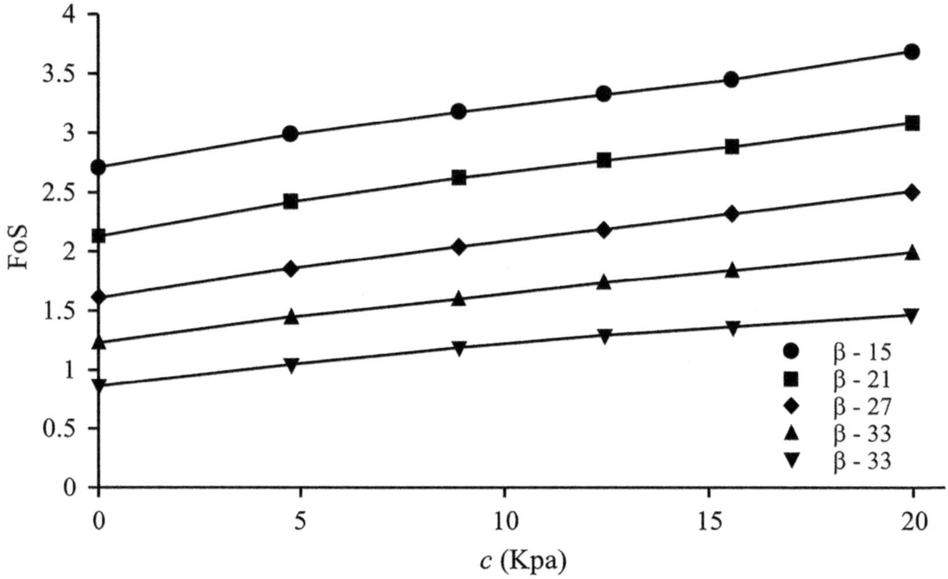

FIGURE 9.5 Variation of stability with cohesion of soils.

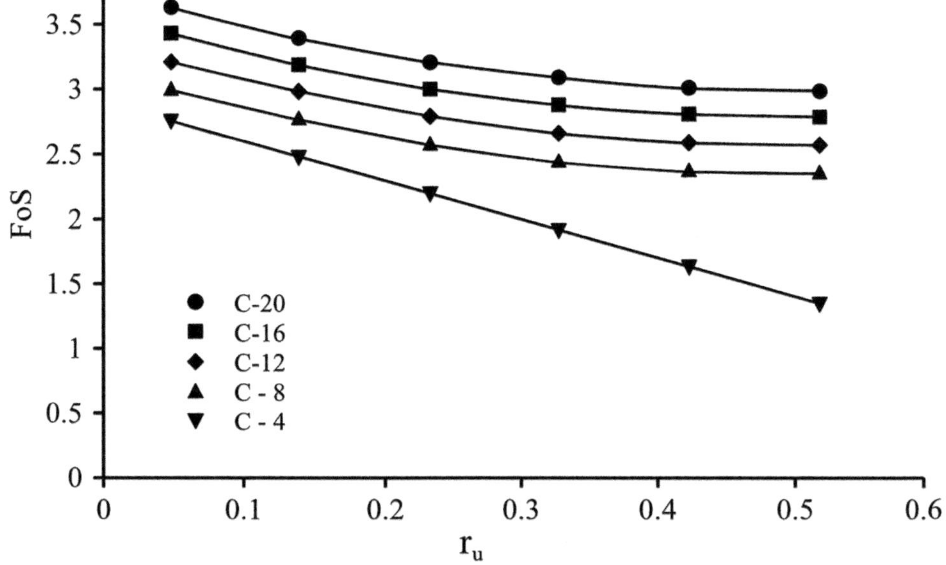

FIGURE 9.6 Variation of stability with Γ_u of soils.

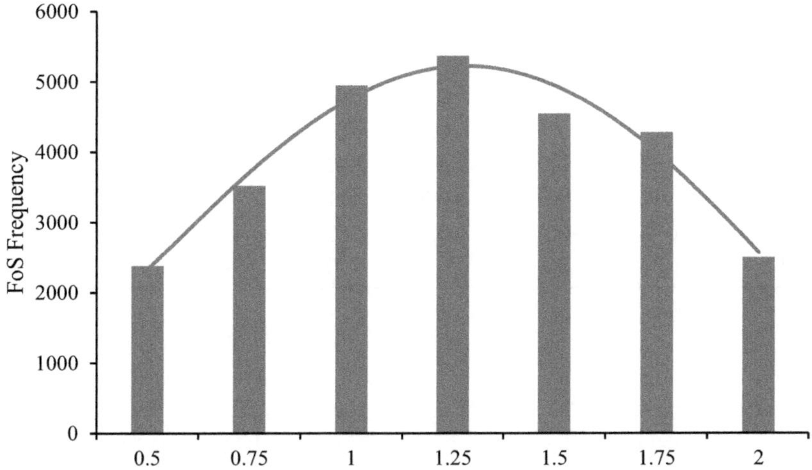

FIGURE 9.7 Normal distribution plot soils (training data).

TABLE 9.2 Details of training data soil–dependent variable descriptive statistics value

DESCRIPTIVE STATISTICS	
Sample variance	0.198
Skewness	–0.001
Kurtosis	–0.882
Mode	0.250
Standard error	0.002
Mean	1.146
Standard deviation	0.445
Median	1.140
Range	1.743
Minimum	0.256
Maximum	1.999
Count	27,000

After excluding 20,000 cases for validating the second assumption, remaining 27,000 cases of learning data is verified for multicollinearity assumption. The presence of multicollinearity in the dataset (i.e., the third assumption) was verified using the Pearson correlation matrix. The correlation coefficients among all the independent variables were calculated and are presented in Table 9.3. From this it can be observed that the correlation values indicate negligible multicollinearity among the independent variables, confirming that they are largely independent of each other.

TABLE 9.3 Multicollinearity of independent variables

	Γ	Φ	β	*H*	*C*	Γ*U*
γ	1	−0.0341	0.164	−0.100	−0.014	−0.007
ϕ	−0.0341	1	0.048	0.001	−0.004	−0.027
β	0.164	0.048	1	−0.370	0.014	0.025
h	−0.100	0.001	−0.370	1	−0.008	0.009
c	−0.014	−0.004	0.014	−0.008	1	−0.008
Γ*u*	−0.007	−0.027	0.025	0.009	−0.008	1

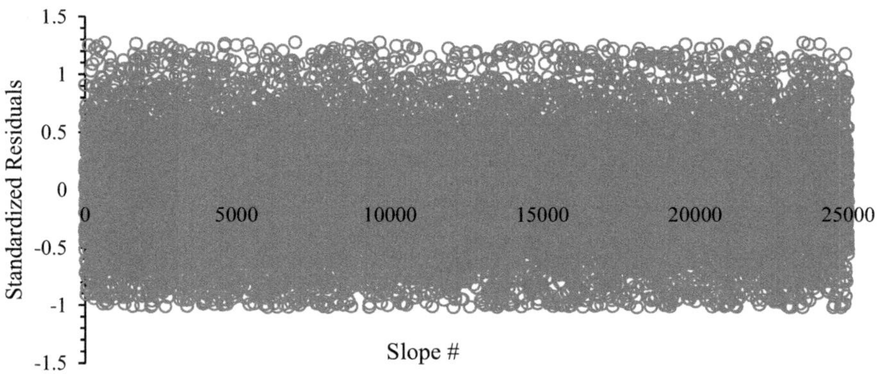

FIGURE 9.8 Homoscedasticity residual plot for soils.

The fourth assumption, known as homoscedasticity, refers to the condition where residual of variance remains constant across all levels of the independent variables. Figure 9.8 clearly presents the scatterplot for the analyzed dry cases, revealing uniform distribution of residuals around zero, thereby satisfying the assumption of homoscedasticity.

After verifying all the required assumptions, the multiple regression (MR) model is developed using SPSS software and presented in Equation (9.1). To evaluate the accuracy of the developed MLR model, it is applied to field data collected from the Jorabat–Shillong Expressway. The predicted results versus the results are obtained from the LEM using metrics such as the mean squared error (MSE) and determination coefficient (R^2). The comparative results are summarized in Table 9.4.

$$\text{FOS} = \tan(\phi) * 1.767 + c * 0.016 - \tan(\beta) * 2.02 - \gamma * 0.03 - H * 0.02 - \Gamma u * 0.89 + 2.75 \tag{9.1}$$

The developed MLR model is validated by applying it to the real field data cases, and FOS for each slope is determined and presented in Table 9.4. MSE and regression coefficient (R^2) are calculated to compare the FOS obtained from MLR model and LEM using Equations (9.2) and (9.3), respectively. These values are also shown in Table 9.4.

TABLE 9.4 Comparison of field data results with predictions from the developed multiple linear regression (MLR) model

S. NO	*H (M)*	*β°*	*C (KPA)*	*Φ°*	*Γ (KN/M³)*	*ΓU*	*FOS LEM*	*FOS MLR*
1	56	33	96	15	15	0.23	1.307	1.580
2	46	39	0	35	14	0.14	0.825	0.928
3	26	39	0	20	17	0.05	0.476	0.708
4	16	45	20	35	14	0.05	1.501	1.445
5	36	21	80	25	17	0.5	2.376	2.298
6	56	39	8	40	15	0.14	1.141	1.065
7	46	21	0	30	15	0.5	0.825	1.243
8	46	15	64	25	13	0.14	2.838	2.529
9	26	15	64	15	15	0.41	2.312	2.280
10	56	33	0	25	15	0.41	0.386	0.42

$$\text{MSE} = \frac{1}{N}\sum_{N}^{1}(y - y')^2 \tag{9.2}$$

$$R^2 = \left[\frac{N\sum y * y' - \left(\sum y\right)\left(\sum y'\right)}{\sqrt{\left[N\sum y^2 - \left(\sum y\right)^2\right]\left[N\sum y'^2 - \left(\sum y'\right)^2\right]}}\right]^2 \tag{9.3}$$

where y = FOS obtained from developed MLR model, y' = reference FOS; and N = total number of cases.

9.3 RELATIVE IMPORTANCE OF FACTORS AFFECTING SLOPE STABILITY

An attempt has been made here to obtain relative importance (RI) of various factors contributing to slope stability using the results of developed MLR model. For this purpose, a separate MR equation has been developed using normalized learning data. Normalization is carried out using Equation (9.4) such that the range varies between [0 and 1].

$$\text{Normalization} = \frac{(X_i - X_{\text{Min}})}{(X_{\text{Max}} - X_{\text{Min}})} \tag{9.4}$$

where X_i is the considered parameter for normalization, $X_{\min}$ is the considered minimum parameter for normalization, and $X_{\min}$ is the considered maximum parameter for normalization.

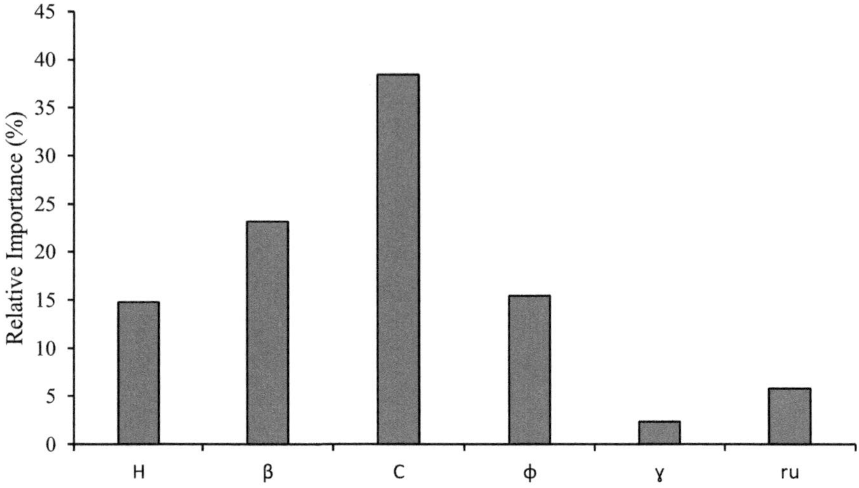

FIGURE 9.9 Relative importance observed for soil under saturated conditions.

The MLR model is developed again with normalizing data using IBM SPSS software and the developed MLR model is shown in Equation (9.5). This normalization of parameters is very important as the data variation of each parameter is different. For example, h value ranges between 6 and 56 m, while γ ranges between 16 and 21 kN/m^3. If the absolute values are used, the coefficients in the developed MLR model are influenced by ranges and they don't really show the relative contribution of these parameters. Developed MLR model using normalized data is shown in Equation (9.5). The coefficients of the respective parameter in the equation are used to calculate the RI of the respective parameters.

$$\text{FoS} = \tan(\phi) * 1.682 + c * 0.014 - \tan(\beta) * 1.973 - \gamma * 0.029 - H * 0.018 - \Gamma u * 0.806 + 2.695 \quad (9.5)$$

The estimated RI of various parameters for soils is shown in Figure 9.9. It is clearly seen from the figure that cohesion c is the most significant parameter affecting the stability of the slope. After cohesion, slope angle, angle of shearing resistance, and height influence the stability of slope.

9.4 RESULTS AND CONCLUSION

After developing the MLR model, its performance was evaluated using two statistical metrics: the coefficient of determination (R^2) and the MSE. The obtained R^2 value was 0.934, indicating a strong correlation between the FOS values predicted by the MLR model and those obtained from the LEM. A value of R^2 close to 1.0 indicates a high degree of accuracy, and it can be seen that the two models give almost equal results. A good fit is indicated when the value of R^2 is above 0.8, as mentioned by Smith (1986). The value of R^2 in this research is 0.934, which clearly indicates that the outcome of the MLR model is highly reliable.

MSE is found to be 0.042, which is a very small number and clearly shows that the average squared error obtained between the FOS predicted by the MLR model and the values obtained using the LEM is very small, almost close to zero. It clearly indicates that the accuracy of the results in prediction is very high. The small value of MSE in this study clearly indicates that the model is performing very well and this developed model can be used in any region.

The significance of each parameter is also tested, and it is found that the cohesion (c) parameter is more significant than others. This clearly indicates that a small change in the value of cohesion can result in a large change in the value of the FOS. Therefore, it is very important to determine the value of cohesion before performing the slope stability analysis.

On the basis of the above statistical outcomes, the proposed MLR model is an efficient and optimized alternative to the conventional LEM of slope stability analysis. Moreover, the efficiency of the proposed model is comparable to that of the ANN model, but without the interpretability issues that come with black-box models. Thus, the proposed MLR model can be applied effectively for rapid slope stability analysis in engineering applications, especially in hilly regions where a large number of slopes have to be analyzed quickly.

The successful application of the proposed model to the Jorabat–Shillong Expressway project clearly indicates that the proposed model can also be applied to other critical infrastructure projects.

REFERENCES

Aswathi, M., & Chandrakaran, S. (2020). Multiple regression (MR) and artificial neural networks (ANN) for predicting factor of safety in slope stability analysis. In *Proceedings of Indian Geotechnical Conference 2020*, Andhra University, Visakhapatnam, India (pp. 32–43).

Bijan, A., & Eftekhari, M. (2022). Prediction of mode I fracture toughness of rock using linear multiple regression and gene expression programming. *Journal of Rock Mechanics and Geotechnical Engineering*, 14(5), 1421–1432.

Bonett, D. G., & Wright, T. A. (2000). Sample size requirements for estimating Pearson, Kendall and Spearman correlations. *Psychometrika*, 65(1), 23–28.

Cetina, T. (2014). The prediction of the critical factor of safety of homogeneous finite slopes subjected to earthquake forces using neural networks and multiple regressions. *Geomechanics and Engineering*, 6(1), 1–15.

Chakraborty, A., & Goswami, D. (2017). Prediction of slope stability using multiple linear regression (MLR) and artificial neural network (ANN). *Arabian Journal of Geosciences*, 10(11), 385.

Erzin, Y., & Cetin, T. (2013). The prediction of the critical factor of safety of homogeneous finite slopes using neural networks and multiple regressions. *Computers & Geosciences*, 51, 305–313.

Erzin, Y., & Cetin, T. (2014). The prediction of the critical factor of safety of homogeneous finite slopes subjected to earthquake forces using neural networks and multiple regressions. *Geomechanics and Engineering*, 6(1), 1–15.

Harle, S. M. (2024). Advancements and challenges in the application of artificial intelligence in civil engineering: A comprehensive review. *Asian Journal of Civil Engineering*, 25(1), 1061–1078.

Hong-In, P., Takahashi, A., & Likitlersuang, S. (2024). Engineering and environmental assessment of soilbag-based slope stabilisation for sustainable landslide mitigation in mountainous area. *Journal of Environmental Management*, 359, 120970.

Kim, D., Kim, T., Kim, Y., Byun, Y. H., & Yun, T. S. (2024). A ChatGPT-MATLAB framework for numerical modeling in geotechnical engineering applications. *Computers and Geotechnics*, 169, 106237.

Kumar, B. R., Vardhan, H., Govindaraj, M., & Vijay, G. S. (2013). Regression analysis and ANN models to predict rock properties from sound levels produced during drilling. *International Journal of Rock Mechanics and Mining Sciences*, 58, 61–72.

Kurnaz, T. F., Erden, C., Dağdeviren, U., Demir, A. S., & Kökçam, A. H. (2024). Comparison of machine learning algorithms for slope stability prediction using an automated machine learning approach. *Natural Hazards*, 120(8), 6991–7014.

Onyelowe, K. C., Mojtahedi, F. F., Ebid, A. M., Rezaei, A., Osinubi, K. J., Eberemu, A. O., & Rehman, Z. U. (2022). Selected AI optimization techniques and applications in geotechnical engineering. *Cogent Engineering*, 10(1), 2153419.

Rao, S. J. (2003). Regression modeling strategies: With applications to linear models, logistic regression, and survival analysis. *Journal of the American Statistical Association*, 98, 257–258.

Sah, N. K., Sheorey, P. R., & Upadhyaya, L. N. (1994). Maximum likelihood estimation of slope stability. *International Journal of Rock Mechanics and Mining Sciences*, 31(1), 47–53.

Sayadi, A., Monjezi, M., Talebi, N., & Khandelwal, M. (2013). A comparative study on the application of various artificial neural networks to simultaneous prediction of rock fragmentation and backbreak. *Journal of Rock Mechanics and Geotechnical Engineering*, 5(4), 318–324.

Selamat, S. N., Majid, N. A., & Taha, M. R. (2025). Multicollinearity and spatial correlation analysis of landslide conditioning factors in Langat River Basin, Selangor. *Natural Hazards*, 121(3), 2665–2684.

Smith, G. N. (1986). *Probability and Statistics in Civil Engineering*. Collins Professional and Technical Books.

Wengang, Z., Hanlong, L., Lin, W., Xing, Z., & Yanmei, Z. (2023). *Application of Machine Learning in Slope Stability Assessment*. Springer Nature.

Won, M., Sadiq, S., Wang, J., & Gao, Y. (2025). Predicting slope stability potential failure surface using machine learning algorithms. *Arabian Journal of Geosciences*, 18(1), 24.

Zhang, G., Tan, J., Zhang, L., & Xiang, Y. (2015). Linear regression analysis for factors influencing displacement of high-filled embankment slopes. *Geomechanics and Engineering*, 8(4), 511–521.

10 Identification of Landslide Susceptibility Areas in Pettimudi Hills Using Frequency Ratio and Logistic Regression Models

Surendar Natarajan and Kamalanandhini Mohan

10.1 INTRODUCTION

The 2016 report from the UN High Commissioner for Refugees indicated that natural disasters have forced 36 million people to leave their homes around the world. By 2050, the number of people affected could reach 50 million. Landslides are one of the natural geological risks that kill hundreds of people every year in mountainous areas (Pourghasemi & Rahmati, 2018). Landslides must be stopped so that they don't cause

DOI: 10.1201/9781003714125-10

damage to infrastructure, affect the environment or kill people. Landslides happen because of a lot of different and complicated factors. Disasters hurt people and property in some way. Landslides can happen on their own or because people are reckless. Climate change makes it rain harder in hills and mountains, which makes landslides happen more often on steep slopes.

The conditioning factors such as topography, soil and vegetation precisely reflect the impact of extreme rainfall on landslides (Xingfu & Erdi, 2025). Infiltration of too much rainwater raises pore water pressure, which makes the soil unstable and causes landslides (Moung-Jin et al., 2014). The landslides also contribute significantly to soil erosion, prompting some important institutions like European Union to prioritize the identification and management of high-susceptibility areas (Panagos et al., 2024).

The geology and geotechnical engineering of slope collapse influence the stability of slopes and landslides (Kotluri et al., 2025; Tien Bui et al., 2012). Environmental changes can cause landslides in a large or small area over short and extended periods of time (N. Wang et al., 2025). Landslides can happen anywhere and at any moment, and there will be different causes for a landslide (Glade & Crozier, 2010). Physical, statistics-based correlation analysis and soft computing are utilized to model how likely a landslide may happen. There are advantages and disadvantages of each forecasting model (Sachdeva et al., 2020).

To be able to estimate how likely a landslide is to happen, physical models need a lot of data. These modeling techniques are prohibitively costly for extensive regions. Previous studies by Meena et al. (2022), Qiu et al. (2022) and Thanh and De Smedt (2014) have thoroughly examined the physically deterministic model for global landslide disaster analysis. Conventional statistical models utilize assumed parameters, impacting the quality of the modeling. Data-driven algorithms are needed to figure out the occurrence of a landslide. In recent times Machine Learning (ML) models such as Random Forest (RF), Stacking Ensemble models such as Artificial Neural Networks (ANNs), Support Vector Machines (SVMs), Gradient Boosting Decision Trees (GBDTs) and UAV image matching based on a deep learning LoFTR algorithm (Zong et al., 2023) are used to generate landslide susceptibility maps (LSMs), but these models require enormous amounts of field data (Nawazuzzoha et al., 2025). The hybrid models are used to generate LSMs in the South Sikkim Himalaya region of India, a data-scarce region that could significantly help in reducing the risk.

Geospatial tools and data-driven models are used to prepare maps that show the occurrence of landslides. Prior research has demonstrated that remote sensing data can be utilized to determine landslide hazard zones and triggering factors (Liu & Shih, 2013). Sameen et al. (2020) utilized training sample selection and incomplete dataset issue resolution, and Stefan Steger et al. (2018) used novel input and conditioning parameters for optimization. Samia et al. (2018) emphasized the conditional factors and employed standard input for landslide analysis. Digital Elevation Models (DEMs) and their derivatives for LSM employ the min–max normalization factor to enhance spatial resolution (Pradhan, 2017). Oh et al. (2012) have investigated how well 3D models with DEM resolutions of 2.5, 5, 10, 20 and 30 m can predict a landslide. Geospatial mapping employing remote sensing and Geographic Information System (GIS) is often used to figure out the occurrence of a landslide (Merghadi et al., 2020).

In 2008, a 7.9 magnitude earthquake hit Wenchuan, China, and caused more than 20,000 landslides over an area of 1160 km^2 (Xu et al., 2013). Ethiopia employs GIS-based statistical stochastic modeling to guess when landslides will happen (Wubalem, 2020). Morocco's landslide investigations employed model training and validation (El-fengour et al., 2021). Naceur et al. (2022) used Logistic Regression (LR) and Support Vector Regression (SVR) to make the LSM of N'fis river basin. Xia et al. (2023) found that ML models were better at predicting landslides in Xiaojin County, China, than deep learning models. Nhu et al. (2020) made the landslide map for Cameron Highlands, Malaysia, by combining GIS and remote sensing data with AHP (analytic hierarchy process), WLC (weighted linear combination) and SMCE (ṣspatial multi-criteria evaluation) maps.

In the eastern Himalayas, superposable neural network (SNN) optimization was used to look at the situation and guess where landslides would happen (Youssef et al., 2023). SNN did a better job of predicting landslides than models based on statistics or physics. Landslides that harm property and kill people are increasing, often in the Western Ghats because more people are moving to cities and constructing buildings (Van Westen et al., 2012). To forecast landslide likelihood and pinpoint susceptible areas, an appropriate modeling methodology must be employed to formulate mitigation strategies. In a recent study by Achu et al. (2020), Dempster–Shafer, Bayesian probability and LR were utilized to make LSMs in the Southwestern Western Ghats of Kerala, South India. Heavy rain and a cloud burst caused many big landslides near Kedarnath (Rao et al., 2014). Saha et al. (2023) used physical, anthropogenic and ML models to accurately study landslides in Arunachal Pradesh's Tawang and West Kameng.

Different modeling methods create different kinds of LSMs. To investigate ways to stop landslides, researchers have made LSMs using a number of different methodologies. Arabameri et al. (2022), Di Napoli et al. (2021) and Novellino et al. (2021) have improved LSM by applying adaptive models with good prediction capability in geospatial and ML approaches. This project uses geospatial approaches and stochastic bivariate and co-variate models to prepare landslide maps. By integrating susceptibility models with risk analysis, landslide-susceptible maps can provide perceptions for regional planning, disaster management and policymaking, stressing targeted mitigation for vulnerable basins.

10.2 MOTIVATION OF THE RESEARCH STUDY

The Kerala state disaster management authority and landslide hazard zonation put Idukki in the high-to-moderate range (Jeggle & Boggero. M., 2018; Sajinkumar et al., 2017; Weidner et al., 2018). On August 6, 2020, the latest landslide happened in Pettimudi Hills in Idukki. The landslides killed people and ruined property. The rapid growth of tea gardens and haphazard development on these slopes made the soil less able to hold water, which caused deadly landslides. This study produces accurate geographic LSMs for Pettimudi Hills by statistical stochastic bivariate and co-variate methodologies, in contrast to previous research that employed management strategies.

A limited number of research initiatives have detected landslides in the designated study region. This study employs LR and frequency ratio (FR) modeling to pinpoint susceptible regions. Many scholars have employed the FR model, a probabilistic method, for more than a year owing to its simplicity. It is known that LR is more accurate. This study showed that mathematical models can be used to check slope stability, find places where landslides are likely to happen and find the variables that cause them to happen (Das & Lepcha, 2019; Moung-Jin et al., 2014).

10.3 OBJECTIVES OF THE STUDY

- To study the landslides that occurred in Pettimudi Hills in Idukki district.
- To generate the spatial landslide triggering factor maps for the study area.
- To identify the landslide susceptibility areas through the two statistical methods of LR model and FR model.

10.4 MATERIALS AND METHODOLOGY

10.4.1 Study Area Details

Pettimudi Hills are in the Western Ghats in Kerala's Idukki district. Figure 10.1 indicates how likely it is to have a landslide. The hills are located between N75° W and S75° E, and between N45° E and S45° W. Idukki district and Pettimudi Hills get 5000 mm of rain every year. The weather here ranges from 16°C to 32°C, thus it's freezing. There are crystalline rocks from the Precambrian era in the southern granulite landscape. Granite, granitic gneiss and migmatite are the most common types of rocks. Most of its hills are denudational, and they go up and down around 2100 m above sea level. The upper hillslopes are slanted, whereas the lower slopes are either mild or steep. The main sources of employment are tea plantations and factories, and the land cover has changed since tea farms have grown a lot. Heavy rain on hills with changing land cover has made landslides more likely to occur.

Figure 10.2 shows the methodology adopted for the study. This study utilized Landsat data from the USGS website. Landslides can happen because of elevation, slope, aspect, soil, rainfall, land use and cover, geology, distance to a lineament, stream, road network, stream power index (SPI) and TWI. ArcGIS 10.4.4 is used to map these features. To quantify the risk of landslides in the study area, FR and LR models are used to find the components of landslides. Figure 10.3a shows DEM data from the Shuttle Radar Topography Mission (SRTM) website.

The main reason for landslides in the mountains is the slope. Models for landslide probability and rainfall runoff should incorporate slopes that are steeper than 3°. Heavy

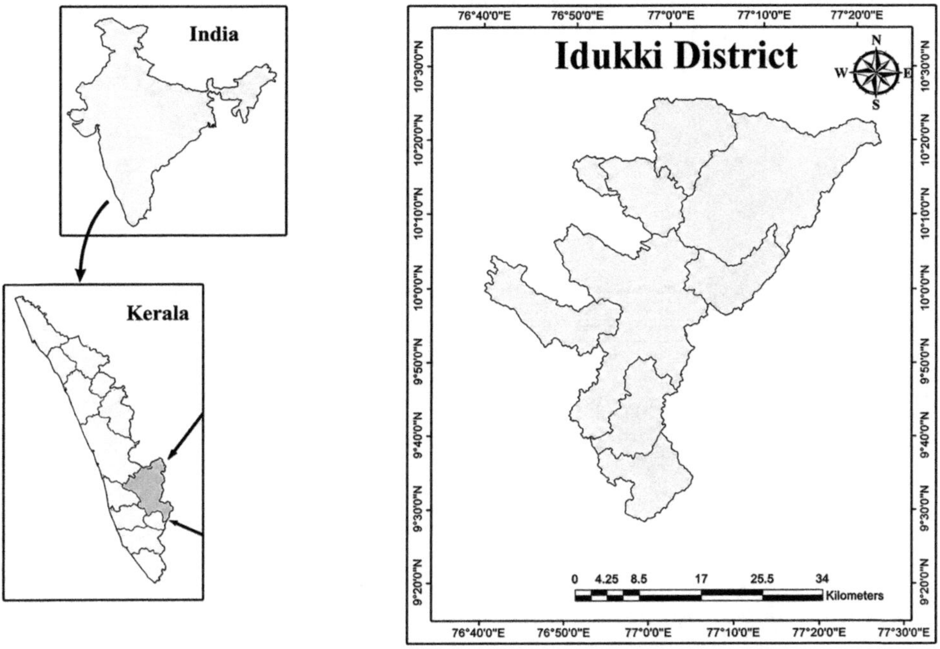

FIGURE 10.1 Study area map of Pettimudi Hills.

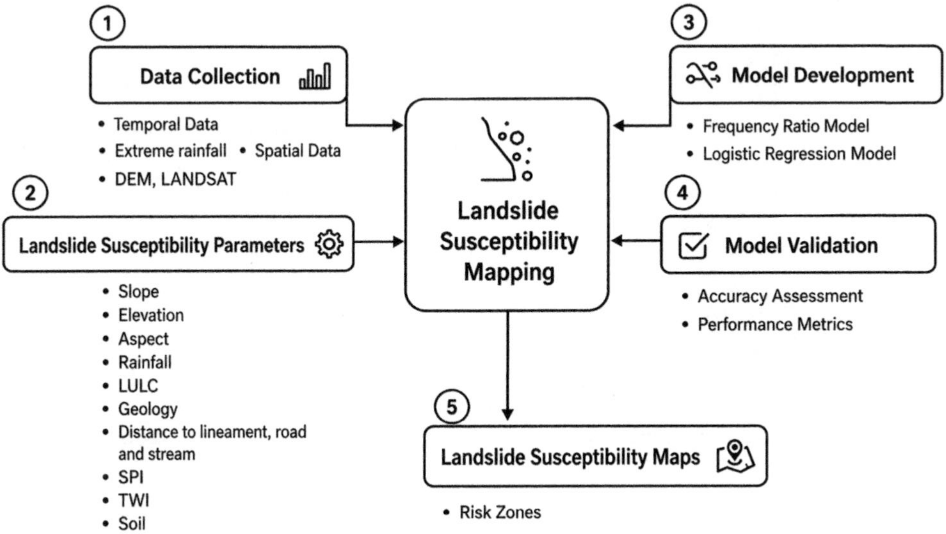

FIGURE 10.2 Methodology.

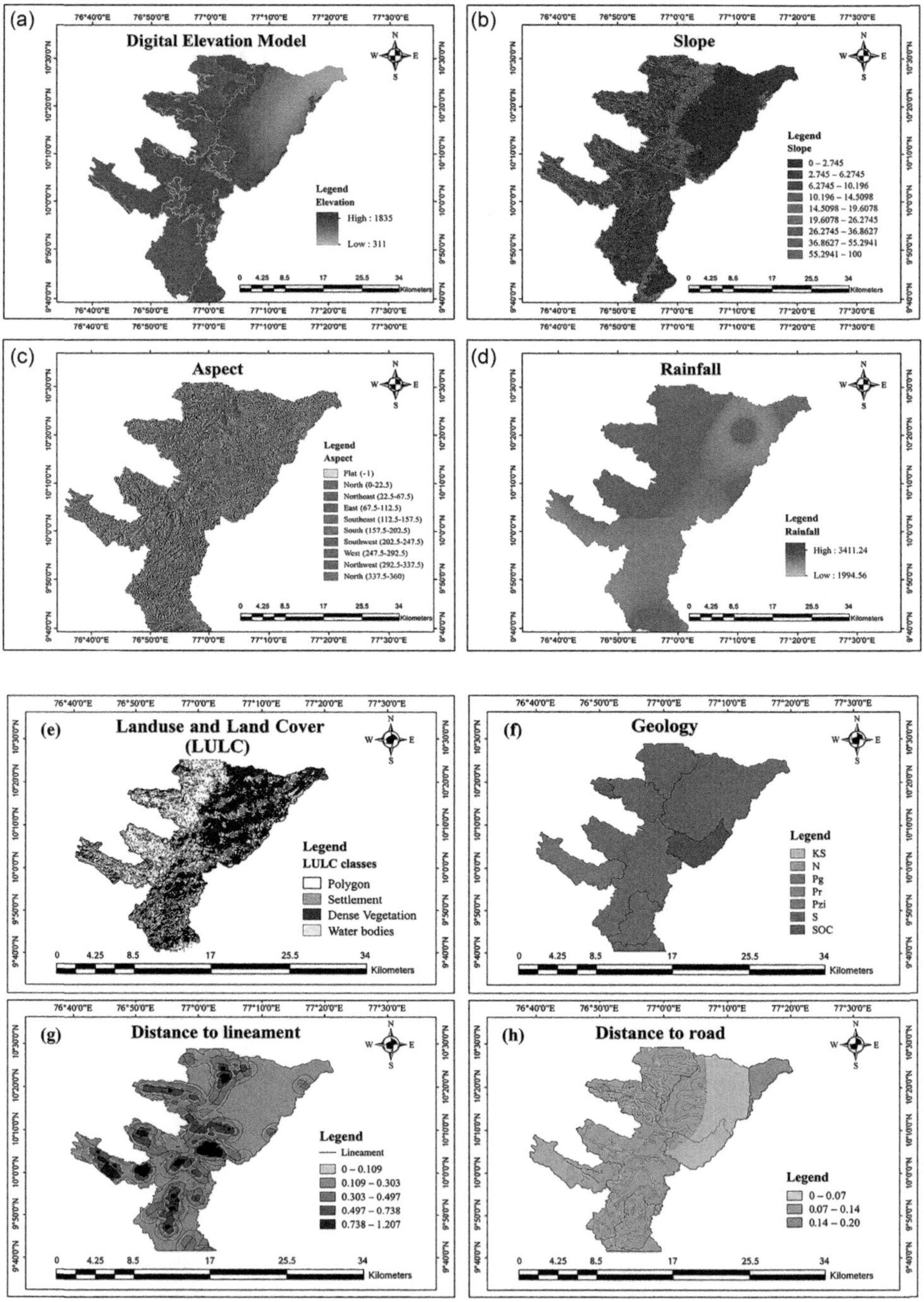

FIGURE 10.3 (a) Digital elevation model (DEM). (b) Slope map. (c) Aspect map. (d) Rainfall map. (e) Land use and land cover (LULC) map. (f) Geology map. (g) Lineament distance map. (h) Distance to road map. (i) Distance to stream map. (j) Stream power index map. (k) Topographic wetness index (TWI) map. (l) Soil map.

(Continued)

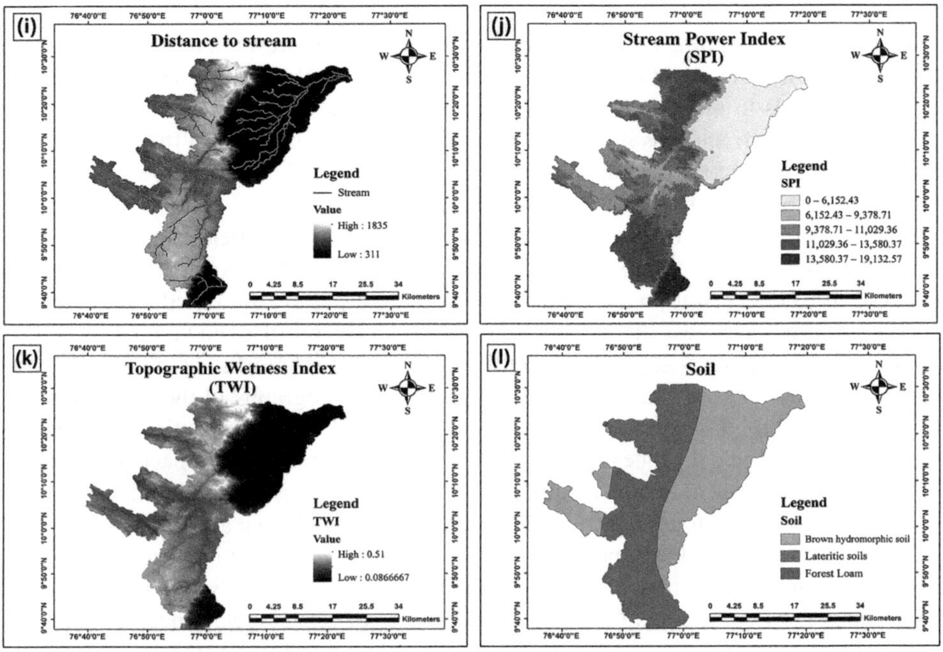

FIGURE 10.3 *(Continued)* (a) Digital elevation model (DEM). (b) Slope map. (c) Aspect map. (d) Rainfall map. (e) Land use and land cover (LULC) map. (f) Geology map. (g) Lineament distance map. (h) Distance to road map. (i) Distance to stream map. (j) Stream power index map. (k) Topographic wetness index (TWI) map. (l) Soil map.

rain can change the stability of slopes and the friction between soil particles. To find unstable and dangerous slopes in the research area and guess how they will act in the future, we need to look at soil mechanics and the steepness of the slopes. The slope map for the research region in Figure 10.3b shows values from 0° to 55.59° in some places. The direction of the terrain is its aspect. Studies of landslides show that most of them happen in one direction. The aspect ratio has an effect on the sun, wind and rain. Figure 10.3c indicates that this study looks at all eight directions. There is no aspect ratio for flat land.

Idukki got 3000 mm of rain on the day of the landslide. There was a lot of rain in a short amount of time in Pettimudi Hills. Heavy rain made the water pressure in the soil pores higher, made the slope less stable and caused the research area to collapse. The average amount of rain that falls in Pettimudi Hills is between 1994.56 and 3411.24 mm. ArcGIS's Inverse Distance Weight (IDW) method is used to create a spatial rainfall map for the research region as shown in Figure 10.3d. Land use land cover (LULC), or the shift in land cover from natural to man-made, is a big reason why landslides happen, especially on steep slopes. This study categorizes land cover into settlement zones, dense vegetation and water bodies.

Using Maximum Likelihood Estimation (MLE), supervised classification in ArcGIS creates LULC classes for the research region. LULC analysis is done for the years before and after the landslide, in 2017 and 2021. Land use has altered, and more communities are popping up in Pettimudi Hills. The large forests in the research region have been turned into tea plantations, which has changed the LULC. The expansion of tea estates leads to more spaces for people to live and less permeable surfaces, which causes more runoff and less infiltration. The slope and soil stability of the research area are affected by more runoff and less infiltration. Figure 10.3e displays the LULC maps of the research area before and after the landslide.

Pettimudi Hills are made up of Precambrian metamorphic shield rocks, basic rocks, the peninsular gneissic complex, charnockite and high-grade meta sedimentary rocks (Soman, 1987). The study area has structural hills, a pediplain, a piedmont zone and rock exposures. The rest of the research area is made up of denudational structural hills. A geological study says that extreme rainfall events change the pore pressure and stability of hills. Figure 10.3f illustrates the geological geography of the area where the research was done. The geomorphology of the examined area is determined by geological and meteorological processes. The study region examines the impacts of physical, chemical and biological factors on landslides. Satellite images are used in the lineament analysis. The study is to understand tectonic plates and faults that are below the surface. Faults break rocks, making them weaker. In lineament, the fractures, rock intrusions and dykes can be identified. Lineaments show areas that are likely to have landslides. Landslides get less often as the lineament distance (n) gets bigger, and the other way around. The five classes of the lineament map are shown in Figure 10.3g.

To make room for new tea estates, infrastructure needs to be extended and road networks need to be built around hill slopes. Building roads on mountains generates tensional cracks that make the slope weaker and hurt the toe and heel of the hill. More roads make slopes less stable, but fewer roads have less of an effect. Figure 10.3h shows how ArcGIS sorted the distance between roads in Pettimudi Hills into five groups. Rivers and streams help keep soil on hill slopes and keep them wet. More water in the soil makes the slope less stable because of pore water pressure. Subsurface runoff from rivers and streams wears away steep soil. Soil erosion caused by streams impacts the stability of hill slopes. Construction debris that blocks the flow of streams affects the soil's slope stability. Streams that are close to home cause more soil erosion than those that are far away. The river distance map for the study area is shown in Figure 10.3i.

The SPI is based on the local slope gradient and the area of the stream's catchment. SPI is directly related to soil erosion and defines how streams erode. A steeper catchment slope makes the water flow faster and more water is available in the upslope area. Equation (10.1) shows the SPI for the area being studied.

$$\text{SPI} = \ln\left(A \cdot \tan i\right) \tag{10.1}$$

where A indicates the area and i denotes the slope gradient. The five SPI classes generated for the study area are shown in Figure 10.3j.

The WI displays the topography, saturation and runoff sources of hydrological processes. High WI values mean that more water runs off the surface, which disrupts and breaks up the soil on slopes. The TWI shows the wetness of soil. A high range means that the soil is moving, while a low range means that rocks are sliding or slumping.

Figure 10.3k illustrates the spatial pattern of TWI in the research area. A higher TWI means that the landslide plate is draining. The WI of the study area is determined using Equation (10.2).

$$\mathrm{WI} = \ln(\alpha / \tan\beta) \tag{10.2}$$

where α implies the catchment area and β denotes the slope gradient.

The soil map for the research area is shown in Figure 10.3l. The soil map is made from FAO raster data. There is forest loam, lateritic and brown hydromorphic soils in the area under investigation. On steep slopes, shallow soil is more likely to wash away or slide down during heavy rain.

10.5 IDENTIFICATION OF LANDSLIDE-SUSCEPTIBLE ZONES

10.5.1 FR Model

Researchers often use the bivariate statistical FR model to look at how likely landslides are to happen in certain areas. It depends on relationships between dependent and independent variables. The FR is the number of pixels in the studied area that cause landslides. Equation (10.3) finds the FR.

$$\mathrm{FR} = \frac{\mathrm{LS}/A}{\mathrm{CP}/\mathrm{TP}} \tag{10.3}$$

FR is the frequency ratio of the class factor, LS is the number of pixels of landslide locations for each class of thematic maps, *A* is the total number of pixels in the research region, CP is the number of pixels in each thematic class and TP is the total number of pixels. To find landslide-risk zones, the FR model employs the FR value of each class in the thematic layer as its weight in the thematic parameters. We use the landslide parameters from each thematic layer to figure out the FR values for training the data set in Equation (10.3). The FR model's accuracy is shown by Receiver Operating Characteristic (ROC) curves for the observed true rate and the expected false rate. Area Under the Curve (AUC) values tell you how accurate your predictions are. The AUC in this study is determined by evaluating the sequential rate of landslides using the training set of parameters while retaining the remainder as test data.

Table 10.1 shows how the FR model connects landslides to training factors in space. Landslides happen in FR model areas with an average value of 1. If the FR value is more than 1, there is a high danger of landslides. If the FR value is less than 1, there is a low risk. In ArcGIS, FR model thematic layers are used to make LSMs. Figure 10.4 depicts the areas of Pettimudi Hills that are most and least likely to slide.

TABLE 10.1 Landslide parameters and frequency ratio model

ADOPTED LANDSLIDE PARAMETERS	*CLASSES*	*TOTAL NO OF PIXELS IN CLASSES*	*PERCENTAGE OF PIXELS IN %*	*NUMBER OF LANDSLIDE PIXELS*	*% OF LANDSLIDE PIXELS*	*FREQUENCY RATIO*
Aspect	Flat	16,508	1.05728204	98	0.312290877	3.385568131
	North	1,00,212	6.418242534	4562	14.53745897	0.441496863
	Northeast	2,01,601	12.91186797	3256	10.37570504	1.244432828
	East	1,98,105	12.6879609	3234	10.30559893	1.231171617
	Southeast	2,56,104	16.40260234	3200	10.19725311	1.60853145
	South	1,58,156	10.12936142	4010	12.77843281	0.792691997
	Southwest	2,00,111	12.81643847	3921	12.49482171	1.025740004
	West	1,78,123	11.40818081	3100	9.878588955	1.154839103
	Northwest	1,51,128	9.679241585	1500	4.779962398	2.024961868
	North	1,01,314	6.488821939	4500	14.33988719	0.452501603
	Overall	15,61,362	100	31381	100	1
Slope	0–2.745	62,357	1.828753926	3160	5.151697941	0.354980813
	2.745–6.2745	85,384	6.716530562	9,105	14.84373726	0.452482447
	6.2745–10.196	530,415	19.30055379	6660	10.8576925	1.777592596
	10.196–14.509	620,142	17.78021802	2611	4.256671938	4.177023337
	14.509–19.607	510,135	16.29924849	501	0.816772363	19.9556807
	19.607–26.274	516,215	2.399712301	11001	17.93475603	0.133802339
	26.274–36.862	316,785	2.484068515	11200	18.25918258	0.136044892
	36.862–55.294	219,152	17.6337046	6001	9.783335235	1.802422607
	55.294–100	105,121	15.55720979	11,100	18.09615416	0.859697019
	Overall	6,75,706	100	61339	100	1

(Continued)

TABLE 10.1 (*Continued*) Landslide parameters and frequency ratio model

ADOPTED LANDSLIDE PARAMETERS	*CLASSES*	*TOTAL NO OF PIXELS IN CLASSES*	*PERCENTAGE OF PIXELS IN %*	*NUMBER OF LANDSLIDE PIXELS*	*% OF LANDSLIDE PIXELS*	*FREQUENCY RATIO*
Geology	Ks	1052672	76.93975268	18776	86.90178654	0.885364453
	N	34352	2.510786251	2111	9.770434139	0.256977962
	Pg	131112	9.582970624	346	1.601407017	5.984094315
	Pr	9408	0.687630329	115	0.532259558	1.291907902
	Pzi	121105	8.851559411	102	0.472091086	18.74968555
	S	10408	0.760720287	98	0.45357771	1.677155359
	SOC	9120	0.666580421	58	0.268443951	2.483126994
	Overall	1368177	100	21606	100	1
Rainfall	3411.24	5,73,310	82.91704391	7222	91.7895272	0.903338828
	1994.56	1,18,116	17.08295609	646	8.210472801	2.080630008
	Overall	6,91,426	100	7868	100	1
SPI	0–6.152	419105	27.34173169	4482	15.36141481	1.77989671
	6.152–9.378	469447	30.62596227	11550	39.58597525	0.773656884
	9.378–11.029	348774	22.75345111	9146	31.34660863	0.725866437
	11.029–13.580	201505	13.14585997	2444	8.376460911	1.569381164
	13.580–19.132	94009	6.132994964	1555	5.329540391	1.150754946
	Overall	1532840	100	29177	100	1
TWI	5.837	1,22,301	15.19448282	11110	52.94005527	0.287012976
	4.474	6,82,603	84.80551718	9,876	47.05994473	1.802074305
	Overall	8,04,904	100	20986	100	1
LULC	Settlement	96,003	11.67408434	2342	63.40010828	0.184133508
	Dense Vegetation	7,11,243	86.48803444	1227	33.21602599	2.603804395
	Water bodies	15,114	1.83788122	125	3.383865728	0.543130658
	Overall	8,22,360	100	3694	100	1

(*Continued*)

TABLE 10.1 (*Continued*) Landslide parameters and frequency ratio model

ADOPTED LANDSLIDE PARAMETERS	*CLASSES*	*TOTAL NO OF PIXELS IN CLASSES*	*PERCENTAGE OF PIXELS IN %*	*NUMBER OF LANDSLIDE PIXELS*	*% OF LANDSLIDE PIXELS*	*FREQUENCY RATIO*
Distance to stream	100	3,23,042	37.76625842	10,244	45.16157475	0.836247598
	200	2,41,201	28.19837451	5,111	22.53229291	1.251464936
	300	1,01,122	11.8219909	4,102	18.08402769	0.653725547
	400	1,00,002	11.69105372	2,221	9.791473791	1.194003473
	500	90,005	10.52232245	1,005	4.430630869	2.374903882
	Overall	8,55,372	100		100	1
Distance to lineament	0–109	4,42,110	35.33710968	10,212	43.1943152	0.293741453
	109–303	4,01,112	32.06020841	8,112	34.31181795	0.934378017
	303–497	2,08,473	16.66289671	4103	17.35470772	0.960136983
	497–738	1,02,212	8.169633473	801	3.388038237	2.411316786
	738–1207	97,214	7.770151728	414	1.751120887	4.437244617
	Overall	12,51,121	100	23,642	100	1
Elevation	311	1,10,124	1.20588092	1214	36.23880597	0.033275956
	1835	90,22,121	98.79411908	2136	63.76119403	1.549439602
	Overall	91,32,245	100	3350	100	1
Soil	Brown hydromorphic soil	6,41,111	53.51296442	14,000	71.91657677	0.74409777
	Laterite soils	1,21,214	10.11762467	114	0.585606411	17.27717539
	Forest loam	1,24,322	10.37704666	212	1.089022448	9.528772046
	Overall	3,11,401	25.99236425	5141	26.40879437	0.984231385
		11,98,048	100	19,467	100	1

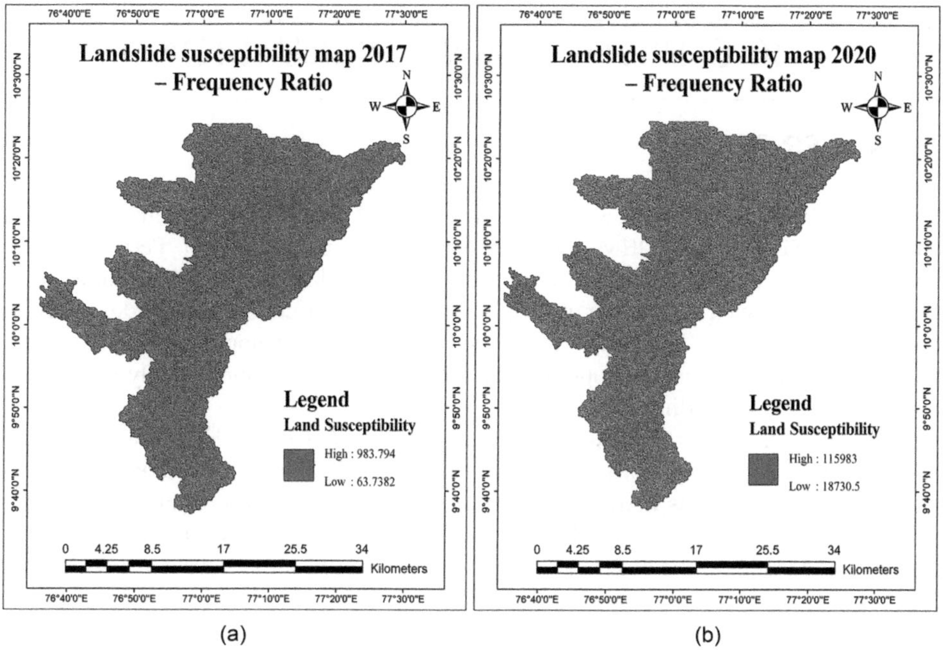

FIGURE 10.4 Land susceptibility map of Pettimudi Hills (frequency ratio). (a) Pre-landslide year 2017. (b) Landslide year 2020.

10.5.2 LR Model

The LR model determines the causes of landslides. This model features nonlinear relationships between the dependent and independent variables (Pal et al., 2025; Tang et al., 2021). The logistic multiple regression model is created using landslide parameters. To make dependent variables in the logistic model, you need a lot of samples. In this study, the pixels that show landslides cover less space than the whole area. Think about all the pixels to make it more accurate. In this study, landslides are the dependent variable, and the LR coefficient predicts the ratios of the model's independent variables. There is either a 0% or 1% chance of landslides. Linear regression is used to figure out how likely a landslide is to happen. Equation (10.4) displays the LR model employed in the investigation. The spatial distribution of landslide occurrence probability in Pettimudi Hills is generated.

$$\begin{aligned} L_{13} = {} & \text{Elevation} + \text{Slope} + \text{Aspect} + \text{Rainfall} + \text{LULC} + \text{Geology} \\ & + \text{Geomorphology} + \text{Distance to lineament} + \text{Distance to road} \\ & + \text{Distance to stream} + \text{SPI} + \text{TWI} + \text{Soil} \end{aligned} \tag{10.4}$$

10.6 RESULTS AND DISCUSSION

10.6.1 FR Model

Pettimudi Hills LSM uses FR and LR models. Spatial data is used to identify landslide causes. The SRTM 30 m DEM generates elevation, slope and aspect. Pettimudi Hills are 311–1835 m high. The western highlands are prone to landslides. Hill altitude affects landslide factors such as weather, rainfall, wind and sun direction. Landslides are more common in western hills than hill folds and peaks due to development. Landslide susceptibility estimates include slope and terrain. Landslides are more likely in the west and center areas, according to FR study.

The analysis shows that excessive rainfall causes Pettimudi Hills landslides during the southwest monsoon. Steep slopes have more drainage than 3°. Water stagnation on non-steep slopes increases infiltration and raises the water table. Infiltration raises soil water pore ratio, which decreases effective mean stress and makes the soil unstable and critical. The analysis demonstrates that low slopes have a larger FR than high slopes.

10.6.2 LR Model

The LR model uses landslide factor coefficients. Landslide factors rank highest. The main cause of landslides in Idukki district is its 3000–5000 mm rainfall. Other things also trigger landslides. Steep slopes in the research area increase runoff and limit soil pore water retention. Land cover change, WI, SPI, slope and rainfall cause landslides. As a less important factor, stream and road distance is given less weight. The LULC) study found that urbanization diminishes vegetation in Pettimudi Hills, reducing plant roots' soil holding ability. The Precambrian crystalline structure of southern granulite terrain rocks makes the area landslip prone. Infiltration increases in this geologic terrain, reducing soil stability.

Hill slides are also more likely in Pettimudi Hills due to their ridges. Lineaments also cause landslides. Landslides are more likely in areas with lineament distances under 500 m. Landslides are more likely near 300–400 m of the road. Landslides are more likely near streams. At 0, the SPI indicates that the location is very prone to landslides. Lower SPIs increase landslide risk. Landslides decrease as the SPI index rises. Landslides are likely on slopes with a high TWI.

Every landslide cause is considered in this study. Both the LR and FR models are simple and help comprehend dependent and independent variables for a better LSM. Figure 10.5 shows how LR model identifies high and low landslide susceptibility in the research area. The FR model's landslide map covers 6%–20% of Pettimudi Hills as low to high-susceptibility zones. However, the LR model suggests that 18%–28% of Pettimudi Hills are in low–high landslide vulnerability zones.

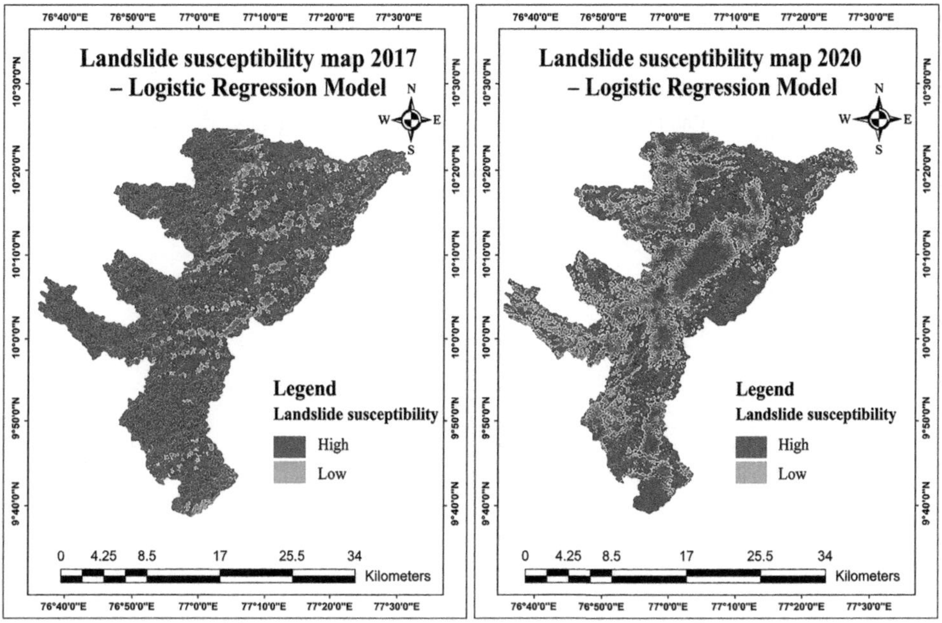

(a) (b)

FIGURE 10.5 Land susceptibility map of Pettimudi Hills (logistic regression model). (a) Pre-landslide year 2017. (b) Landslide year 2020.

10.7 VALIDATION

This study uses ROC curves to verify model performance. ROC characteristics evaluate model performance. The ROC component is used to determine model accuracy (Giri et al., 2025). Figure 10.6 shows the study region ROC curve. Positive rates on the Y-axis and negative rates on the X-axis define the ROC curve. The optimal negative rate is zero while positive rate is one. A model is better if its AUC is closer to one. This study uses the ROC curve to determine model accuracy using five AUC categories: very high (0.9–1), high (0.8–0.9), medium (0.7–0.8), low (0.6–0.7) and very low (0.5–0.6). Models with high AUC are used. Pettimudi Hills' landslide-prone areas had AUC values of 0.75 and 0.8 in the FR and LR models, respectively.

10.8 SENSITIVITY ANALYSIS

The sensitivity analysis was adopted to measure the effectiveness of individual conditioning factors responsible for landslide occurrence and to evaluate the strength of

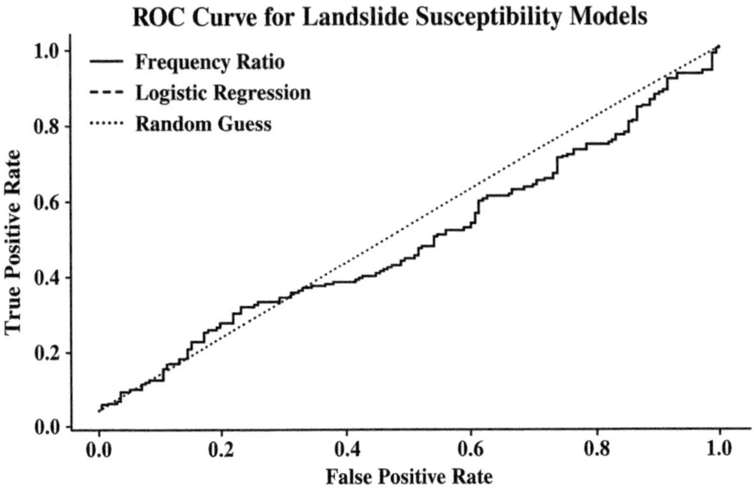

FIGURE 10.6 ROC curve for LS modeling of Pettimudi Hills.

FR and LR models used in this study (J. Wang et al., 2025). This analysis helps to understand how changing input parameters affect the final landslide susceptibility output and identifying the most dominant controlling factors reliable for slope instability in Pettimudi Hills.

In the case of FR model, sensitivity was analyzed by assessing the spatial distribution of high FR values across the different landslide conditioning factors. Parameters that are constantly showing FR values much higher than unity in multiple classes were considered as highly sensitive. In this study slope, rainfall, LULC, soil type and TWI are the most sensitive factors. The steep to moderate slopes showed extremely higher landslide pixel density when compared to their spatial coverage, which shows strong sensitivity to slope angle. Likewise, zones receiving higher rainfall recorded a notably high frequency of landslides, proving that rainfall is the dominant triggering factor in the study area. The LULC classes in the study area representing settlements and plantation modified terrain also show higher FR values by stressing the role of anthropogenic alteration in increasing slope instability.

Similarly, the sensitivity analysis was done for LR model which is based on magnitude and sign of model coefficients applied to each conditioning parameter. The parameters with high absolute coefficient values were considered as most influential in controlling landslide probability. The parameters like rainfall, slope, LULC, TWI, SPI and distance to lineament showed the more effect on landslide occurrence. The rainfall shows the strong positive contribution, highlighting its important role in triggering landslides.

Slope and TWI also confirmed the strong positive relationships with landslide probability, by showing that moisture accumulation on steep slopes critically governs slope failure in Pettimudi Hills. LULC showed a negative coefficient for dense vegetation and positive coefficient for settlement and plantation classes, which reveals the stabilizing role of natural vegetation and destabilizing effects of land modification.

The model stability was further analyzed by selecting the parameters which are not systematic and removed one at a time. The model performance was reassessed by using AUC values. The rejection of rainfall and slope resulted in the highest reduction in prediction accuracy in both FR and LR models, by proving their influential control on landslide occurrence. The elimination of secondary factors such as distance to road and distance to stream caused reasonably minor variations in model performance, by proving their supplementary role in landslide conditioning.

The sensitivity analysis confirms that landslides in Pettimudi Hills are mostly governed by the associated interaction of intense rainfall, terrain steepness, moisture accumulation and land cover transformation. The strong relationship between the sensitivity behavior of the FR and LR models further strengthens the reliability of the susceptibility modeling results. This assessment also reveals that any future landslide mitigation strategy in the region must spotlight slope stabilization, surface and subsurface drainage control, and rigorous regulation of land use change on steep hill slopes.

10.9 CONCLUSIONS

In the last few years, tea estates and communities have quickly changed the way the land on Pettimudi Hills looks. A landslide in Pettimudi Hills killed more than 65 people because of severe rain and changes in land usage and land cover. So, it's important to use a competent modeling method to find hills that are likely to slide. This research employs FR and LR bivariate and multivariate statistical models to identify areas susceptible to landslides. It's easy to use these models to figure out how likely a landslide is and to make maps of the areas where they are likely to happen. This modeling study quantifies landslides by employing relevant parameters and corroborates the findings through ROC curves.

The sensitivity carried in this study strengthens the model's reliability by detecting the most influential factors leading to slope failure in the study area. Rainfall arose as the most dominant triggering parameter, followed by slope, LULC and TWI. These factors continuously show high sensitivity in both FR and LR models, by confirming that intense rainfall occurring on steep, moisture-prone and anthropogenically modified slopes is the main driver of landslides in Pettimudi Hills. The other factors such as soil type, SPI and distance to lineaments also contributed to the slope instability, while distance to roads, streams, elevation and aspect ratio play a minor role. The convergence of results from both sensitivity and validation analysis confirms the robustness of the adopted model.

Geospatial techniques provide landslide maps for Pettimudi Hills. Most landslides happen in the center and northern parts of Pettimudi Hills. The study's south and other locations have only a small effect on landslides. If the same trend of rapid, uncontrolled growth and settlement expansion continues, the slopes and landslides in the southern study area may become unstable, making them unsafe for people who live there. You can use the spatial maps to find mountainous areas with unstable slopes and a high risk of landslides so that you can take action to prevent them from happening. The study also tells the district government not to build on Pettimudi Hills to lower the pressure on the hill slopes.

Future study scope:

- The real time monitoring of rainfall, soil moisture and movement of slope by using sensors and early warning systems can notably improve landslide preparedness in Pettimudi Hills.
- Upcoming studies may combine high-resolution LIDAR data, UAV-based terrain mapping and advance ML models and can be implemented for early detection of landslides, so that precautionary measures can be taken at earlier stage.
- The future studies can improve mapping and modeling of landslide susceptibility to reduce risks and stop rapid urbanization in steep areas.

ACKNOWLEDGMENTS

The author would like to convey his gratitude to USGS Earth Explorer for readily providing the needed satellite data. The author also extends his sincere thanks to IMD Thiruvananthapuram for providing access to rainfall data. Finally, a thanks to the anonymous reviewer of this manuscript.

REFERENCES

Achu, A. L., Aju, C. D., & Reghunath, R. (2020). Spatial modelling of shallow landslide susceptibility: a study from the southern Western Ghats region of Kerala, India. *Annals of GIS*, *26*(2), 113–131. https://doi.org/10.1080/19475683.2020.1758207

Arabameri, A., Chandra Pal, S., Rezaie, F., Chakrabortty, R., Saha, A., Blaschke, T., Di Napoli, M., Ghorbanzadeh, O., & Thi Ngo, P. T. (2022). Decision tree based ensemble machine learning approaches for landslide susceptibility mapping. *Geocarto International*, *37*(16), 4594–4627. https://doi.org/10.1080/10106049.2021.1892210

Das, G., & Lepcha, K. (2019). Application of logistic regression (LR) and frequency ratio (FR) models for landslide susceptibility mapping in Relli Khola river basin of Darjeeling Himalaya, India. *SN Applied Sciences*, *1*(11), 1–22.

Di Napoli, M., Di Martire, D., Bausilio, G., Calcaterra, D., Confuorto, P., Firpo, M., Pepe, G., & Cevasco, A. (2021). Rainfall-induced shallow landslide detachment, transit and runout susceptibility mapping by integrating machine learning techniques and GIS-based approaches. *Water, 13*(4), 488. https://doi.org/10.3390/W13040488

El-fengour, M., Motaki, H. El, & Bouzidi, A. El. (2021). Landslides susceptibility modelling using Multivariate Logistic Regression Model in the Sahla Watershed in Northern Morocco. *Sociedade & Natureza (Online)*, *33*. https://doi.org/10.14393/SN-V33-2021-59124

Giri, S., Tripathi, V. K., Thenmozhi, M., Panda, K. C., Sharma, R., Rakshit, A., & Panda, S. (2025). Capability assessment of machine learning classifiers for spatial landslide susceptibility mapping in Malappuram district, Kerala, India. *Physical Geography*. https://doi.org/10.1080/02723646.2025.2552290

Glade, T., & Crozier, M. J. (2010). Landslide geomorphology in a changing environment. *Geomorphology*, *120*(1–2), 1–2. https://doi.org/10.1016/J.GEOMORPH.2009.09.018

Jeggle, T., & Boggero, M. (2018). *Post-Disaster Needs Assessment (PDNA): Lessons from a Decade of Experience*. https://documents.worldbank.org/en/publication/documents-reports/documentdetail/481761539848031116

Kotluri, S. K., Kumar, T. S., Prajith, A., Eldhose, K., Shivapriya, S., Kaliraj, S., Nandakumar, V., & Chalapathi Rao, N. V. (2025). Geomorphological evolution of the multi-chain cascading disaster of Wayanad (30 July 2024), Kerala, India. *Journal of the Geological Society of India*, *101*(8), 1247–1260. https://doi.org/10.17491/JGSI/2025/174230

Liu, J. K., & Shih, P. T. Y. (2013). Topographic correction of wind-driven rainfall for landslide analysis in Central Taiwan with validation from aerial and satellite optical images. *Remote Sensing*, *5*(6), 2571–2589. https://doi.org/10.3390/RS5062571

Meena, V., Kumari, S., & Shankar, V. (2022). Physically based modelling techniques for landslide susceptibility analysis: A comparison. *IOP Conference Series: Earth and Environmental Science*, *1032*(1), 012033. https://doi.org/10.1088/1755-1315/1032/1/012033

Merghadi, A., Yunus, A. P., Dou, J., Whiteley, J., ThaiPham, B., Bui, D. T., Avtar, R., & Abderrahmane, B. (2020). Machine learning methods for landslide susceptibility studies: A comparative overview of algorithm performance. *Earth-Science Reviews*, *207*, 103225. https://doi.org/10.1016/J.EARSCIREV.2020.103225

Moung-Jin, L., Won-Kyong, S., Joong-Sun, W., Inhye, P., & Saro, L. (2014). Spatial and temporal change in landslide hazard by future climate change scenarios using probabilistic-based frequency ratio model. *Geocarto International*, *29*(6), 639–662. https://doi.org/10.1080/10106049.2013.826739

Naceur, H. A., Abdo, H. G., Igmoullan, B., Namous, M., Almohamad, H., Al Dughairi, A. A., & Al-Mutiry, M. (2022). Performance assessment of the landslide susceptibility modelling using the support vector machine, radial basis function network, and weight of evidence models in the N'fis river basin, Morocco. *Geoscience Letters*, *9*(1), 1–20.

Nawazuzzoha, M., Rashid, M. M., Iqbal, F., Suheb, Bhardwaj, A., Shakeel, A., Arabameri, A., & Naqvi, H. R. (2025). Advanced machine learning models for landslide susceptibility mapping in South Sikkim Himalayas, India. *Geological Journal*.

Nhu, V. H., Mohammadi, A., Shahabi, H., Ahmad, B. Bin, Al-Ansari, N., Shirzadi, A., Geertsema, M., Kress, V. R., Karimzadeh, S., Kamran, K. V., Chen, W., & Nguyen, H. (2020). Landslide detection and susceptibility modeling on Cameron Highlands (Malaysia): A comparison between random forest, logistic regression and logistic model tree algorithms. *Forests*, *11*(8), 830. https://doi.org/10.3390/F11080830

Novellino, A., Cesarano, M., Cappelletti, P., Di Martire, D., Di Napoli, M., Ramondini, M., Sowter, A., & Calcaterra, D. (2021). Slow-moving landslide risk assessment combining machine learning and InSAR techniques. *CATENA*, *203*, 105317. https://doi.org/10.1016/J.CATENA.2021.105317

Oh, H. J., Park, N. W., Lee, S. S., & Lee, S. (2012). Extraction of landslide-related factors from ASTER imagery and its application to landslide susceptibility mapping. *International Journal of Remote Sensing*, *33*(10), 3211–3231. https://doi.org/10.1080/01431161.2010.545084

Panagos, P., Matthews, F., Patault, E., De Michele, C., Quaranta, E., Bezak, N., Kaffas, K., Patro, E. R., Auel, C., Schleiss, A. J., Fendrich, A., Liakos, L., Van Eynde, E., Vieira, D., & Borrelli, P. (2024). Understanding the cost of soil erosion: An assessment of the sediment removal costs from the reservoirs of the European Union. *Journal of Cleaner Production*, *434*, 140183. https://doi.org/10.1016/J.JCLEPRO.2023.140183

Pal, S. C., Biswas, T., Ghorai, S., Pande, C. B., Islam, A., Abu Reza Md. Towfiqul Islam, & Alam, E. (2025). Landslide vulnerability mapping using GIS-based statistical model for sustainable ecosystem management in the Himalayan Region of Teesta River Basin, India. *Environmental Sciences Europe*, 37. https://doi.org/10.1186/s12302-025-01107-8

Pourghasemi, H. R., & Rahmati, O. (2018). Prediction of the landslide susceptibility: Which algorithm, which precision? *CATENA*, *162*, 177–192. https://doi.org/10.1016/J.CATENA.2017.11.022

Pradhan, B. (2017). *Laser Scanning Applications in Landslide Assessment*, 1–359. https://link.springer.com/book/10.1007/978-3-319-55342-9

Qiu, H., Zhu, Y., Zhou, W., Sun, H., He, J., & Liu, Z. (2022). Influence of DEM resolution on landslide simulation performance based on the Scoops3D model. *Geomatics, Natural Hazards and Risk*, *13*(1), 1663–1681. https://doi.org/10.1080/19475705.2022.2097451

Rao, K. V., Rao, V. V., Dadhwal, V., & Diwakar, P. (2014). Kedarnath flash floods: A hydrological and hydraulic simulation study. *Current Science*, *106*(4), 598–603.

Sachdeva, S., Bhatia, T., & Verma, A. K. (2020). A novel voting ensemble model for spatial prediction of landslides using GIS. *International Journal of Remote Sensing*, *41*(3), 929–952. https://doi.org/10.1080/01431161.2019.1654141

Saha, S., Bera, B., Shit, P. K., Sengupta, D., Bhattacharjee, S., Sengupta, N., Majumdar, P., & Adhikary, P. P. (2023). Modelling and predicting of landslide in Western Arunachal Himalaya, India. *Geosystems and Geoenvironment*, *2*(2), 100158. https://doi.org/10.1016/J.GEOGEO.2022.100158

Sajinkumar, K. S., Asokakumar, M. R., Sajeev, R., & Venkatraman, N. V. (2017). A potential headward retreat landslide site at Munnar, Kerala. *Journal of the Geological Society of India*, *89*(2), 183–191.

Sameen, M. I., Pradhan, B., Bui, D. T., & Alamri, A. M. (2020). Systematic sample subdividing strategy for training landslide susceptibility models. *CATENA*, *187*, 104358. https://doi.org/10.1016/J.CATENA.2019.104358

Samia, J., Temme, A., Bregt, A. K., Wallinga, J., Stuiver, J., Guzzetti, F., Ardizzone, F., & Rossi, M. (2018). Implementing landslide path dependency in landslide susceptibility modelling. *Landslides*, *15*(11), 2129–2144. https://doi.org/10.1007/S10346-018-1024-Y

Soman, K. (1987). *Geology of Kerala PDF | PDF | Granite | Rocks* (1st ed., Vol. 2). https://www.scribd.com/document/342438076/Geology-of-Kerala-pdf

Steger, S., Brenning, A., Bell, R., & Glade, T. (2018). Incompleteness matters-An approach to counteract inventory-based biases in statistical landslide susceptibility modelling. *EGU General Assembly Conference Abstracts,* Vienna, Austria.

Tang, R. X., Yan, E. C., Wen, T., Yin, X. M., & Tang, W. (2021). Comparison of logistic regression, information value, and comprehensive evaluating model for landslide susceptibility mapping. *Sustainability*, *13*(7), 3803.

Thanh, L. N., & De Smedt, F. (2014). Slope stability analysis using a physically based model: a case study from A Luoi district in Thua Thien-Hue Province, Vietnam. *Landslides*, *11*(5), 897–907.

Tien Bui, D., Pradhan, B., Lofman, O., Revhaug, I., & Dick, O. B. (2012). Landslide susceptibility mapping at Hoa Binh province (Vietnam) using an adaptive neuro-fuzzy inference system and GIS. *Computers & Geosciences*, *45*, 199–211. https://doi.org/10.1016/J.CAGEO.2011.10.031

Van Westen, C. J., Jaiswal, P., Ghosh, S., Martha, T. R., & Kuriakose, S. L. (2012). Landslide inventory, hazard and risk assessment in India. In B. Pradhan and M. Buchroithner (eds.), *Terrigenous Mass Movements: Detection, Modelling, Early Warning and Mitigation Using Geoinformation Technology*, Springer, Berlin Heidelberg, *9783642254956*, 239–282. https://doi.org/10.1007/978-3-642-25495-6_9

Wang, J., Zhu, H.-H., Zhang, W., Tan, D.-Y., & Pasuto, A. (2025). Enhancing landslide displacement prediction using a spatio-temporal deep learning model with interpretable features. *Journal of Geophysical Research: Machine Learning and Computation*, *2*(3), e2025JH000592. https://doi.org/10.1029/2025JH000592

Wang, N., Wang, M., & Zhang, J. (2025). A hybrid VMD-LSTM-SVR model for landslide prediction. *Frontiers in Built Environment*, *11*, 1630201.

Weidner, L., Oommen, T., Escobar-Wolf, R., Sajinkumar, K. S., & Samuel, R. A. (2018). Regional-scale back-analysis using TRIGRS: An approach to advance landslide hazard modeling and prediction in sparse data regions. *Landslides*, *15*(12), 2343–2356.

Wubalem, A. (2020). Modeling of landslide susceptibility in a part of Abay Basin, northwestern Ethiopia. *Open Geosciences*, *12*(1), 1440–1467. https://doi.org/10.1515/GEO-2020-0206

Xia, L., Shen, J., Zhang, T., Dang, G., & Wang, T. (2023). GIS-based landslide susceptibility modeling using data mining techniques. *Frontiers in Earth Science*, *11*, 1187384.

Xingfu, Z., & Erdi, A. B. I. (2025). Research on rainfall-induced landslide susceptibility prediction considering spatial heterogeneity. *Earthquake Research Advances*, 100400. https://doi.org/10.1016/J.EQREA.2025.100400

Xu, C., Xu, X., Dai, F., Wu, Z., He, H., Shi, F., Wu, X., Xu, S., Xu, C., Xu, Á. X., He, Á. H., Shi, Á. F., Wu, Á. X., Dai, Á. F., Wu, Z., & Xu, S. (2013). Application of an incomplete landslide inventory, logistic regression model and its validation for landslide susceptibility mapping related to the May 12, 2008 Wenchuan earthquake of China. *Natural Hazards*, *68*, 883–900. https://doi.org/10.1007/s11069-013-0661-7

Youssef, K., Shao, K., Moon, S., & Bouchard, L. S. (2023). Landslide susceptibility modeling by interpretable neural network. *Communications Earth and Environment*, *4*(1), 1–16.

Zong, H., Yuan, X., Gan, S., Zhang, X., Yang, M., & Lv, J. (2023). UAV image matching of mountainous terrain using the LoFTR deep learning model. *Frontiers in Earth Science*, *11*, 1203078.

AI- and IoT-Driven Prediction Models for Stone Column Performance under Liquefaction

11

Priyanka Ahirwar and Pradeep Kumar Jain

11.1 INTRODUCTION

It has long been known that one of the most damaging ground-failure mechanisms in earthquake-prone areas is seismic liquefaction. Embankments, bridge foundations, and lifeline infrastructure have frequently failed catastrophically due to abrupt loss of soil strength combined with significant deformations such settlement and lateral spreading. Despite decades of research and practice, this hazard persists, as demonstrated by historic case studies ranging from the 1964 Niigata and Alaska earthquakes to the more recent Christchurch (2011) and Palu Bay (2018) occurrences (Cubrinovski et al., 2011; Adalier & Elgamal, 2004).

One of the most popular ground improvement methods for liquefaction mitigation is the use of stone columns (SCs), both ordinary (OSC) and geosynthetic-encased (ESC). They work by spreading stresses, densifying loose soil, and—above all—by offering

DOI: 10.1201/9781003714125-11

drainage channels with high permeability that hasten the dissipation of pore pressure during shaking (Ahirwar & Jain, 2025; Pal & Deb, 2021). They have been shown to stabilize saturated sandy foundations in the United States through field data from the Loma Prieta earthquake and dam rehabilitation efforts (Hausler & Sitar, 2001; Luehring et al., 2001). Table 11.1 synthesizes key literature on SCs in liquefiable soils, showing that while OSC and ESC have been widely studied, clogging-aware monitoring and artificial intelligence (AI)–Internet of Things (IoT) integration remain largely unexplored. The real performance of SCs, however, is heavily impacted by time-dependent mechanisms and site conditions. Clogging from fines migration is still a major problem among them since it gradually lowers column permeability and compromises the drainage benefit of SCs (Onoue et al., 1987; Pal & Deb, 2021). The drainage efficiency contrast between unclogged and clogged SCs is illustrated in Figure 11.1 where clogging-induced permeability reduction delays excess pore pressure dissipation (Onoue et al., 1987; Pal & Deb, 2021).

Despite the fact that numerical models such as PLAXIS 3D and FLAC have been widely used to simulate SC behavior under seismic pressure (Kumari et al., 2015, these models struggle to capture stochastic, nonlinear, and dynamic phenomena like clogging.

TABLE 11.1 Summary of literature on stone columns in liquefiable soils

AUTHOR(S)	*YEAR*	*METHOD (FIELD/ EXP./NUMERICAL)*	*TREATMENT (OSC/ESC)*	*KEY FINDINGS*
Onoue et al.	1987	Analytical + Field	OSC	Identified clogging mechanism; pore water pressure dissipation delayed due to fines infiltration.
Hausler and Sitar	2001	Field	OSC	Documented SC performance in dam rehabilitation; effectiveness depends on site conditions.
Kumari et al.	2015	Numerical (PLAXIS 3D)	OSC	SCs reduce liquefaction potential but clogging not modeled.
Sawant and Sharma	2018	Numerical (PLAXIS 3D)	OSC	Seismic performance improved; clogging neglected.
Pal and Deb	2019	Numerical/ Experimental	OSC	Showed clogging reduces drainage capacity under cyclic loads.
Pal and Deb	2021	Experimental + Modeling	ESC	Geotextile encasement reduces clogging, improves drainage efficiency.
Zhou et al.	2021	Centrifuge	OSC	Demonstrated time-dependent pore pressure dissipation in liquefiable soils.
Ahirwar and Jain	2025	Review	OSC/ESC	Highlighted clogging-aware SC performance gaps motivating AI–IoT integration.

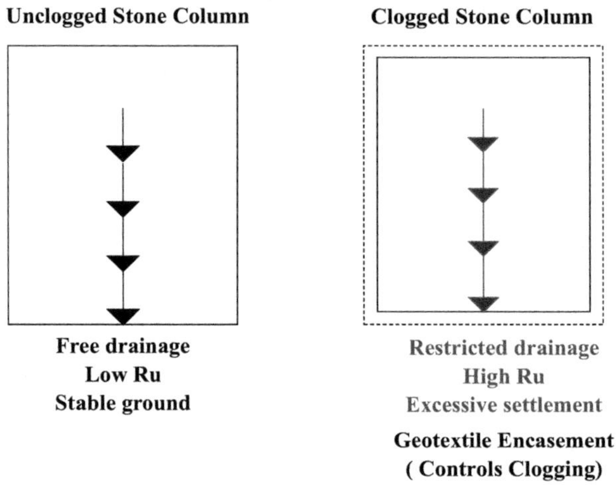

FIGURE 11.1 Comparative mechanism of unclogged and clogged stone columns under liquefaction. Unclogged columns allow free pore water drainage, reducing Ru and settlement, whereas clogged columns experience restricted drainage, excessive Ru build-up, and higher deformation. Geotextile encasement mitigates clogging by maintaining filtration stability.

Similarly, field instrumentation provides delayed input and is essential but often deficient. The discrepancy between simulation accuracy and real-time soil response monitoring is what motivates the incorporation of AI and the IoT into geotechnical engineering. Predicting soil properties, settlement, and liquefaction potential from complex datasets has already demonstrated the usefulness of AI techniques. These techniques include deep learning architectures, tree-based ensembles, support vector machines (SVMs), and artificial neural networks (ANNs) (Zou et al., 2020). Concurrently, IoT-enabled sensor networks that offer continuous in situ measurement of vibration, deformation, and pore pressure include piezometers, accelerometers, strain gauges, and smart geotextiles (Bhatta & Dang, 2024; Jara et al., 2014). SC performance during earthquakes can be predicted by processing the enormous, real-time datasets generated by these technologies using AI algorithms.

Building on the body of established geotechnical research (Onoue et al., 1987; Pal & Deb, 2021; Ahirwar & Jain, 2025) and new developments in AI/IoT applications for infrastructure monitoring (Bhatta & Dang, 2024; Jara et al., 2014), this chapter occupies this frontier and investigates how intelligent frameworks can improve the sustainability and dependability of SCs in seismic locales. Thus, the intersection of IoT and AI presents a revolutionary avenue: prediction of SC performance under liquefaction in real time with clogging awareness. Key elements include physics-guided features, multi-tier validation, and an end-to-end AI–IoT decision flow. This study introduces the first clogging-aware AI–IoT predictive framework for assessing SC performance under liquefaction, bridging the gap between field monitoring, numerical simulation, and intelligent decision-making.

11.2 METHODOLOGY

This study's methodology is designed to forecast the seismic performance of OSC/ESC in liquefiable soils by combining clogging-aware geotechnical knowledge with AI and IoT technology. The framework progresses through sequential stages—data curation, IoT monitoring, feature engineering, AI modeling, validation, and deployment—each designed to ensure that advanced computational tools remain firmly anchored to geotechnical mechanisms.

11.2.1 Problem Definition

Liquefaction-induced failures are controlled by the interaction of seismic demand, SC drainage capacity, and soil resistance. Conventional numerical evaluations (e.g., PLAXIS 3D; Kumari et al., 2015) mimic these processes, but they typically assume constant permeability and fail to account for the growing efficiency loss caused by clogging (Onoue et al., 1987; Pal & Deb, 2021). This study specifically uses time-dependent permeability degradation and real-time observational data to estimate pore pressure evolution, settlements, and lateral deformation in soils treated with SC. The prediction problem is reformulated in this way.

11.2.2 Data Curation and Knowledge Integration

A comprehensive dataset was produced by merging:

1. Field case studies, such as Christchurch (2011) and Palu Bay (2018), in which ground improvement was evaluated during real earthquakes (Cubrinovski et al., 2011).
2. Centrifuge and shaking table studies provide experimental evidence of pore pressure growth and dissipation in SC-treated soils (Zhou et al., 2021).
3. Numerical models, particularly validated PLAXIS 3D simulations (Kumari et al., 2015; Sawant et al., 2018).

Each record was enriched with:

- **Soil Parameters**: relative density, fines content, shear wave velocity, and corrected Standard Penetration Test (SPT)/Cone Penetration Test (CPT) values (Youd et al., 2003; Boulanger & Idriss, 2008).
- **Seismic Parameters**: peak ground acceleration, Arias intensity, shaking duration.

- The parameters of the column include stiffness, length, encasement modulus, area replacement ratio, diameter, and spacing (Priebe, 1998; Rayamajhi et al., 2014).
- Descriptors of clogging include well-resistance indices and relative permeability decrease factors (Onoue et al., 1987; Pal & Deb, 2021).

Incorporating experimental, analytical, and empirical expertise, this hybrid dataset ensures that AI algorithms identify patterns consistent with soil mechanics.

11.2.3 IoT-Enabled Monitoring Framework

An IoT-based field monitoring system is envisioned to supplement static datasets. The system uses inclinometer or settlement sensors to monitor deformations, Micro-Electro-Mechanical System (MEMS) accelerometers to record seismic input, and piezometers to measure transient pore water pressures. According to the layered Structural Health Monitoring (SHM) architecture comprising physical, communication, cloud, and application layers, these devices continuously stream data into a cloud environment over secure narrowband-IoT or Long-Term Evolution for Machines (LTE-M) channels (Bhatta & Dang, 2024). The representative IoT monitoring layout in SC-improved ground is shown in Figure 11.2, integrating piezometers, accelerometers, settlement gauges, and cloud-based data networks. The real-time inputs needed for predictive AI models are provided by such a system, which permits direct monitoring of drainage behavior and clogging effects during and after earthquake stress (Table 11.2).

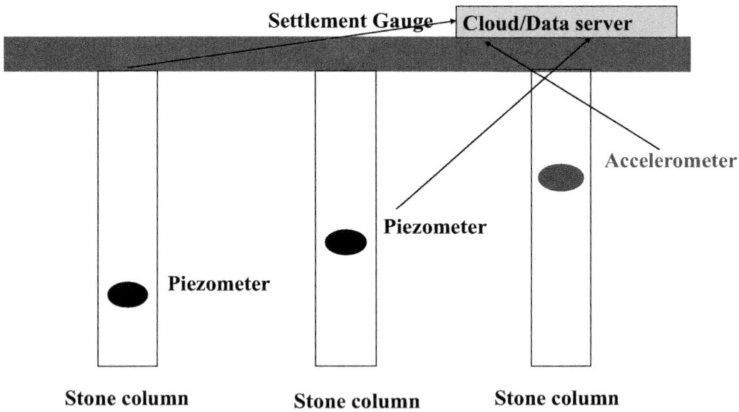

FIGURE 11.2 IoT-based monitoring layout in stone column improved ground. Sensors including piezometers, accelerometers, and settlement gauges transmit real-time data to a cloud server, enabling clogging-aware liquefaction assessment.

TABLE 11.2 IoT sensors for clogging-aware liquefaction assessment

SENSOR TYPE	*MEASURED PARAMETER*	*APPLICATION IN SC/LIQUEFACTION*	*STRENGTHS*	*CHALLENGES*
Piezometer	Pore water pressure	Tracks Ru evolution during shaking	Direct measure of liquefaction	Sensitive to clogging and saturation
Accelerometer (MEMS)	Seismic acceleration	Captures input motion and soil response	High-frequency response	Installation durability under cycles
Settlement gauge	Surface settlement	Measures Uz in SC-treated ground	Simple, reliable	Limited to surface displacements
Inclinometer	Lateral deformation	Detects slope/ embankment spreading	Good for stability monitoring	Expensive, long-term drift
Smart geotextile	Strain + permeability	Monitors clogging and encasement response	Innovative, multi-parameter	Still experimental; field validation needed

11.2.4 Physics-Guided Feature Development

AI model input features are carefully chosen to strike a compromise between scientific interpretability and predictive performance:

- **Seismic Demand**: duration of violent motion, peak acceleration, and Arias intensity
- **Soil Resistance**: shear wave velocity, fines content, and clean-sand equivalent SPT/CPT values
- **Column Efficiency**: encasement tensile modulus, stiffness contrast, area replacement ratio, and spacing ratio
- Indicators of clogging include drainage resistance indices and permeability decay factors (Onoue et al., 1987)

While specifically mentioning clogging as a deterioration mechanism, these characteristics demonstrate the dual function of SCs as drainage channels and reinforcing inclusions.

11.2.5 AI Modeling Framework

The predictive engine combines traditional and advanced approaches:

- Nonlinear soil–structure interactions are captured via gradient boosting and SVM models.

- Temporal dependencies in pore pressure build-up and dissipation are learned by recurrent neural networks (Long Short-Term Memory, LSTM).
- Cyclic Stress Ratio (CSR)–Cyclic Resistance Ratio (CRR) factor-of-safety tests and other baseline techniques are kept for reference (Seed & Idriss, 1971; Boulanger & Idriss, 2008).

This team ensures that predictions remain both innovative and consistent with accepted geotechnical practice. SHapley Additive exPlanations (SHAP) analysis improves the interpretability of the model by showing that trends like a higher replacement ratio that results in lower excess pore pressures are compatible with observed soil behavior. Table 11.3 provides an overview of AI algorithms that have been applied in geotechnical engineering, summarizing their input features, predictive capabilities, and limitations.

11.2.6 Validation Strategy

To ensure dependability, three stages of validation are performed:

1. Retroactive validation with case histories and centrifuge data
2. External validation is done using independent field recordings that were not used for training
3. In contrast to simplified techniques for liquefaction assessment (Youd et al., 2003; Boulanger & Idriss, 2008)

This multi-tiered validation keeps the results trustworthy for engineers by avoiding algorithmic overfitting.

11.2.7 Development as a Decision-Support System

Finally, the framework is operationalized to create a real-time decision-support tool. A conceptual framework (Figure 11.3) is put forth to show how AI-driven predictive models (Zou et al., 2020; Bui et al., 2017) and numerical simulations (Kumari et al.,

TABLE 11.3 Artificial intelligence models applied in geotechnical engineering with their input variables, predicted outputs, advantages, and limitations

ALGORITHM	*INPUT FEATURES*	*OUTPUT*	*ADVANTAGE*	*LIMITATION*
ANN	Seismic + soil + SC + clogging indices	Ru, Uz, FS	Nonlinear mapping	Needs large data
LSTM	Time-series pore pressure	Ru(t), Uz(t)	Temporal learning	Complex tuning
SVM	Static geotech features	FS	Robust small data	Less interpretability
XGBoost	Hybrid features	Clogging Index, FS	High accuracy	Can overfit

2015; Sawant et al., 2018) can be combined with IoT-enabled monitoring (Bhatta & Dang, 2024; Jara et al., 2014) for clogging-aware liquefaction assessment of SCs (Onoue et al., 1987; Pal & Deb, 2021). Incoming IoT data processing edge devices set off alarms when pore pressure ratios exceed critical thresholds. Predictions are updated continuously by cloud-based AI models and reported with safety margins based on physics. The system's dual outputs ensure transparency by providing both AI predictions and geotechnical benchmarks, which are useful for field engineers tasked with monitoring and managing liquefaction-mitigated sites. The results representation framework is outlined in Figure 11.4, linking IoT monitoring, numerical simulations, and AI predictions into a unified decision-support cycle.

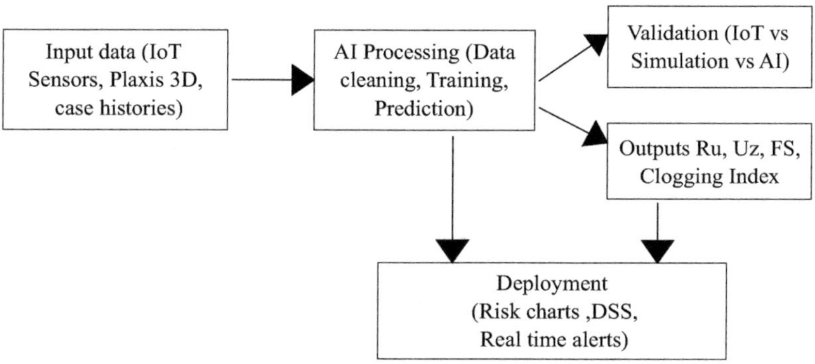

FIGURE 11.3 Results representation framework integrating IoT monitoring, PLAXIS 3D simulations, and AI-based predictions. The framework ensures multi-source data fusion, validation, and deployment into practice via decision-support tools and real-time alerts.

TABLE 11.4 AI–IoT Integrated Framework for Clogging-Aware Liquefaction Assessment

STAGE	*DESCRIPTION*	*EXPECTED OUTPUT*	*LINK TO PRACTICE*
Data Curation	Combine field, centrifuge, numerical results with clogging indices	Hybrid datasets with soil, seismic, SC properties	Standardized input for AI
IoT Monitoring	Deploy piezometers, accelerometers, smart geotextiles	Real-time Ru, Uz, deformation data	Continuous site monitoring
Feature Engineering	Physics-guided selection of seismic, soil, SC, clogging variables	Balanced interpretable AI inputs	Ensures geotechnical realism
AI Modeling	Use ANN, SVM, XGBoost, LSTM	Predicted Ru, Uz, FS, clogging index	Real-time performance forecasts
Validation	Benchmark with case histories and centrifuge tests	Accuracy and reliability metrics	Engineer trust in outputs
Deployment	DSS integrating IoT + AI outputs	Alerts, design charts, decision tools	Practical engineering use

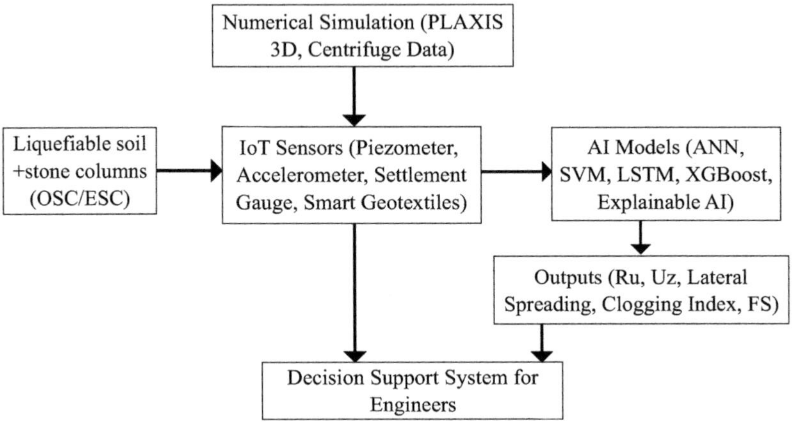

FIGURE 11.4 Conceptual framework integrating IoT-enabled monitoring, AI-driven predictive models, and numerical simulations (PLAXIS 3D) for clogging-aware liquefaction assessment and stone column performance.

11.3 DISCUSSION

The proposed methodology shows how the assessment of SC performance during liquefaction may be revolutionized by merging IoT-based sensors, AI modeling, and clogging-aware soil mechanics. Although traditional CSR–CRR methods (Seed & Idriss, 1971; Boulanger & Idriss, 2008) and finite element analyses (Kumari et al., 2015; Sawant et al., 2018) provide valuable insights, they often assume constant permeability and fail to account for the progressive drainage loss caused by clogging (Onoue et al., 1987; Pal & Deb, 2021). By including permeability decay factors and well-resistance indices in prediction models, this study closes a big gap in current practice. The inclusion of IoT-enabled monitoring allows for real-time surveillance of clogging-related changes in pore pressure and settlement. Unlike traditional sparse instrumentation (Hausler & Sitar, 2001), layered IoT networks (Bhatta & Dang, 2024) enable continuous, high-frequency data streams that directly feed AI algorithms. Combined with physics-guided features and tested against actual data, AI becomes a useful tool rather than a mystery. The strategy maintains interpretability while producing predictions more quickly and adaptably than conventional techniques (Zou et al., 2020). The proposed platform represents the first integrated framework combining clogging indices, explainable AI, and IoT monitoring for liquefaction mitigation.

11.4 CHALLENGES AND RESEARCH GAPS

Notwithstanding its potential, the proposed AI–IoT system for clogging-aware SC performance still faces several challenges. The absence of data is the primary limitation.

While centrifuge experiments (Zhou et al., 2021) and case studies provide helpful benchmarks, few high-quality datasets combine clogging effects and real-time monitoring. AI models are at risk of overfitting and losing their generalizability in the absence of more samples. Second, there is still no standard method for quantifying obstruction. Although drainage resistance theory (Onoue et al., 1987) and current studies on geotextile encasement (Pal & Deb, 2021) provide conceptual indicators, there is currently a lack of progress in the field calibration of these theories. Third, when IoT is implemented in geotechnical scenarios, problems with data security, signal stability, and durability occur. Subsurface sensors commonly fail under cyclic loads and high pore pressures, yet reliable long-term communication in seismic regions remains challenging (Bhatta & Dang, 2024). Finally, the integration of AI with geotechnical practice needs to be more widely accepted by practitioners. Many engineers remain cautious about black-box forecasts; therefore, frameworks that combine interpretability with physics-based checks are needed (Zou et al., 2020). These gaps highlight the need for (i) larger datasets that incorporate clogging; (ii) field validation to improve permeability decay models; (iii) developing trustworthy IoT sensor systems that are appropriate for subsurface conditions; and (iv) ensuring AI outputs are transparent and trustworthy for engineering practice.

11.5 CONCLUSIONS AND FUTURE SCOPE

This study develops a clogging-aware AI–IoT framework to predict the seismic performance of SCs in liquefiable soils. By combining conventional geotechnical processes with modern digital tools, the method closes a major gap in current practice—the neglect of time-dependent drainage loss due to clogging. Using knowledge-driven datasets, IoT-enabled sensing, physics-guided feature engineering, and explainable AI models, the framework demonstrates how real-time predictions can complement established CSR–CRR procedures (Seed & Idriss, 1971; Boulanger & Idriss, 2008) and numerical analyses (Kumari et al., 2015; Pal & Deb, 2021).

The primary contribution is the first combination of congestion indices, AI forecasts, and IoT monitoring into a single decision-support system. This approach moves geotechnical practice closer to digital resilience by enhancing the reliability, transparency, and adaptability of SC design.

Future studies ought to concentrate on the following topics:

1. Produce thorough, superior datasets that particularly document dynamic performance and clogs.
2. Enhance the permeability decay models' field calibration for different soil–column types.
3. Examine hybrid frameworks that integrate AI, IoT, and numerical modeling in digital twin environments for liquefaction mitigation.
4. Develop dependable IoT sensors that can operate for a long time in difficult subsurface and seismic situations.

The proposed framework can become a practical, scalable tool for engineers that offers real-time, clogging-aware prediction and, ultimately, improves the resilience of critical infrastructure in seismic regions by working toward these objectives.

ACKNOWLEDGMENTS

The authors acknowledge institutional support from MANIT Bhopal.

REFERENCES

Adalier, K., & Elgamal, A. (2004). Mitigation of liquefaction and associated ground deformations by stone columns. *Engineering Geology*, *72*(3–4), 275–291.

Ahirwar, P., & Jain, P. K. (2025). Performance and challenges of stone columns for liquefaction mitigation: a comprehensive review. *Geotechnical Research. 12*(4), 172–185.

Bhatta, S., & Dang, J. (2024). Use of IoT for structural health monitoring of civil engineering structures: a state-of-the-art review. *Urban Lifeline*, *2*(1), 17.

Boulanger, R. W., & Idriss, I. M. (2008). Closure to "liquefaction susceptibility criteria for silts and clays" by ross W. Boulanger and IM Idriss. *Journal of Geotechnical and Geoenvironmental Engineering*, *134*(7), 1027–1028.

Cubrinovski, M., Bradley, B., Wotherspoon, L., Green, R., Bray, J., Wood, C., ... & Wells, D. (2011). Geotechnical aspects of the 22 February 2011 Christchurch earthquake. *Bulletin of the New Zealand Society for Earthquake Engineering*, *44*(4), 205–226.

Hausler, E. A., & Sitar, N. (2001). Performance of soil improvement techniques in earthquakes. In *Proceedings of the Fourth International Conference on Recent Advances in Geotechnical Earthquake Engineering and Soil Dynamics* (Paper No. 10.15), Missouri University of Science and Technology, USA.Jara, A. J., Parra, M. C., & Skarmeta, A. F. (2014). Participative marketing: extending social media marketing through the identification and interaction capabilities from the Internet of things. *Personal and Ubiquitous Computing*, *18*(4), 997–1011.

Kumari, N., Sawant, V. A., & Mehndiratta, H. C. (2015). Effectiveness of stone columns in liquefaction mitigation: a numerical study using PLAXIS 3D. *Geotechnical and Geological Engineering, 33*(4), 1105–1117.

Luehring, R., Snorteland, N., Stevens, M., & Mejia, L. (2001). Liquefaction mitigation of a silty dam foundation using vibro-stone columns and drainage wicks: A case history at Salmon Lake Dam. *Proceedings of the 21st USSD Annual Meeting and Lecture*, Denver, Colorado, USA.

Onoue, A., Mori, N., & Takano, J. U. N. (1987). In-situ experiment and analysis on well resistance of gravel drains. *Soils and Foundations*, *27*(2), 42–60.

Pal, S., & Deb, K. (2019). Effect of clogging of stone column on drainage capacity during soil liquefaction. *Soils and Foundations*, *59*(1), 196–207.

Pal, S., & Deb, K. (2021). Filtration performance of geotextile encasement to minimize the clogging of stone column during soil liquefaction. *Geotextiles and Geomembranes*, *49*(3), 771–788.

Priebe, H. J. (1998). Vibro replacement to prevent earthquake induced liquefaction. *Ground Engineering*, *31*(9), 30–33.

Rayamajhi, D., Nguyen, T. V., Ashford, S. A., Boulanger, R. W., Lu, J., Elgamal, A., & Shao, L. (2014). Numerical study of shear stress distribution for discrete columns in liquefiable soils. *Journal of Geotechnical and Geoenvironmental Engineering*, *140*(3), 04013034.

Seed, H. B., & Idriss, I. M. (1971). Simplified procedure for evaluating soil liquefaction potential. *Journal of the Soil Mechanics and Foundations division*, *97*(9), 1249–1273.

Youd, T. L., Idriss, I. M., Andrus, R. D., Arango, I., Castro, G., Christian, J. T., ... & Stokoe, K. H. (2003). Closure to "Liquefaction Resistance of Soils: Summary Report from the 1996 NCEER and 1998 NCEER/NSF Workshops on Evaluation of Liquefaction Resistance of Soils" by TL Youd, IM Idriss, Ronald D. Andrus, Ignacio Arango, Gonzalo Castro, John T. Christian, Richardo Dobry, WD Liam Finn, Leslie F. Harder Jr., Mary Ellen Hynes, Kenji Ishihara, Joseph P. Koester, Sam SC Liao, William F. Marcuson III, Geoffrey R. Martin, James K. Mitchell, Yoshiharu Moriwaki, Maurice S. Power, Peter K. Robertson, Raymond B. Seed ... *Journal of Geotechnical and Geoenvironmental Engineering*, *129*(3), 284–286.

Zhou, Y. G., Liu, K., Sun, Z. B., & Chen, Y. M. (2021). Liquefaction mitigation mechanisms of stone column-improved ground by dynamic centrifuge model tests. *Soil Dynamics and Earthquake Engineering*, *150*, 106946.

Zou, Y. X., Zhang, J. M., & Wang, R. (2020). Seismic analysis of stone column improved liquefiable ground using a plasticity model for coarse-grained soil. *Computers and Geotechnics*, *125*, 103690.

12 Development of Mobile Application for Calculating the Bearing Capacity of a Pile

Rakesh Kumar Dutta,
Tammineni Gnananandarao,
Amit Kumar, and V. Gayathri

12.1 INTRODUCTION

The increasing significance of deep foundation systems in the past few years has been caused by the rapid urbanization and the increasing requirement to build high-rise buildings. Among them, however, the most significant are the pile foundations as they are more capable of shifting the structural loads to deeper and more stable layers. This is particularly necessary in places where there are weak, compressible, or highly variable soils. The conventional approaches to estimating the bearing capacity of piles require intensive site investigations, penetration tests, laboratory findings, and manualized calculations that rely on the use of empirical–semi empirical relationships. Despite several

DOI: 10.1201/9781003714125-12

years in geotechnical practice, these methods may be both laborious, time-consuming, and prone to human error—especially when engineers need to be able to make rapid decisions in the difficult field conditions.

Geotechnical design is becoming increasingly tough because engineers must deal with unpredictable ground conditions, complex soil behaviors under varied loads, and modern requirements such as Load and Resistance Factor Design (LRFD). These complexities have made fast and reliable analysis tools essential, with many practitioners turning to computational and data-driven solutions to reduce uncertainty and handle bigger data sets in their decision-making workflows.

In the last few years, civil engineering researchers have been leaning a lot more toward machine learning (ML) and deep learning for all sorts of prediction-type problems. It's kind of everywhere now. People have used these methods to estimate the compressive strength of modified recycled aggregate concrete (Shaban et al. 2021) and to figure out the strength of pressed bricks (Elbaz et al. 2023a). And there's a whole bunch of studies showing that ML works pretty well for predicting soil-bearing capacity too—there's a long list of authors on that (Ahmad et al. 2021; Dutta et al. 2018, 2019c; 2019a; 2019b; Gnananandarao et al. 2019; Kalinli et al. 2011; Kashani et al. 2019; Kumar Dutta et al. 2019; Moayedi and Hayati 2018; Padmini et al. 2008; Pakdel et al. 2019; Shahnazari and Tutunchian 2012; Gnananandarao et al. 2022; Tsai et al. 2012). There are also ML models being used for predicting different air-quality parameters (Elbaz et al. 2023b) and even in tunneling work (Elbaz et al. 2023c; Shen et al. 2022). It's basically spreading across most areas of civil engineering now.

Padmini et al. (2008) basically showed that a Neuro-Fuzzy model did a better job predicting the bearing capacity of shallow foundations on cohesionless soils than using Fuzzy or Artificial Neural network (ANN) alone. But then Kalinli et al. (2011) came up with results saying ANN actually outperformed the Adaptive Neuro-Fuzzy Inference System when it came to estimating ultimate bearing capacity, so the findings don't all point in the same direction. Later, Shahnazari and Tutunchian (2012) used Multigene Genetic Programming (GP) and found that it could give pretty dependable estimates for foundations on non-cohesive soils. And Tsai et al. (2012) suggested a hybrid GP setup that mixed weighted GP with some soft-computing polynomial ideas, and that one also seemed to work quite well for predicting shallow foundation bearing capacity. So overall, different people have found different models that work better depending on what exactly they were testing.

A variety of nonlinear ML and soft-computing models were compared by Moayedi and Hayati (2018), which showed that the feedforward neural network gave an almost similar result to what was obtained through the analysis with finite element methods, a fact that is remarkable when it comes to geotechnical models. Pakdel et al. (2019) tested the method of foundation design using the concept of differential evolution, evolutionary strategy, and biogeography-based optimization, finding that there is no unequivocal scenario where any of these methods outperforms the rest. According to Kashani et al. (2019), the same observation was made in which no one single optimization method would be rated as superior in all the cases analyzed. Ahmad et al. (2021) have made another addition to the literature by creating a Gaussian-process regression model, which is able to predict the ultimate bearing capacity of shallow foundations on cohesionless soils with high level of accuracy. Recent studies like Fattahi et al. (2024), Khanmohammadi et al. (2022), Kumar et al. (2024), Onyelowe et al. (2025), and Ren and Sun (2023) keep on showing how artificial intelligence and probabilistic methods and digital tools are becoming central to the development of geotechnical engineering.

Despite recent progress, there is still little effort made into adopting mobile applications in normal foundation engineering. Most of the available applications are devoted to similar yet different activities like slope stability analysis, initial soil classification using ASTM requirements, or field logging (Aydın et al. 2023; Moreno-Maroto et al. 2021; Srivastava et al. 2021). Very few of these tools integrate the information about Standard Penetration Test (SPT) and soil parameters as well as code-based design processes through a single interface. Besides, the number of applications that offer calculations with adequate reliability to be used on-site immediately and meet the traditional engineering standards is very limited. This is especially lacking with remote or otherwise resource-constrained environments, as engineers still need to resort to manual chart interpolation, paper-based record keeping, and hand computation. Although effective, it has the drawbacks of being very slow, prone to errors, and not predictable.

The gap in the literature related to the estimation of safe bearing capacity of piles has long existed and the development of a mobile application devoted to this issue is becoming urgent. It is not just an auxiliary calculator but a significant move toward the digitalization of geotechnical practice in combining the time-tested principles of soil mechanics with modern technological interventions. With the ability to offer a portable, reliable tool that eliminates the necessity of consulting reference handbooks or making tentative calculations on paper, the application will help to make quicker, more confident decisions within the profession. It, therefore, generates efficiency in the time and increases safety to boost the efficiency of the overall processes.

Riding on this assumption, this chapter introduces an Android application that can be used to calculate the bearing capacity of the piles as defined in the Indian Standard Code (IS 2911 (Part 1/Sec 1) 2010) as safe. User inputs to the application (e.g., pile dimensions, stratigraphic data, SPT N-values, and adhesion/friction parameters) are subjected to the established calculations of the code. The IS code formulae and verification controls provide the accuracy of the established workflow, and the obtained results are similar to the traditional hand-based designing processes. The first benefit is that much of the manual numerical effort is removed in this way, making the process faster, cleaner, and by far less tedious than using the traditional methods.

The conceptual background of this digital tool can be based on the fundamental geotechnical engineering techniques. The construction industry players are still relying on underground information a lot, since the behavior of soil is one of the main determinants of the performance of the structure and the choice of foundations. The use of pile foundations is of further importance when the surface strata are either of weak or highly compressible nature and this is highlighted in IS 2911 (Part 1/Sec 3) 2010. The traditional method of determining safe bearing capacity requires a lot of site research, hand calculations, and full code compliance, all of which are also dependable but time-intensive and require a skilled individual to manage. In this way, it is not an optimum efficient process.

With the growing integration of digital tools into the world of civil engineering, mobile apps become relevant as viable options to achieve complicated calculations, eliminating mistakes, and faster decision-making processes. The Android program herein is in line with this trend and it serves different types of piles, such as bored piles and cast-in-situ concrete piles, besides providing a high input validation, stratigraphic detailing, and comprising of IS-code-compliant calculations. All of these aspects ensure that the outputs are credible and consistent with the standard engineering activities, thus facilitating the design process without the need to sacrifice the methodological rigor.

The historical approach to the analysis of this development explains its importance. The understanding of the pile behavior has developed through the generations as the pioneering work on soil mechanics by Terzaghi (1973), the analytical formulations by Meyerhof (n.d.), and the next stage of knowledge development of multiple scholarly publications. Modern design requirements (IS 2911 (Part 1/Sec 1) 2010; IS 2911 (Part 1/Sec 3) 2010) are heavily based on these fundamental theories to assess the condition of cohesion, friction angle, pile size, and other involved parameters. However, manual processing of such analyses, especially in the field, is still burdensome with stratified soil profiles, repetitive computation pathways, and highly strict code compliance. These burdens can be relieved by digital tools, which will lead to more accurate and efficient results.

The earlier efforts of implementing mobile technology in geotechnical engineering were mainly focused on the auxiliary functions like slope stability measurements (Rakesh et al. 2019), ASTM-like soil classification applications, or multi-criterion soil identification systems (Kumar et al. 2015). The exceptions include SPT results used to estimate allowable loads used in shallow foundations in the work by Kumar et al. (2015) and the study by Khatri et al. (2023). Regarding these endeavors, there is no single mobile tool that can be used to properly calculate the safe bearing capacity of the deep foundations by combining SPT measurements with the process of calculating the IS codes. This gap that remains is sufficient to support the current research. This gap is, thus, addressed by the study that proposes an Android app that combines the concepts of the design, based on SPT, with the interpretation of layered soils and the calculation of the IS code in a single, convenient app. By mixing the usual geotechnical concepts with newer digital tools, the app improves the accuracy of the calculations and cuts down a lot of the manual effort engineers normally put in. It also gives field engineers a bit more analytical support right where they need it. Overall, this feels like a meaningful step toward modernizing geotechnical work. It helps make designs safer, speeds up the whole process, and fits well with the movement toward more efficient, tech-driven workflows in today's fast-changing construction industry.

12.2 BACKGROUND ON MOBILE APPLICATION ARCHITECTURE, CODING FRAMEWORK, AND EMBEDDED GEOTECHNICAL EQUATIONS

To clarify the reason of having designed this mobile application, it is important to outline the technical foundations, which are the logical architecture, the programming framework, and the incorporation of geotechnical equations in the system. The app itself was built on the Android Studio, which is the de facto standard integrated development environment of native Android development. The main calculation engine is written in Java, and the user interface software, such as input fields, dropdown lists, selection boxes, and expressing output displays, was written in XML. Such a mixture gives stable and modular architecture which is essential to the field engineers who need a lightweight and reliable tool to perform fast calculations. The application scheme consists of several activities, and each of them deals with specific stage of work process: scanning piles features, setting up soil layers, authenticating parameter input, and demonstrating the ultimate outcome. In order to have visual consistency on heterogeneous mobile devices, Constraint Layout is used to position UI elements. Subsequently, navigating between activities is achieved through explicit intents, and contextual feedback is achieved through the short pop-up notifications and validation prompts that guide users during the progress. One of the main technical strengths is that the geotechnical equations were directly incorporated into the calculating machine. The program complies with the IS2911 standards of calculating end-bearing and skin friction capacities of cohesive soils, thus making sure that the results generated are in line with the accepted engineering practice.

$$Q_u = A_p N_c C_u + \alpha C_a A_s \tag{12.1}$$

In this context, the app automatically calculates the important parameters, such as the adhesion factor, the average cohesion value, and the pile shaft area, directly out of the information provided by the user. In the event that the soil type is considered as non-cohesive, the manual bearing capacity equation is used to calculate its application. As a result, the user will not have to execute each step of the computation manually and perform each assessment.

$$Q_u = cN_c A_c + qN_q A_q + 0.5\gamma B N_\gamma A_\gamma \tag{12.2}$$

The bearing capacity factors, namely N_c, N_q and the N_g, are automatically calculated based on the IS 2911 guidelines. Other procedural components like the factor of safety determination, ultimate capacity estimation, and the combination of multiple strata contributions in the soil are solved by brief iterative loops written in Java. The application also includes some logic to reason out field measurements and SPT N-values typed in by the user, thus allowing it to categorize the soil as either cohesive or non-cohesive.

The equations to determine the adhesion factor (a) and the frictional resistance are obtained in the reliable sources, where Ranjan and Rao (2022) focus on the fact that all calculations follow the accepted approaches to design. Comprising of Java-based computations, an XML-built interface, and formulae that follow the codes of IS, the application achieves a strong technical background and can be used by the working engineer. Such quantitative foundation not only confirms the accuracy of the calculations but also shows that the application can be actually used by the field.

12.3 ALGORITHM AND APP DEVELOPMENT

The advent of computing instruments and the state-of-the-art software techniques have greatly influenced the development of geotechnical engineering in the digital age. The modern practice is more dependent on mobile applications, numerical model, and data-driven workflow, which combined together allow gaining a better insight into complicated subsurface conditions and resolving the problems that are harder to tackle without it. With these digital tools coming into everyday use, the results of designs have turned out to be more precise, consistent, and in line with the modern sustainable goals. The mobile app shown conforms to this trend as it provides two major features, namely, recording SPT field measurements and also estimating the safe load-bearing capacity of pile foundations. These options are offered with a lean interface that can facilitate the quick opening of the necessary procedure even as the engineer is on the ground. The flowchart in Figure 12.1 presents a summary of the steps to be followed, which eventually lead to the calculation of the pile capacity. It is aimed at contrasting the current digital convenience with the traditional geotechnical principles, thus keeping the accuracy and the ease of use at the same time.

After choosing the type of pile to be used, which could be a precast concrete pile or a cast-in-situ driven pile, the application will request some of the necessary inputs, such as the diameter of the piles, their overall length, concrete grade, and some additional parameters.

In order to make sure that the analysis relates to site conditions in actual situations, the user specifies basic soil properties, such as unit weight and coefficient of earth pressure at rest. Subsequent workflow decisions depend on whether the substrate has cohesive behavior or has a cohesionless character. The application follows the IS Guidelines for the type of soil selected and in turn ensures that the calculations are accurate and reliable. For cohesive soils, IS equations are used to calculate bearing capacity while for non-cohesive soils the applicability of an alternative of IS based formulae of that category is adopted. Upon completing all of the computations, the results are presented in a concise, readable format that allows engineers to make quick, informed decisions and, in effect, create safer and more efficient pile designs.

On the main screen of the application, users can choose between some of the different types of piles, such as cast-in-situ bored piles, precast piles installed in pre-bored holes, and cast-in-situ driven piles. Once a pile type is selected, the app asks for all the necessary inputs: pile length and diameter, grade of concrete, applied load, soil unit weight, coefficient of earth pressure at rest, soil class, SPT N-value and data for each soil strata such as depth, cohesion, and adhesion factor.

For cohesive soils, the adhesion factor is then evaluated in accordance with the guidance of Ranjan and Rao (2022), and the SPT values are taken as they are as N-value. For non-cohesive soils the internal friction angle has to be given because the calculations are dependent on the parameter mentioned above.

If more than one soil layer is present, the user can include them one after the other, by pressing the "+" button, and thereby enable building of the layered profile. In

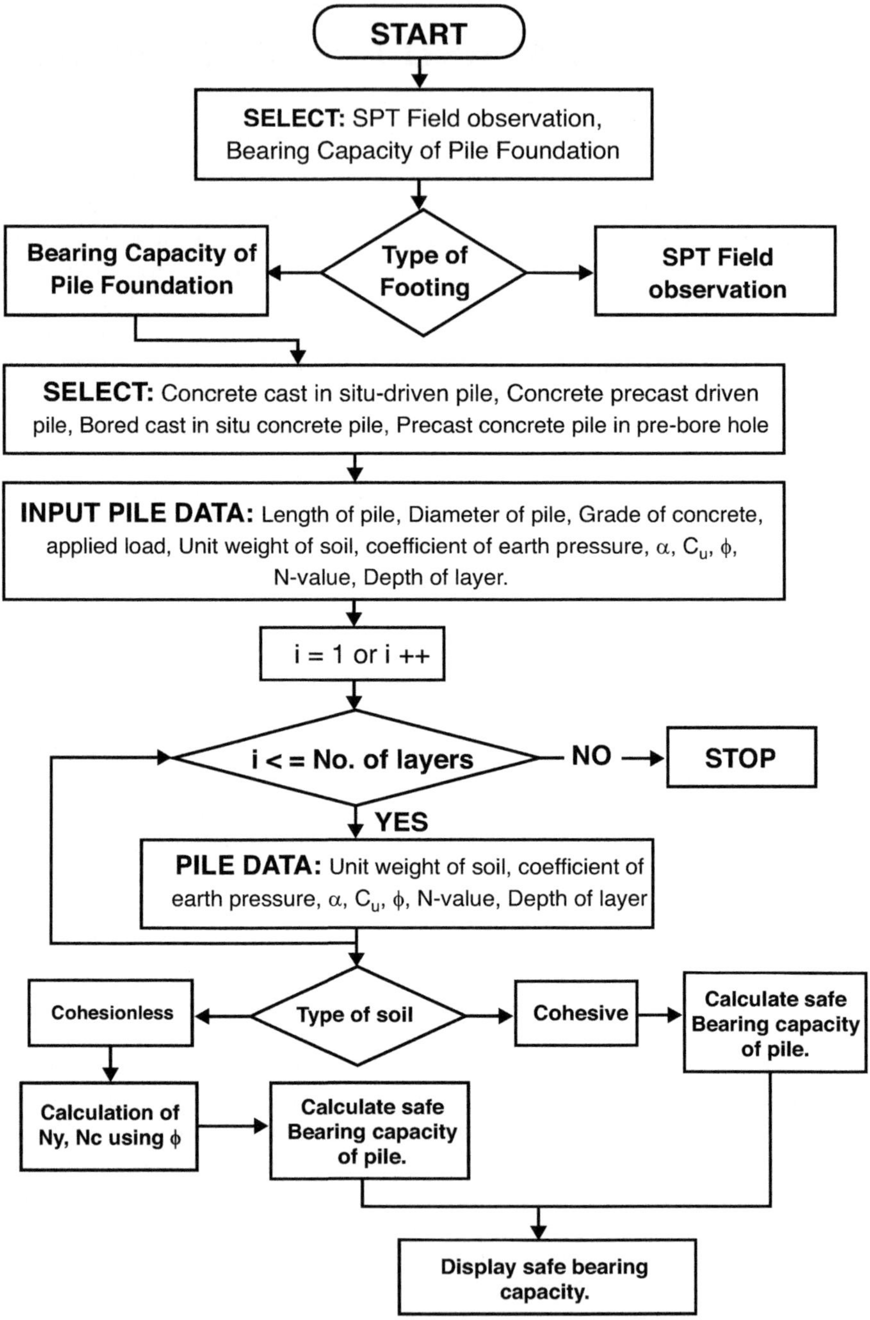

FIGURE 12.1 Flow diagram for entering data and determining the pile's safe bearing capacity.

order to ensure that the system is clear and error-free, all necessary available piles are represented by number fields, and context-dependent questions are showed according to the chosen type of pile. This step-by-step configuration makes the process simple as possible without compromising engineering accuracy. This makes the application a pragmatic digital assistant which is completely code-based and simple. An effective input validation system verifies the data during entry. Immediate pop-up warnings alert the user to any omissions or inconsistencies and non-applicable values can be input as zero. Everything is implemented with Java, Android Studio, and XML; predefined XML components perfectly fit in the structure of the activity and add more cleanliness to the code and consistency to the interface. Each activity screen would be laid out with a hierarchical structure of UI elements with a central constraint layout that is the main container holding the main elements (button, text fields, images, recyclers, etc.).

Visual consistency from one screen to the next (Figure 12.2a–c) is therefore achieved which, in particular, is a great advantage, because many buttons and inputs with numbers use the same layout. When someone is opening the app for the first time they would first select pile foundation type from the main menu. That opens another screen in which all the relevant inputs can be filled in. Additional layers can be added easily with the "+" button as shown in Figure 12.3. Once everything is entered, when the "bearing capacity of pile foundation" option is tapped, the calculations run immediately and the results are instantly shown. The whole workflow is easy, fast, and efficient for on-site use.

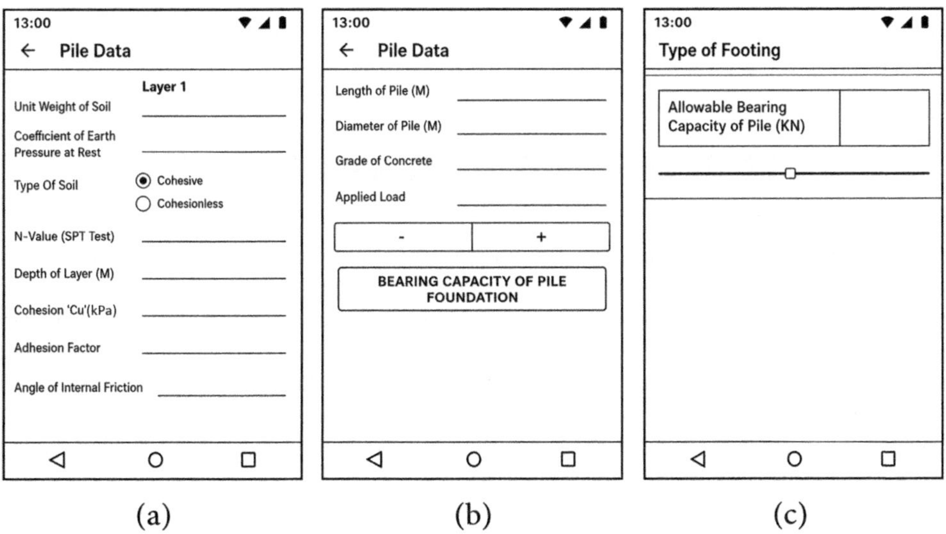

FIGURE 12.2 Activities in the application (a) Addition of new layer, (b) screen for input data, and (c) result display.

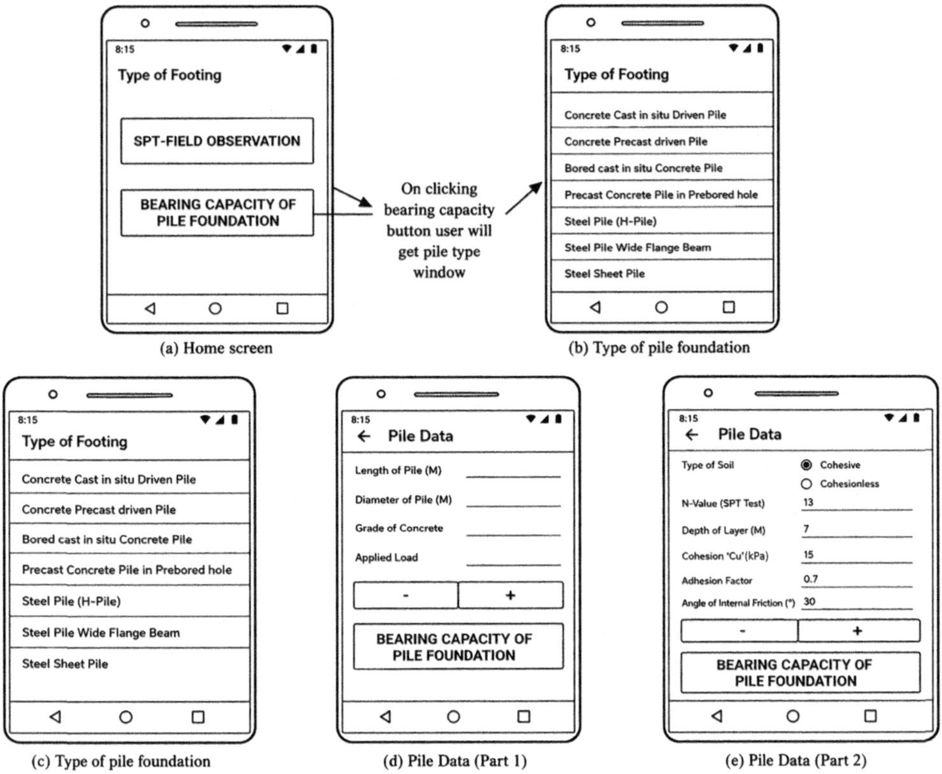

FIGURE 12.3 Home screen, type of pile, and input window of concrete-driven pile.

12.4 TESTING AND VALIDATION

Testing and validation of the application was done in a systematic way based on three representative pile foundation problems that are frequently encountered in geotechnical practice. These examples were chosen to represent a variety of soil types and pile dimensions, in order to determine if results obtained with the Android application could replicate results obtained with conventional manual calculations. For each case, the pile capacity was determined first manually by using standard equations and the same set of input was entered in the application. Side-by-side comparison of the outputs gave a quantitative measure of the accuracy of the tool.

The first example was a 400 mm circular pile that was driven 10 m into a profile of two layers of cohesive soil. The upper 7 m consisted of soft clay (unit weight 18 kN/m^3, cohesion 15 kN/m^2), using the SPT N-value for the calibration of the shaft friction and end-bearing contributions. Starting from 7 to 10 m, there was a medium stiff clay layer with identical unit weight, but with a higher cohesion of 50 kN/m^2, thus contributing a disproportionate part to both the skin friction and base resistance. Manual calculations included satisfactory factors for adhesion and IS end-bearing equations. Comparison of the output of the application with the hand-worked figures resulted in a close agreement between the two figures for both the case of the layered cohesive soils, leading to the conclusion that the tool is accurate in handling cohesive soils which are layered.

The second case was the pile in a uniform c–φ medium, a situation common in the alluvial deposits where the soils have both cohesion and friction. The pile, which is 500 mm diameter, was driven 10 m deep. The main properties of the soil used were the cohesion of 50 kPa, the unit weight of 18 kN/m^3, and the friction angle of 30°. Both shaft friction and the end-bearing were calculated by using c–φ conventional calculations such as the beta-method and the formulation based on an effective stress analysis. Manual analysis consisted of the determination of earth pressure coefficients, overburden stresses, and frictional resistances along the pile. When the same inputs were fed into the application, the results were nearly the same, and thus confirmed the reliability of the software when both cohesion and friction are factors in influencing pile capacity.

The third validation example was for a pile buried in deep, uniform soft clay, which is typical of coastal or deltaic environments. A pile 450 mm diameter was driven 12 m into clay of unit weight 16.5 kN/m^3 and cohesion 35 kN/m^2. Given the soft and low strength of the clay, this was a severe test for the application's ability to handle difficult conditions. The SPT N-value was informative for the empirical correlations used for consistency. The output from the application was similar to the figures derived by hand, indicating sufficient performance in these difficult circumstances.

The manual calculations mainly used the methods of undrained shear strength, in which both the resistance of the shaft and the base are directly related to the cohesion and adhesion. Accordingly, the results of the application showed an excellent agreement with the hand-derived results, proving the effectiveness of such application on even weak and homogeneous clays. Across all three scenarios investigated (layered cohesive soil, mixed c–φ conditions, and deep soft clay), the application was able to reproduce the manual computations with considerable accuracy, thus making the Android tool a reliable and interesting tool for preliminary design, pedagogical purposes, and neglected on-site decision-making in geotechnical engineering.

12.5 RESULTS AND DISCUSSION

In order to assess the real performance of the application to pile foundations, three real-world examples with different soil configurations and pile typology were chosen. These cases have been selected for a representative spectrum of purposive geotechnical situations in order to give a thorough evaluation of the accuracy, reliability, and practical utility of this app. The first case examined a circular pile that was installed in a cyclic soil profile, with soft clay overlying a "medium stiff clay soil"—very common circumstances found by an engineer. Given that the strength and consistency of clay are affected in a substantial way with depth, this heterogeneity has a strong influence on the behavior of the pile. For validation, intensive calculations were manually made, based on classical methods regarding an estimation of the end-bearing resistance and skin friction in cohesive soils. These step-by-step manual computations are displayed in Table 12.1, in order to make the computations transparent and easy to scrutinize. Subsequent comparison with results of the app illustrated in Figure 12.4 showed that the correspondence is extremely close. This concordance has proven the abilities of the app to correctly estimate the pile capacity for layered cohesive profiles, a fact that shows that this app is useful in the field and also for instructive or training purposes.

The second validation case used a larger diameter circular pile to test the app in different soil conditions. Herein, the pile was located in a c–φ layer with given values of cohesion, unit weight, and friction angle—parameters that play a major role on both end-bearing and skin friction components. Such c–φ soils, with a combination of cohesive and frictional behavior, give good testing of the app's ability to deal with complex strength regimes. Detailed manual calculations were made using standard design formulas for design of piles in c–φ soils, safe bearing capacity being compiled in Table 12.2. Comparative analysis against the mobile app results, illustrated in Figure 12.5, showed a nearly exact match, meaning that the tool is accurate for managing soil conditions where both cohesion and friction are dominant.

The third validation example was performed considering a circular pile located wholly inside a uniform soft clay layer; in this way, it was possible to evaluate the app under purely cohesive conditions. Soft clays usually have low shear strength and high compressibility, so it is critical to accurately estimate capacity of these clays to ensure safe design. The manual calculations were based either on methods established by cohesive soils rather than those proposed for this shaft friction or in terms of end-bearing resistance, in which all procedure steps of these processes are listed in Table 12.3.

TABLE 12.1 Calculations of bearing capacity of pile for example 1

EQUATIONS	*DESCRIPTIONS*
$Q_u = Q_E + Q_S$	Q_E is end-bearing resistance of pile shaft and Q_S is the skin fiction resistance of pile
$Q_u = (A_p{*}N_c{*}C_u) + (\alpha{*}C_a{*}A_s)$	where A_p = area of pile at tip, C_u = cohesion, α = adhesion factor, C_a = average cohesion, A_s = area of pile shaft
$Q_E = \frac{\pi}{4} \times d^2 \times N_c \times C_u$	End-bearing resistance of pile
$Q_E = \frac{\pi}{4} \times 0.4^2 \times 9 \times 15$	
$Q_E = 56.54$ kPa	
$Q_S = \alpha{*}C_a{*}A_s$	Skin friction resistance of pile
$Q_S = \alpha \times \left(\frac{C_{u1} \cdot h_1 + C_{u2} \cdot h_1}{h_1 + h_2}\right) \times (\pi \times d \times L)$	
$Q_S = 0.7 \times \left(\frac{15 \times 7 + 50 \times 3}{7+3}\right) \times (\pi \times 0.4 \times 10)$	
$Q_S = 224.30$ kPa	
$q_u = \left(\frac{Q_E + Q_S}{2.5}\right) = \left(\frac{56.54 + 224.30}{2.5}\right)$	The allowable bearing capacity of pile calculated by taking factor of safety 2.5, So allowable bearing capacity of pile is q_u
$q_u = 112.3$ kPa	

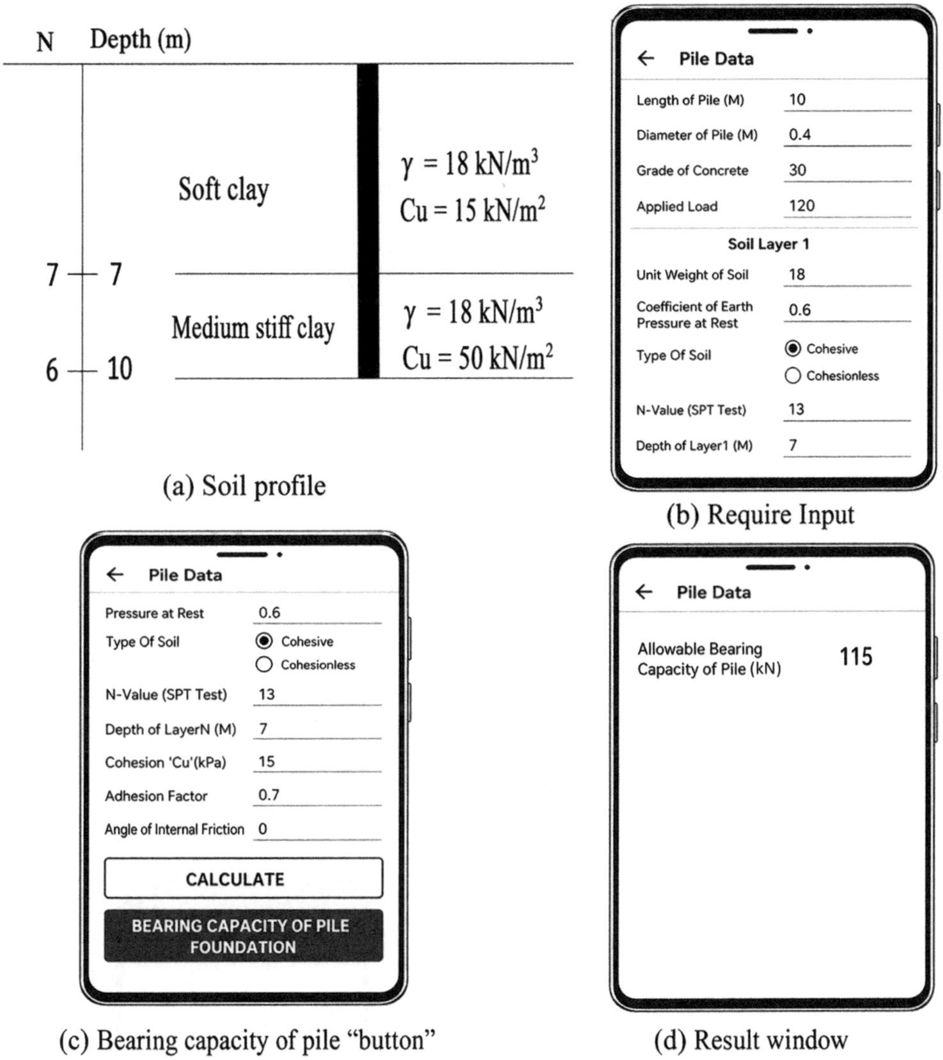

FIGURE 12.4 Soil profile and result of example 1 calculated using mobile app.

Comparison of these manually obtained values to the results of the app, in Figure 12.6, again showed a good agreement, confirming the efficacy of the app even in environments of weak, compressible soils. The uniformity of predictions from all three cases gives a strong demonstration of the reliability and strength of the tool for practical use in geotechnical design.

Overall, the results of the validation prove that the developed mobile application is reliable and provides adequate estimates of the pile-bearing capacity in various soil conditions.

The close match between hand calculations and the app's results helps to substantively validate the correctness of the results and in so doing to endorse its trustworthiness

TABLE 12.2 Calculations of bearing capacity of pile for example 2

EQUATIONS	*DESCRIPTIONS*
$Q_u = cN_cA_c + qN_qA_q + 0.5\gamma BN_\gamma A_\gamma$	Q_u is the ultimate bearing capacity of pile, N_c, N_q, and N_γ are bearing capacity factors, A_c, A_q, and A_γ are the areas of the pile shaft, pile base, and pile tip, respectively. *B* is the width of the pile base for a single pile, *q* is the effective overburden pressure at the pile base
$N_c = 9 + 0.4\frac{B}{d}$, $N_q = 9 + \frac{B}{d}\left(\frac{1}{3}\tan\varnothing\right)$, $N_\gamma = 0$	Bearing capacity factors
$N_c = 9 + 0.4\frac{0.5}{10}$, $N_q = 9 + \frac{0.5}{10}\left(\frac{1}{3}\tan 30°\right)$, $N_c = 9.02$, $N_q = 1.17$, $N_\gamma = 0$	
$A_c = \pi DL = \pi \times 0.5 \times 10 = 15.71\ m^2$ $A_q = \pi(4D) = \pi\ (40.52) = 6.28\ m^2$ $A_\gamma = 0$	Areas of the pile shaft, pile base, and pile tip, respectively
$Q_u = 50 \times 9.02 \times 15.71 + 200 \times 1.17 \times 0.2 + 0.5 \times 20 \times 0.02 \times 0\ \text{kPa}$ $Q_u = 1413.6\ \text{kPa}$	Ultimate bearing capacity of pile
$q_u = \frac{1413.6}{2.5} = 565.44\ \text{kPa}$	The allowable bearing capacity of pile calculated by taking factor of safety 2.5, so allowable bearing capacity of pile is q_u

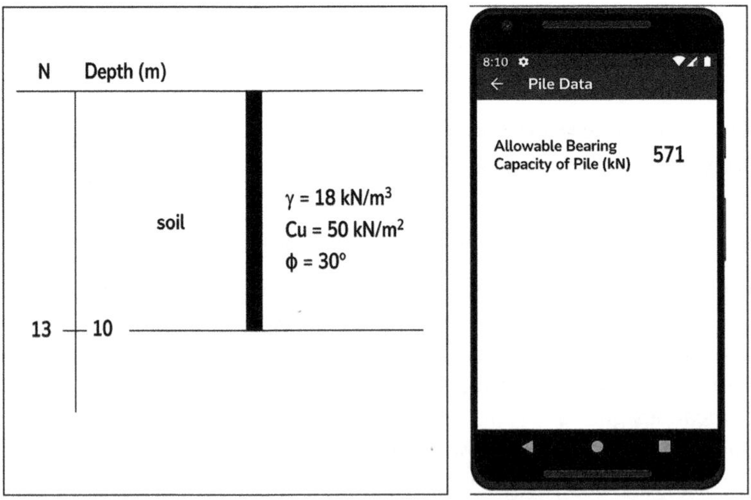

(a) Soil Profile (b) Result window

FIGURE 12.5 Soil profile and result of example 2 calculated using mobile app.

TABLE 12.3 Calculations of bearing capacity of pile for example 3

EQUATIONS	*DESCRIPTIONS*
$Q_u = (A_p * N_c * C_u) + (\alpha * C_a * A_s)$	Ultimate bearing capacity of pile
$Q_u = \left[\frac{\pi}{4} \times 0.45^2 \times 9 \times 35\right] + \left[0.7 \times \left(\frac{35 \times 12}{12}\right) \times (\pi \times 0.45 \times 12)\right]$	
$Q_u = 465.75$ kN	
$q_u = \frac{465.75}{2.5}$ kN	Safe bearing capacity of pile
$q_u = 186.3$ kPa	

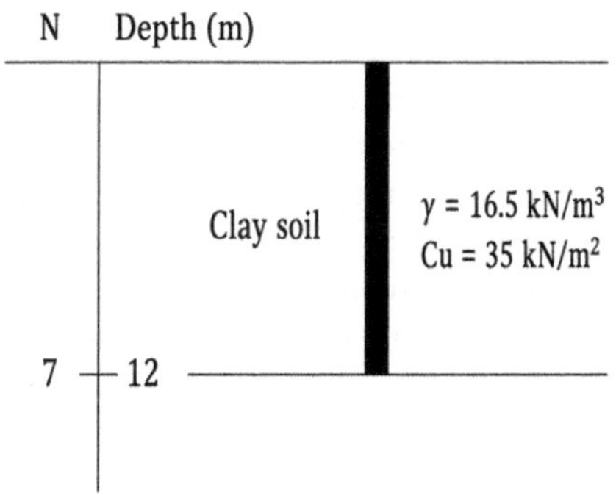

(a) Soil profile

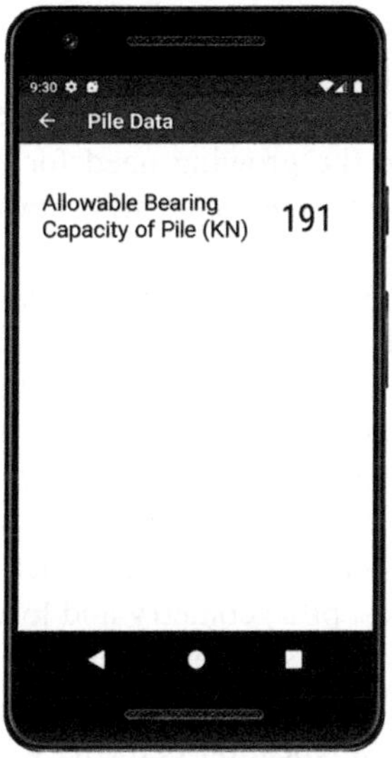

(b) Result window

FIGURE 12.6 Soil profile and result of example 3 calculated using mobile app.

in a wide array of geotechnical contexts in its own right. The uniformity of performance between the examples examined indicates that the tool is not restricted to elementary soil setups, but is also able to control more complex scenarios involving layered or mixed behavior soils.

The truthfulness of the application is further enhanced by the performance in actual case studies with different types of soils and pile designs. By following the standard analytic frameworks, and having a commitment toward preserving the fidelity to field observations, the app reflects the traditional design approach with an unexpected precision to have engineered capacities of particular specific value for engineers, researchers, and practitioners who need expedited and reliable capacity estimates in their design decision-making.

Beyond technical accordance, the application is for greater efficiency of operation by significantly reducing the time involved in making manual calculations for pile design work. Its intuitive interface and powerful computational architecture make it possible for users to get accurate results without having to perform long hand calculations. By rapidly advancing processes and supporting the informed use of engineering choices, the application becomes a sensible tool to geotechnical engineering practice today.

12.6 CONCLUSION

The general objective of this research was to create a simple and user-friendly mobile application that would be able to calculate the safe bearing capacity of piles quickly with little effort. The effort was aimed to overcome the gap between time-consuming and monotonous calculations performed with traditional instruments (by hand) on one side, and the growing need for efficient digital tools on the other, especially among geotechnical engineers who normally work at construction sites and may not always have access to extensive charts and code books. The application was developed using Android Studio and written in Java based on its stability and compatibility with a wide range of Android devices. This technological platform had been designed to supply engineers with a peacetime working instrument without compromising the accuracy of some routine design tasks. The impetus for the development of the application comes from the importance of pile foundations in modern construction practice. As structures get heavier and more stressed, distributing loads to deeper levels of stronger soil layers becomes critical. Designing such foundation conventionally requires evaluation of soil properties, pile geometry and loads, then the application of code-conforming formulae and handbooks—a procedure which is totally right, time-consuming, and error-prone, especially in case of tight deadlines or busy site conditions. By digitalizing the workflow, the application is trying to improve the whole procedure, but still achieving the precision that is needed in geotechnical work.

The tool is aimed at field engineers, practicing geotechnical experts, and students alike. Its interface is also easy to use: users type in soil parameters such as cohesion, unit weight, or friction angle, along with the dimensions of the pile and details about the loading. Upon submission, the application runs standard analytical formulas and the safe bearing capacity is almost instantaneous. This functionality is useful in situations where it is necessary to do preliminary design, site visits, or quick feasibility checks. An important merit of the application is the good agreement of the results obtained with the manual computations; in Table 12.3 it is evident that any differences are very small and are in good agreement, most of the time they can be attributed to a rounding error or to small values of the numerical precision. Since hand calculations serve as the benchmark of accuracy, this close match confirms the hand calculations' application methodology acceptance.

Reliability of the tool was evaluated using several case studies with a variety of soil types, pile sizes, and loading situations. Whether the subsurface was made of cohesive, granular, or multi-layered materials, the application always had an accurate outcome, thus emphasizing its versatility for use in a range of different engineering contexts.

Beyond precision, the work is greatly reduced with the app. Conventional pile design forces one to look through tables, code clauses, and charts and make repetitive design calculations, but integrating all of these tasks in one application significantly reduces the workload in such routine steps. Consequently, engineers may devote more attention to interpretation of the site and decision processes and less time to mere mathematics.

Another excellent advantage is the portability of the application. Field engineers often work in a remote or difficult working environment where they may not have easy access to reference books or computers, and a pocket-sized, offline ready tool is extremely beneficial to have. Besides, the application has educational purposes because students can learn about the effect of the soil parameters and the geometry of the pile on the capacity of the bearing. Changing the inputs and then seeing the updated outputs helps to get a better insight into underlying concepts. In a much more general sense, this application ties in with the current trend toward digitalization that is taking place in civil and geotechnical engineering. As the field becomes more heavily based around automation, data-driven design, and even real-time monitoring, fields like these have a role in helping to drive overall efficiency and to reduce the number of errors. This project continues that trend by realizing a real-time, motivating, field-ready tool that facilitates modern working practices.

In summary, the study describes a significant attempt to improve geotechnical practice with the use of digital innovation. The mobile application addresses a need that has existed for a long time for a fast, accurate, and readily accessible pile design tool. With its easy-to-use interface and accurate calculations, it complements the speed, accuracy, and overall effectiveness of pile capacity analysis, making it an invaluable tool for geotechnical engineers.

REFERENCES

Ahmad, M., F. Ahmad, P. Wróblewski, R. A. Al-Mansob, P. Olczak, P. Kamiński, M. Safdar, and P. Rai. 2021. "Prediction of ultimate bearing capacity of shallow foundations on cohesionless soils: A Gaussian process regression approach." *Applied Sciences* 11 (21): 10317. Multidisciplinary Digital Publishing Institute. https://doi.org/10.3390/APP112110317.

Aydın, Y., Ü. Işıkdağ, G. Bekdaş, S. M. Nigdeli, and Z. W. Geem. 2023. "Use of machine learning techniques in soil classification." *Sustainability* 15 (3): 2374. Multidisciplinary Digital Publishing Institute. https://doi.org/10.3390/SU15032374.

Dutta, R. K., T. Gnananandarao, and V. N. Khatri. 2019a. "Application of soft computing techniques in predicting the ultimate bearing capacity of strip footing subjected to eccentric inclined load and resting on sand." *Journal of Soft Computing in Civil Engineering* 3 (2): 30–42. Pouyan Press. https://doi.org/10.22115/scce.2019.144535.1088.

Dutta, R. K., T. Gnananandarao, and A. Sharma. 2019b. "Application of random forest regression in the prediction of ultimate bearing capacity of strip footing resting on dense sand overlying loose sand deposit." *Journal of Soft Computing in Civil Engineering* 3 (4): 28–40. Pouyan Press. https://doi.org/10.22115/scce.2019.137910.1080.

Dutta, R. K., R. Rani, and T. Gnananandarao. 2018. "Prediction of ultimate bearing capacity of skirted footing resting on sand using artificial neural networks." *Journal of Soft Computing in Civil Engineering* 2 (4): 34–46. https://doi.org/10.22115/scce.2018.133742.1066.

Dutta, R. K., A. Singh, and T. Gnananandarao. 2019c. "Prediction of free swell index for the expansive soil using artificial neural networks." *Journal of Soft Computing in Civil Engineering* 3 (1): 47–62. https://doi.org/10.22115/scce.2018.135575.1071.

Elbaz, K., I. Hoteit, W. M. Shaban, and S. L. Shen. 2023a. "Spatiotemporal air quality forecasting and health risk assessment over smart city of NEOM." *Chemosphere* 313: 137636. Pergamon. https://doi.org/10.1016/J.CHEMOSPHERE.2022.137636.

Elbaz, K., W. M. Shaban, A. Zhou, and S. L. Shen. 2023b. "Real time image-based air quality forecasts using a 3D-CNN approach with an attention mechanism." *Chemosphere* 333: 138867. Pergamon. https://doi.org/10.1016/J.CHEMOSPHERE.2023.138867.

Elbaz, K., A. Zhou, and S. L. Shen. 2023c. "Deep reinforcement learning approach to optimize the driving performance of shield tunnelling machines." *Tunnelling and Underground Space Technology* 136: 105104. Pergamon. https://doi.org/10.1016/J.TUST.2023.105104.

Fattahi, H., H. Ghaedi, F. Malekmahmoodi, and D. J. Armaghani. 2024. "Optimizing pile bearing capacity prediction: Insights from dynamic testing and smart algorithms in geotechnical engineering." *Measurement* 230: 114563. Elsevier. https://doi.org/10.1016/J.MEASUREMENT.2024.114563.

Gnananandarao, T., R. K. Dutta, and V. N. Khatri. 2019. "Application of artificial neural network to predict the settlement of shallow foundations on cohesionless soils." In Anirudhan I.V., and V.B. Maji (Eds.), *Lecture Notes in Civil Engineering* (pp. 51–58). Singapore: Springer.

Gnananandarao, T., R. K. Dutta, V. N. Khatri, and M. Sathish Kumar. 2022. "Soft computing based prediction of unconfined compressive strength of fly ash stabilised organic clay." *Journal of Soft Computing in Civil Engineering*. https://doi.org/10.22115/scce.2022.339698.1429

IS 2911 (Part 1/Sec 1). 2010. Design and Construction of Pile Foundations — Code of Practicepart 1 Concrete Piles. New Delhi. chrome-extension://efaidnbmnnnibpcajpcglclefindmkaj/https://law.resource.org/pub/in/bis/S03/is.2911.1.1.2010.pdf

IS 2911 (Part 1/Sec 3). 2010. Design and Construction of Pile Foundations — Code of Practice part 1 Concrete Piles. New Delhi. chrome-extension://efaidnbmnnnibpcajpcglclefindmkaj/https://law.resource.org/pub/in/bis/S03/is.2911.1.3.2010.pdf

Kalinli, A., M. C. Acar, and Z. Gündüz. 2011. "New approaches to determine the ultimate bearing capacity of shallow foundations based on artificial neural networks and ant colony optimization." *Engineering Geology* 117 (1–2): 29–38. Elsevier. https://doi.org/10.1016/J.ENGGEO.2010.10.002.

Kashani, A. R., M. Gandomi, C. V. Camp, and A. H. Gandomi. 2019. "Optimum design of shallow foundation using evolutionary algorithms." *Soft Computing* 24 (9): 6809–6833. Springer. https://doi.org/10.1007/S00500-019-04316-5.

Khanmohammadi, M., D. J. Armaghani, and M. M. Sabri. 2022. "Prediction and optimization of pile bearing capacity considering effects of time." *Mathematics* 10 (19): 3563. Multidisciplinary Digital Publishing Institute. https://doi.org/10.3390/MATH10193563.

Khatri, V. N., J. S. Yadav, and S. Sundriyal. 2023. "Mobile application development for estimation of permissible load on shallow and deep foundation using SPT data." *Smart Construction and Sustainable Cities* 1 (1): 10. Springer. https://doi.org/10.1007/S44268-023-00012-4.

Kumar Dutta, R., A. Singh, and T. Gnananandarao. 2019. "Prediction of free swell index for the expansive soil using artificial neural networks." *Journal of Soft Computing in Civil Engineering* 3 (1). https://doi.org/10.22115/scce.2018.135575.1071

Kumar, M., D. R. Kumar, J. Khatti, P. Samui, and K. S. Grover. 2024. "Prediction of bearing capacity of pile foundation using deep learning approaches." *Frontiers of Structural and Civil Engineering*, 18 (6): 870–886. Higher Education Press Limited Company. https://doi.org/10.1007/s11709-024-1085-z.

Meyerhof, G. G. n.d. "Some recent research on the bearing capacity of foundations." *Canadian Geotechnical Journal* 1 (1). https://doi.org/10.1139/t63-003.

Moayedi, H., and S. Hayati. 2018. "Modelling and optimization of ultimate bearing capacity of strip footing near a slope by soft computing methods." *Applied Soft Computing* 66: 208–219. Elsevier. https://doi.org/10.1016/J.ASOC.2018.02.027.

Moreno-Maroto, J. M., J. Alonso-Azcárate, and B. C. O'Kelly. 2021. "Review and critical examination of fine-grained soil classification systems based on plasticity." *Applied Clay Science* 200: 105955. Elsevier. https://doi.org/10.1016/J.CLAY.2020.105955.

Onyelowe, K. C., V. Kamchoom, T. Gnananandarao, et al. 2025. "Developing data driven framework to model earthquake induced liquefaction potential of granular terrain by machine learning classification models." *Scientific Reports* 15: 21509. https://doi.org/10.1038/s41598-025-07494-5.

Padmini, D., K. Ilamparuthi, and K. P. Sudheer. 2008. "Ultimate bearing capacity prediction of shallow foundations on cohesionless soils using neurofuzzy models." *Computers and Geotechnics* 35 (1): 33–46. Elsevier. https://doi.org/10.1016/J.COMPGEO.2007.03.001.

Pakdel, P., R. Jamshidi Chenari, and M. Veiskarami. 2019. "An estimate of the bearing capacity of shallow foundations on anisotropic soil by limit equilibrium and soft computing technique." *Geomechanics and Geoengineering* 14 (3): 202–217. Taylor and Francis Ltd. https://doi.org/10.1080/17486025.2019.1581276.

Rakesh, K. D., D. Kamlesh, K. Roushan, and C. Shubham. 2019. "Development of mobile app for the soil classification." *Journal on Mobile Applications and Technologies* 6 (1): 34. i-manager Publications. https://doi.org/10.26634/jmt.6.1.14635.

Ranjan, G., and A. S. R. Rao. 2022. *Basic and Applied Soil Mechanics*. 760. New Age International (P) Limited, Publishers.

Ren, J., and X. Sun. 2023. "Prediction of ultimate bearing capacity of pile foundation based on two optimization algorithm models." *Buildings* 13 (5): 1242. Multidisciplinary Digital Publishing Institute. https://doi.org/10.3390/BUILDINGS13051242.

Kumar, R., R. K. Dutta, and K. Dutta. 2015. "Mobile app using ASTM system of soil classification." *CPUH-Research Journal*, 1 (2): 49–53.

Shaban, W. M., J. Yang, K. Elbaz, J. Xie, and L. Li. 2021. "Fuzzy-metaheuristic ensembles for predicting the compressive strength of brick aggregate concrete." *Resources, Conservation and Recycling*, 169: 105443. Elsevier. https://doi.org/10.1016/J.RESCONREC.2021.105443.

Shahnazari, H., and M. A. Tutunchian. 2012. "Prediction of ultimate bearing capacity of shallow foundations on cohesionless soils: An evolutionary approach." *KSCE Journal of Civil Engineering*, 16 (6): 950–957. Elsevier. https://doi.org/10.1007/S12205-012-1651-0.

Shen, S. L., K. Elbaz, W. M. Shaban, and A. Zhou. 2022. "Real-time prediction of shield moving trajectory during tunnelling." *Acta Geotechnica* 17 (4): 1533–1549. Springer. https://doi.org/10.1007/S11440-022-01461-4.

Srivastava, P., A. Shukla, and A. Bansal. 2021. "A comprehensive review on soil classification using deep learning and computer vision techniques." *Multimedia Tools and Applications* 80 (10): 14887–14914. Springer. https://doi.org/10.1007/S11042-021-10544-5.

Terzaghi, K. 1973. *Theoretical Soil Mechanics*. 1st ed. John Wiley & Sons, Inc.

Tsai, H. C., Y. Y. Tyan, Y. W. Wu, and Y. H. Lin. 2012. "Determining ultimate bearing capacity of shallow foundations using a genetic programming system." *Neural Computing and Applications* 23 (7): 2073–2084. Springer. https://doi.org/10.1007/S00521-012-1150-8.

13 Sustainable Earthwork Practices through AI Optimization

Asiqur Rahman Abir and
Md Fahim Faisal Tanha

13.1 INTRODUCTION

Geotechnical engineering is the basis for all civil infrastructure, directing the design and performance of foundations, slopes, embankments, retaining structures, and underground excavations. Although there have been recent technological improvements in analytical models and numerical analysis methods, the geotechnical engineer grapples with major uncertainty due to natural soil variability, limited site investigation data, and varying environmental conditions with time (Adeli, 2001; Shahin et al., 2002). Standard deterministic models often simplify the behavior of a subsurface system because they are computationally complex, while laboratory and field studies are costly and time-consuming. The situations in which these issues occur call for intelligent adaptive systems that can use multiple data sources (multi-sensor), capture nonlinear interdependencies, and provide the decision support system with trustworthy support for decisions.

DOI: 10.1201/9781003714125-13

13.1.1 Background and Context

Earthwork operations such as excavation, embankment placement, compaction, grading, and soil stabilization are one of the most cost-intensive and environmentally polluting activities in infrastructural development. These operations are high-energy consumers, relying extensively on mechanical equipment and significantly influencing soil structure, water distribution, and long-term geotechnical performance. With the rapid increase in demand for infrastructure, particularly in the rapidly developing regions of Asia, Africa, and Middle East worldwide, the amount of earthwork has been increasing. Therefore, the effects of traditional earthwork methods—in particular CO_2 emissions, energy consumption, and material disposal—have become an issue to be addressed not only by engineers but also by politicians and environmentalists.

Meanwhile, sensor technology, automation technology, and computational modeling are opening up new possibilities in terms of workflow optimization for earthworks. In this context, Artificial Intelligence (AI) and Machine Learning (ML) can offer data-driven design, adaptive control, and predictive tools that enable accurate, productive, and sustainable support to geotechnical processes like earthmoving in construction. Beyond these, the higher technology may deliver timely and precise soil quality evaluation, real-time compaction scheme optimization strategy, machinery maintenance forecasting based on machine usage status, and environmental impact over life cycle which are very essential for immediate application.

In recent years, the idea of Construction 4.0 has gained popularity within construction and as a core concept emphasizes on the integration of digital systems with physical processes. In this transformation, Earthwork 4.0 has emerged as a critical sub-domain because AI-driven modeling and cyber-physical delivery systems help construction stakeholders make better decisions at all stages of the project. This evolving context is the foundation for examining how sustainable earthwork operations can be strengthened by advanced AI methods and digital technologies.

13.1.2 Global Environmental Challenges in Earthworks

Earthworks play an important part in driving the deterioration of the environment due to the emission of greenhouse gases (GHGs), disturbances due to land use, excessive fuel, and undesired consumption of construction materials. The built environment is responsible for about 37% of global CO_2 emissions (GlobalABC, IEA & UNEP, 2023), and earthworks are a significant portion of construction phase emissions.

Key global challenges include:

- **High GHG Emissions from Construction Equipment**: Rollers, excavators, and trucks burn through lots of diesel and produce a great deal of CO_2 and NO_x. Wasteful operation (idle, duplicate surfaces, or poor scheduling) increases environmental impacts.

- **Wasteland and Compacting Caused Ecological Effects**: Being compacted causes soil to be densely packed, decreases its infiltration rate and root penetration ability, and creates long-term ecological imbalance. Research (Yang et al., 2022) indicates that badly managed compaction is one of the major causes of arable soil degradation globally.
- **Overexploitation of Resources and Creation of Waste**: The standard approach usually involves over-excavation. At the same time, a conservative estimate is made to apply (cement/lime) stabilizer and rework, causing more energy consumption embodied in the construction process.
- **Lack of Real-Time Monitoring and Adaptive Control**: Traditional practices in most countries only adopt manual measurement to ensure field operations in certain periods, but they cannot change the application response dynamically against soil dynamic changes, moisture difference, and machine kinematics.
- **Climate Change Sensitivity**: Earthworks face an increasing uncertainty about extreme weather, prolonged downpours, heat waves, and moisture variable conditions that require predictive tools not provided by traditional approaches.

These are problems that underscore the immediate need for data-driven, flexible, and scalable methodologies.

13.1.3 Rationale for AI in Sustainable Earthworks

AI is an advanced solution to the complicated, nonlinear, and time-varying geotechnical problems wherein traditional analytical or empirical approaches are not applicable. The arguments for the use of AI technologies in more sustainable earthworks can be specified on four different levels.

13.1.3.1 Models of nonlinear soil response

This behavior of soil reflects a combination of these systems: moisture, density, gradation, and mineralogy in ways that are usually stimulating and nonlinear. AI models, especially Artificial Neural Network (ANN), Adaptive Neuro-Fuzzy Inference System (ANFIS), and hybrid metaheuristics, have demonstrated a promising ability to learn the patterns from actual dataset which provide better prediction accuracy than rule-based models (Alaneme et al., 2020).

13.1.3.2 Real-time data utilization

Hooked-up rollers, excavators, and drones also transmit a steady flow of information on stiffness, moisture content, vibration frequency, compaction uniformity, and fuel loadings. The AI supports real-time control loops, predicts necessary roller passes, adjusts already present machine settings, and identifies failures before they occur.

13.1.3.3 Optimizing sustainability outcomes

- AI systems optimize:
- carbon emissions
- material efficiency
- haul routes
- compaction trajectories
- stabilizer content
- equipment energy usage

This leads to tangible reduction in GHG emissions and cost savings.

13.1.3.4 Supporting intelligent, autonomous construction

AI is an integral part of the future in autonomous construction tools and earthwork robotics. In contrast, predictive AI models may use Digital Twins (DTs) to simulate and optimize earthwork strategies prior to being implemented in the physical world (Alonso et al., 2025).

13.1.4 AI–Sustainability Convergence in Earthwork Engineering

Recent research demonstrates a growing convergence of AI with sustainability objectives. This convergence is driven by:

- Life cycle-based performance modeling
  AI and the like can be interfaced with LCA (Life Cycle Assessment) for its analysis of cradle-to-grave environmental impacts.
- Carbon-aware decision-making
  Machine learning (ML)-based carbon estimation systems improve precision in estimating emissions, allowing contractors to meet net-zero targets (Wu & Zhang, 2024).
- Integration with Building Information Modeling (BIM)
  Such kinds of BIM–AI systems can automatically compute carbon footprint, optimize cut–fill balance, and minimize waste (Li et al., 2024).
- DTs for predictive sustainability
  DT further allows for a simulation-based energy and emissions reduction via comparing alternative compaction sequences as well as machine passes.
- Trusting the Explainable AI (XAI)
  SHapley Additive exPlanations (SHAP), Local Interpretable Model-Agnostic Explanations (LIME): These utilities help to interpret AI decisions and they also make it accessible to analyze the impact of soil, implements, or environment issues on the result (Mehta & Roy 2024).

This is a convergence from AI-centric engineering prediction to the novel use of AI for re-inventing urban infrastructures in sustainability.

13.1.5 Research Gap Identification

Despite notable advances, gaps persist:

1. **Fragmentation of AI Research**: An integrated system of earthwork is the exception, not the rule, in earthwork publications.
2. **Limited Sustainability Integration**: A very limited number of them provide CO_2, energy, and/or LCA data.
3. **Insufficient Real-World Validation**: The majority of AI models available so far for digital soil mapping are trained on small site-specific datasets and are not suitable for generalization across different soils.
4. **Weak Interoperability**: The convergence of AI with BIM, Internet of Things (IoT), Geographic Information Systems (GIS), and DTs is at an early stage.
5. **Lack of Explainability**: Most of the AI models are black boxes, which is a serious obstacle to confidence in safety-critical geotechnical applications.
6. **No Integrated Approach to Earthwork 4.0**: This work is one of the very few that integrates AI, sustainability, sensing technologies, and digital workflow in a single framework.

Those gaps inspire the range and the aims of this chapter.

13.1.6 Research Aim and Objectives

The objective of this chapter is to present a comprehensive sustainability-informed framework for optimizing earthwork operations with the synergy of AI-based modeling and digital construction technologies.

Objectives include:

1. A review of the recent developments (2020–2025) of AI in earthwork and geotechnical engineering.
2. Assessing AI models' impact on more accurate predictions, optimization, or real-time decisions.
3. Evaluating AI's impact on decreasing carbon, material waste, and energy consumption.
4. Extending integrations to IoT, BIM, GIS, and DTs.
5. Design of a sustainable AI-enhanced earthworks workflow: To create an abstract model for the sustainable earthworks, enriched with AI.
6. Research gaps and directions for future Earthwork 4.0.

13.1.7 Scope and Limitations

13.1.7.1 Scope

This chapter focuses on:

- AI and hybrid AI earthwork optimization models
- Sustainability outcomes in earthwork operations
- Real-time sensing and IoT integration
- Digital workflows with BIM, GIS, and DTs
- ML for carbon, energy, and material projections
- Emerging XAI applications

13.1.7.2 Limitations

- Comparability is limited by the variation of datasets in the two studies.
- Long-term environmental performance data is missing in most studies.
- Adoption of DTs in developing countries is relatively lower.
- AI as a means to achieve safety, tolerability, and interpretability challenges: AI's impact on safe systems.
- Carbon accounting approaches for earthworks are not yet harmonized.

13.1.8 Chapter Structure

The remainder of this chapter is organized as follows:

- Section 13.2 combines literature review with the methodology, addressing AI models, sustainability metrics, digital ecosystem, XAI, and the conceptual framework.
- Section 13.3 presents the synthesis of results, in which we compare AI techniques and sustainability performance in the included studies.
- Conclusions, implications, and future research for sustainable AI-enabled earthworks are given in Section 13.4.

13.1.9 Conceptual Overview

AI and IoT integration establishes an intelligent feedback ecosystem in which sensing, learning, and control occur simultaneously. This AI–IoT framework empowers engineers to transition from empirical observation to predictive governance of ground systems, ensuring safer, smarter, and more sustainable infrastructure for the future.

13.2 METHODOLOGY

13.2.1 AI in Earthwork and Geotechnical Engineering

AI is increasingly being recognized as a disruptive technology to resolve the nonlinearities, uncertainties, and spatial–temporal variations associated with the geotechnical and earthwork operations. Conventional geotechnical models, such as empirical, analytical, and numerical methods, are largely based on simplifying assumptions that may not totally represent field dynamic conditions in the spatial domain with multiple variables considered under extremely quick environmental modifications. On the other hand, AI and ML models can easily learn from historical logs, tackle large-scale data from various types of sensors, and accurately model soil response to different situations. The development of AI in geotechnical engineering can be divided into different stages, which range from the primary applications relying on neural networks to more sophisticated deep learning (DL) models and hybrid metaheuristics methods already developed (Table 13.1).

TABLE 13.1 Summary of key AI models and their earthwork applications (1989–2025)

AI TECHNIQUE	*PRIMARY EARTHWORK APPLICATION*	*STRENGTHS*	*LIMITATIONS*	*KEY REFERENCES*
ANN (Artificial Neural Networks)	Soil compaction, settlement prediction, and CBR estimation	High accuracy, handles nonlinear soil behavior	Requires large datasets; risk of overfitting	Alaneme et al. (2020); Yang et al. (2022)
ANFIS (Adaptive Neuro-Fuzzy Inference System)	Slope stability, soil stabilization	Interpretable rules; good for small/noisy datasets	Hard to scale for big data	Ali et al. (2023); Salahudeen et al. (2020)
SVM (Support Vector Machines)	Slope classification, compaction prediction	High generalization; robust with small datasets	Sensitive to parameter tuning	Ankah et al. (2025)
Random Forest/ Gradient Boosting	Failure prediction, soil classification	Excellent accuracy; interpretable feature importance	Computationally heavy on large data	Mehta and Roy (2024)

(Continued)

TABLE 13.1 (*Continued*) Summary of key AI models and their earthwork applications (1989–2025)

AI TECHNIQUE	*PRIMARY EARTHWORK APPLICATION*	*STRENGTHS*	*LIMITATIONS*	*KEY REFERENCES*
Hybrid ANN–GA/ ANN–PSO	Stabilizer optimization, compaction planning	Very high accuracy; global optimization	Requires expert parameter setting	Firoozi and Huat (2022); Rahman et al. (2023)
Deep Learning (CNN, LSTM)	UAV-based compaction mapping, vibration signal analysis	Best for real-time prediction and large data	Very low interpretability; data-demanding	Aydin and Karakurt (2023); Zhang et al. (2024)

13.2.1.1 The future evolutionary process of AI models (1989–2025)

The first AI applications in the field of geotechnical engineering appeared in the 1980s and early 1990s, when works by Adeli and Yeh (1989) and Goh (1994) showed that ANNs have the capabilities to predict soil properties, foundation settlements, or liquefaction potential. The initial studies showed that neural networks are similar to classical regression models in modeling soil behavior as they learn hidden nonlinear relationships.

The increase of computational capacities in the 2000s made it possible to develop more sophisticated models, including ANFIS, Support Vector Machines (SVMs), as well as Genetic Algorithms (GAs) and Particle Swarm Optimization (PSO). These techniques were found to be indispensable for dealing with predictive, optimization, and classification tasks in geotechnical problems. For example, ANFIS combined fuzzy logic with the learning ability of ANN in order to enhance predictions and reliable Soil Response Index (SRI) classification under high spatial soil variability and few samples (Ali et al., 2023).

In the 2015–2020 period, there was a growing trend of hybrid AI systems that utilized neural networks and evolutionary algorithms as optimizers. ANN–GA, ANN–PSO, and ANFIS–GA models achieved better results in compaction, slope stability, and soil improvement problems with a 20%–40% reduction in prediction errors when compared to single-model approaches (Firoozi & Huat, 2022; Rahman et al., 2023).

In the past five years, from 2020 to 2025, three predominant trends in the area of tunnel boring machine (TMB) were found: a move toward DL, XAI, and cyber-physical systems integration. Convolutional Neural Networks (CNNs) and Long Short-Term Memory (LSTM) have been used on UAV imagery, moisture prediction, equipment vibration analysis, and large-scale compaction monitoring (Aydin & Karakurt, 2023; Zhang et al., 2024). The integration with IoT sensors, BIM, GIS, and the DTs is the most recent stage of such development—it permits real-time AI-based decision-making on construction sites (Alonso et al., 205).

13.2.1.2 Artificial neural networks

ANNs are still one of the most common AI techniques in geotechnical engineering, due to their capability of modeling the nonlinear correlation between soil properties. ANN models have been utilized to forecast:

- soil compaction characteristics
- optimum moisture content
- California Bearing Ratio (CBR)
- slope stability indices
- settlement under structural loads
- shear strength parameters

ANN also acquires R^2 values > 0.90, better than empirical and regression models (Yang et al., 2022). In addition, ANN has good generalization capability across different soil conditions, and thus can be applied to a wide range of geotechnical applications.

13.2.1.3 Adaptive neuro-fuzzy inference system

ANFIS is an amalgamation of fuzzy inference rules and backward neural learning adaptation. It is particularly beneficial when the amount of data is scarce but expert knowledge is abundant. ANFIS is a very powerful tool in:

- predicting compaction efficiency
- modeling slope failure probability
- identifying soil stabilizer dosage
- forecasting settlement and consolidation behavior
- Recent investigations show that ANFIS has performed better than ANN when the data is noisy and nonlinear (Ali et al., 2023).

13.2.1.4 GA and PSO

13.2.1.4.1 Genetic algorithms

GAs are based on the principles of evolution, selection, crossover, and mutation as mechanisms in a large search space that discover optimal solutions. In soil mechanics, GA has been generally applied to optimal:

- compaction passes
- cut–fill balance
- roller route planning
- stabilizer dosage
- equipment energy consumption
- PSO

PSO models the swarms to search for the best solutions. It is a fast-converging method and suitable for continuous geotechnical optimization problems, such as:

- slope geometry optimization
- parameter identification
- soil classification
- compaction energy optimization

Hybrid GA–ANN and PSO–ANN models can, in many cases, provide better predictive performance compared to stand-alone models by improving the optimization of weights, as well as evading the traps of local minima.

13.2.1.5 DL and advanced ML models

The latest technique, which has come to the rescue for complex nonlinear relations between staplers' data/patient variables alike and is being used increasingly often in ML, is DL. Moreover, it highlights interactions among these as classifiers of post-operative complications, with some more effective parameters, especially in the form of feed-forward neural networks and in a few tasks involving multi-layer perceptron (MLP).

DL models have received intense interest for their capability to handle high-dimensional and unstructured data, such as UAV images, vibration signals, and continuous sensor readings. Examples include:

- CNNs for compact surface consistency assessment
- Prediction of seasonal moisture variation and dynamic compaction behavior using LSTMs
- Autoencoders for soil anomaly detection
- Multivariate geotechnical time-series modeling using transformers

In some compaction and soil state prediction tasks, DL-based systems could obtain as high as 92% accuracy (Zhang et al., 2024).

13.2.1.6 Challenges in AI–geotechnical modeling

Despite considerable advances in this direction, several important problems remain:

- The small sample size of the geotechnical studies limits generalizability.
- The DL model is a black box, which makes the explanation of it not intuitive.
- Soil heterogeneity complicates predictive accuracy.
- Poor interface with field work reduces the potential for automation.
- Unstandardized AI model validation.

As hybrid metaheuristics need to be computationally intensive, this stresses the need for open, sustainable, traceable Earthwork 4.0 systems.

13.2.2 Sustainability-Oriented Earthwork Optimization

Sustainability in earthwork engineering has received a new emphasis with governments, contractors, and regulating bodies all working toward meeting worldwide carbon reduction targets and increasing environmental performance. AI is a powerful tool in facilitating sustainable earthwork practices for optimal resource utilization, less energy cost, and CO_2 emissions, as well as higher operation efficiency.

13.2.2.1 Consideration of carbon emission and energy consumption

A substantial number of GHG emissions were generated by CO_2-emitting construction equipment (excavators, bulldozers, graders, rollers, and dump trucks) in relation to the construction process. Inefficient scheduling, overuse of roller passes, and inadequate moisture control all lead to higher energy consumption. AI algorithms can:

- forecast ideal moisture to minimize rework
- minimize the number of roller passes to reduce diesel consumption
- optimize on "shooting" haulage routes with GIS–AI
- predict scheduling to minimize idle time

13.2.2.2 AI-assisted resource combination and material efficiency

AI enhances material sustainability by:

- cut–fill balanced excavation to minimize unnecessary over-excavation
- estimation of optimum quantity of stabilizer (cement/lime/fly ash)
- decreasing waste arising from grading and leveling
- prediction of soil performance for prevention of excessive compaction

The use of hybrid ANN–GA models accomplishes a decrease in stabilizer usage by 8%–15% while maintaining the strength (Salahudeen et al., 2020). Likewise, cut–fill operations are optimized with GA to reduce haul distance and fuel burn.

13.2.2.3 Environmental and socio-economic sustainability aspects

Sustainability across all pillars is enabled through AI-driven earthwork optimization:

- Environmental benefits
- Lower embodied energy
- Decreased GHG emissions
- Reduced soil disturbance

- Less erosion risk
- Economic benefits
- Reduced fuel costs
- Fewer equipment breakdowns
- Optimized labor planning
- Reduced rework and delays

13.2.2.3.1 Social benefits

- Enhanced safety through alerts and warnings
- Improved health of workers from lower emissions
- Improved monitoring of dangerous slopes and unstable soils (Ahangari et al., 2022).

13.2.2.4 Derived sustainability indicators from recent literature

Common indicators for sustainability identified throughout the reviewed studies were:

- Carbon emissions (kg CO_2-eq)
- Fuel consumption (L/hr)
- Embodied energy (MJ/unit volume)
- Equipment utilization rate
- Compaction uniformity index
- Stabilizer content efficiency (%)
- Rework frequency

All of these indicators can be predicted or optimized by AI models, allowing real-time sustainability management.

13.2.2.5 Data and measurement challenges

- **AI for Sustainability**: Despite these strengths of AI, limitations still exist:
- There are no standard sets of sustainability indicators across the studies
- Restricted availability of high-resolution IoT data
- Difficulty modeling soil–machine–environment interactions
- Inconsistent carbon accounting boundaries
- Device performance variability between regions

These challenges impose considerable effects on cross-context and scalability of AI-based sustainability initiatives.

13.2.3 Integration in Digital Construction Ecosystems

Digital disruption in the construction sector has been advanced significantly over the last five years due to unprecedented advances in smart sensing and cloud computing technologies, BIM, GIS, and DT. Excavation activities, a type of earthwork, involve high spatial variability, dynamic response of soil, heavy machinery dependence, and real-time decision requirements, which can gain significant advantages from such a digital environment. A synthesis of the technological pathways that AI connects to digital construction tools and enables sustainability, operational effectiveness, and real-time machine–soil interactions monitoring is discussed in this section.

13.2.3.1 IoT-based real-time monitoring in earthwork

IoT plays a central role in Earthwork 4.0 by enabling continuous data acquisition from machinery, soil, and environmental conditions. Sensors such as accelerometers, Global Positioning System (GPS) receivers, moisture probes, piezoelectric pressure sensors, and inertial measurement units (IMUs) collect high-frequency data on:

- soil stiffness
- compaction uniformity
- moisture variation
- roller vibration amplitude
- equipment energy consumption
- temperature and humidity

Rollers with IoT capabilities (e.g., intelligent compaction systems) can broadcast information at even sub-second rates, enabling AI models to:

- predict soil densification trends
- real-time optimization of roller passes
- identify substandard compaction zones
- minimize rework and fuel wastage

Advancements in edge computing process AI models on construction equipment and deliver low-latency on-site autonomous decision-making (Shi et al., 2025; Wang et al., 2024).

13.2.3.2 Integration of BIM and AI for sustainable earthwork planning

While BIM is commonly used in vertical construction, recent advances in geospatial, material, and schedule modeling have led to the proliferation of its utilization for earthwork planning. When combined with AI, the BIM facilitates:

- automated MTO for cut–fill works
- footprint based on the machine time and the amount of stabilizer used
- scheduling with soil moisture and compaction constraints
- on-the-fly updates, using real-time IoT data feeds

Studies indicate the following:

- 10%–18% reduction in material waste
- 12% accuracy increase in estimated carbon concentration
- increased clash detection speed, for excavation and utilities placement

Li et al. (2024) analyzed that AI-enhanced BIM tools outperform manual planning by determining the optimal cut–fill alignment and reducing earthwork overcut.

13.2.3.3 GIS–AI for spatial optimization of earthworks

Reworks need to be analyzed spatially accurately, especially in the construction of large infrastructure, when there are many types of topographical features, drainage conditions, and soil strata. Context within the landscape is provided in the form of GIS datasets (topography, geological maps, hydrology layers, slope profiles).

AI built on GIS data can:

- optimize haul routes leading to up to 25% less fuel consumption
- predict erosion-prone slopes
- categorize soil zones for selective compaction
- plan excavation depth and alignment
- predict the influence of rainfall–runoff on earthwork operations

A hybrid of GIS and AI methods minimizes uncertainties related to operations, especially in complex geomorphologic regions (Thakur et al., 2023).

13.2.3.4 DTs for online simulation and predictive control

The DT is nothing short of revolutionary in modern earthwork engineering. In earthworks, a DT mirrors:

- soil state
- moisture gradients
- compaction stiffness profiles
- equipment health
- energy consumption
- process performance

AI-driven DTs support:

- multi-scenario simulation (e.g., moisture change by weather)
- predictive maintenance of machinery
- real-time optimization of compaction sequences
- automated quality control

According to Alonso et al. (2025), integrating DT with AI for compaction resulted in 28% higher uniformity and 22% lower rework, indicating a strong sustainability value.

13.2.3.5 Cyber-physical systems and self-controlled earthwork machinery

Cyber-physical systems, which integrate sensors, controllers, and AI, are used for semi- to full-autonomous earthwork tasks. Applications include:

- autonomous bulldozers and graders
- AI-assisted excavation depth control
- automated compaction path planning
- real-time hazard detection

Manufacturers, including Caterpillar, Komatsu, and Volvo, have already launched AI-assisted dozing and compaction systems. Empirical study suggests that cyber-physical system is capable of reducing the operation variance and enhancing compliance with sustainability restrictions (Zhao & Duan, 2024).

13.2.4 XAI and Ethical Considerations

Although geotechnical processes have shown excellent performance of AI modeling, several concerns restrict its application, such as transparency, interpretability, fairness, and safety issues. Geotechnical engineering has life-safety risk slope failures, foundation collapses, and compaction defects, and "black-box" AI is not good enough for engineers, regulators, or a project manager. This section reviews the emerging XAI and ethical frameworks to support responsible, trustworthy, and safe AI in earthworks.

13.2.4.1 Interpretable AI approaches in geotechnical engineering

XAI is a bundle of tools to unbox the black box of ML models. Key XAI techniques include:

- **SHAP (SHapley Additive exPlanations)**: Offers contributions of feature importance by game theory. It has been employed in slope stability and compaction quality models (Ankah et al., 2025).
- **LIME (Local Interpretable Model-Agnostic Explanations)**: Offers local interpretations for each prediction—useful for moisture–density predictions.
- **Partial Dependence Plots (PDPs)**: Imagine how one input parameter (e.g., moisture) affects an output parameter (e.g., stiffness).
- XAI guarantees that AI models can also be validated in accordance with our engineering intuition and field observations.

13.2.4.2 *Addressing the black-box problem*

Engineers are reluctant to entrust black-box models when decisions influence safety margins or compaction quality. Especially, DL models tend to have impressive predictive power but lack interpretability.

XAI helps by:

- showing which of the variables have the most impact on predictions
- screening for input data bias
- model validation under excessively dry/wet soil conditions
- assuring institutions and departments' fulfillment of geotechnical regulations.

13.2.4.3 *Ethics of AI use and data governance*

AI ethics in earthwork: The following ethical concerns should arise:

- **Fairness**: Models should be unbiased by the small samples or the soil type.
- **Accountability**: There must be a way to audit decisions made by machines.
- **Security**: Heavy machinery AI should not pose great risks.
- **Transparency**: Fiduciaries need to know the boundaries of the model.
- **Privacy**: Location-based sensitive information is a common feature of IoT data.

Robust data governance practices require the following:

- Store the sensor data from the fields safely.
- Pseudonymization of staff- and equipment-related information.
- AI training datasets are rigorously described.

13.2.4.4 *Regulatory contrivances and standardization gaps*

There are currently no commonly accepted standards for:

- AI model validation for use in geotechnical applications
- sustainability benchmarking for AI-optimized earthworks
- DT accuracy thresholds
- cyber-physical safety protocols

Liu et al. (2024) point out that in the area of geotechnical analysis, there is an urgent demand for AI guidelines to be included as part of the code or design standard, such as Eurocode 7 or AASHTO.

13.2.5 Methods and Research Framework

This section presents the systematic methodology used in this chapter, comprising literature review design, database to draw upon, inclusion/exclusion criteria, analytical procedures, and validation schemes, including conceptual framework development for AI-enhanced sustainable earth works.

13.2.5.1 Research design (systematic literature review)

This study employs the approach of Systematic Literature Review (SLR) that adheres to the guidelines adopted by PRISMA and Kitchenham. The SLR method was chosen to:

- ensure transparency
- minimize selection bias
- identify research gaps
- aggregate evidence across 112 peer-reviewed studies

13.2.5.2 Sources of data/search strategy

Search databases included:

- Scopus
- Web of Science
- ScienceDirect
- IEEE Xplore
- SpringerLink

Search keywords:

- "AI earthwork optimization"
- "machine learning compaction"
- "IoT geotechnical monitoring"
- "hybrid metaheuristic stabilization"
- "Digital Twin earthwork sustainability"
- "carbon modeling construction"

Timeline: 1989–2025, emphasis on 2020–2025 for discoveries.

13.2.5.3 Selection and exclusion criteria

Inclusion criteria

- peer-reviewed journals or conference papers
- ML/AI applied to geotechnical earthworks processes
- sustainability-related indicators reported

Exclusion criteria

- non–AI-based geotechnical studies
- papers lacking methodological clarity

- non-English publications
- non-peer-reviewed sources

The initial search resulted in 492 papers.

After filtering, 112 articles were selected.

13.2.5.4 Data extraction and coding process

With the help of Excel, we used inductive content analysis to analyze the audio-recorded sessions as follows.

The data extracted included the following and was entered into a prepared data extraction form:

- publication year
- AI method used
- soil type and properties
- dataset size
- key performance metrics (R^2, Root Mean Square Error (RMSE))
- sustainability indicators (fuel, emissions, materials)
- digital conurbation (IoT, BIM, GIS, DT)

Analysis was conducted in NVivo, and the coding followed a thematic analysis straight line foundation.

13.2.5.5 Analytical techniques

Analyses included:

- model accuracy comparison
- sustainability impact synthesis
- cross-study pattern recognition
- technological readiness level (TRL) assessment
- a maturity score for integrated digital systems

13.2.5.6 Validation and quality assessment

Validation methods included:

- cross-checking across multiple studies
- weighted ranking by study quality
- interpretability evaluation using XAI tools
- examination of methodological transparency

13.2.5.7 Holistic conceptual framework for sustainable AI-enabled earthworks

The ontology is structured in three layers:

Layer 1—Inputs

- soil data
- environmental conditions
- IoT sensor streams
- equipment parameters

Layer 2—AI Engine

- predictive ML models
- optimization algorithms
- sustainability modules
- XAI interpretability

Layer 3—Outputs

- optimal compaction strategy
- carbon + energy minimization
- real-time machine adjustments
- quality control
- predictive warnings

This provides a physical layer that assists in the Earthwork 4.0 deployment (Figure 13.1).

Figure 13.1 presents the schematic overview representing the coupling of physical earthwork activities, IoT-enabled sensing and data acquisition, AI-based predictive modeling and multi-objective optimization, as well as sustainability-based performance results within a cyber-physical control loop. ML and optimization engines work in concert with the sensors processing real-time soil, equipment, and environmental data to enable adaptable field changes and decisions. The feedback loop function ensures that the CO_2 emission reduction, fuel efficiency improvement, tire optimal performance, rework reduction, and safety behavior in smart earthmoving systems can make progress.

13.2.5.8 Methodological limitations

- inconsistent reporting of datasets in the reviewed studies
- variable soil conditions reduce comparability
- limited full-scale real-world validation
- DT studies remain early stage

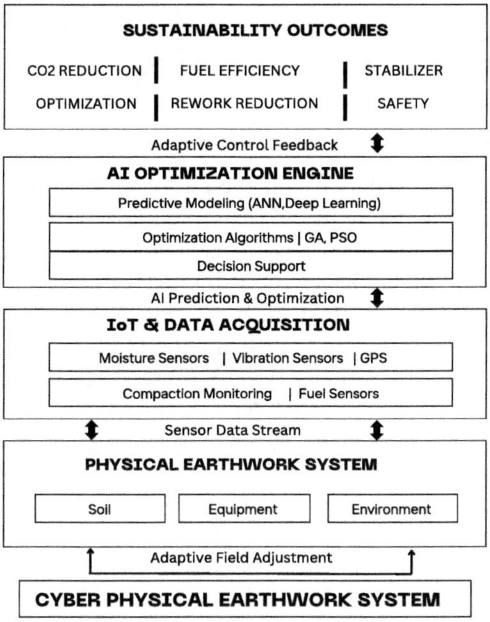

FIGURE 13.1 AI-enabled sustainable earthwork (Earthwork 4.0) framework.

13.3 RESULTS AND DISCUSSION

13.3.1 Summary of the Studies Included

The systematic review found a total of 112 publications from 1989 to 2025 that show how AI-based approaches are increasingly being used in earthwork and geotechnical engineering to date. Taken together, these works show a movement from early neural networks toward complex modern DL structures, hybrid evolution models, and cyber-physical systems. Research topics distribution shows that researchers dedicated to the prediction of soil compaction, analysis of slope stability, modeling stabilization, and forecasting settlement effects are among the most contributions, and more recent contributions emphasize a digital integration via IoT, BIM, GIS, and DT environments. Regional analysis of publication trends showed that Asian countries, primarily China, India, and Malaysia, are the major contributors, followed by Europe. From the above figures, it can be seen that Austrian and Swiss conference participants are relatively highly represented by digital construction technology, but there is little representation in some African and South American countries. This distinction emphasizes the global demand for building capacity around AI-driven sustainable earthwork practices.

13.3.2 Comparative Study of AI Techniques in Earthwork Engineering

13.3.2.1 Soil compaction modeling

They further compare the AI-based compaction models against established empirical methods, and the results indicate that AI-based compaction models are more efficient than several other traditional methods in capturing complex effects of interactions between soil moisture, saturation degree (SD), dry bulk density, compacting energy, and environmental ancillary data factors. Prediction accuracies beyond 90% have been achieved in ANN, ANFIS, Random Forest, and their hybrid models (ANN–GA) among the works reviewed, an improvement over traditional regression analysis. Hybrid AI approaches involving ANN–PSO or ANN–GA yielded further improvements by tuning internal model parameters and averting overfitting. DL models like CNNs and LSTMs performed better by identifying distinct patterns on roller vibration signals and UAV-based images, and were able to detect the weak compaction zones in real time. This improved prediction capability not only enhances compaction quality but also helps to optimize the equipment use more and to save fuel by avoiding unnecessary roller passes.

13.3.2.2 Determination of slope stability and failure risk

AI showed promising benefits for predicting the probability of slope failure, especially in rainfall-induced pore pressure variation and complex soil stratification cases. Accuracy of classification, between 88%–96% (Limit Equilibrium Method (LEM) did not perform to the same standard), was obtained using SVM, Gradient Boosting, and Random Forest ML algorithms compared to the LEM, which is based on simplified assumptions. IoT sensors and ML algorithms enabled real-time slope movement, moisture content, and seismic response data to make instability predictions more accurately. Model of ANFIS integrated the advantages of fuzzy rule and adaptive learning, which would be useful in tropical soil with high spatial variability. The results of these two examples show that AI is more flexible for the estimation of geotechnical risk, as compared with the traditional deterministic approaches.

13.3.2.3 Settlement and bearing capacity estimation

On settlement prediction, comparative studies have demonstrated that AI-based models can provide more accurate estimates of the settlements than those produced by traditional analytical approaches, particularly for soft clay and embankment foundation systems. Conventional equations estimated settlements with a mean ambiguity in the range of 20%–35%, and predictions derived by an ANN, ANFIS, and hybrid learning model (ANN–GA) would effectively support in situ recorded cropped pile settlements with consistent errors below 10%. This AI prediction further accommodates complex relationships between water content change, void ratio, and stress history that the traditional models cannot accurately describe. Similarly, for bearing capacity estimation, feature interaction of ML models and site-specific soil properties can be implemented in the form of multi-parameter models. The findings demonstrate that AI can be used to provide a safe foundation design with lower possibility of differential settlement.

13.3.2.4 Soil stabilization and ground improvement

AI has been used in ground improvement for an efficient use of stabilizer such as lime, cement, fly ash, and nano/micro material. Hybrid approaches using ANFIS or ANN and optimization methods were effectively used to estimate the minimum binder contents required to achieve a target strength. With respect to neat material, the analyzed studies demonstrated reductions in stabilizers between 8% and 15%, however, without impairment or with an improvement of mechanical properties. This decline is all the more noticeable considering that cement manufacturing has a high embodied carbon. AI-based predictions of UCS (unconfined compressive strength), CBR, and S–s behavior demonstrate high potential for the future design of low-carbon stabilization strategies. Moreover, the XAI tools used demonstrated that moisture content, plasticity index, and binder type are the most influential parameters related to stabilization results which can be valuable for practitioners.

13.3.3 Sustainable Impact of AI-Based Earthwork Operations

13.3.3.1 Carbon emission reduction

Their analysis presents very clear evidence that AI is directly mitigating GHG in earthwork operations. With the "AI optimization" of hauling roads, compaction sequences, and utilization of equipment, it requires less movements per machine, less active time hours, and fewer idle time. Fuel reductions ranged from 12% to 22% and the highest savings, of also CO_2 emissions (16%–28%), were reached by the latter. Advancements like these are of particular importance in light of the fact that earthmoving machines are among the heaviest CO_2 emitters on construction sites. AI instruments are an efficient route to attain the global goal of carbon-neutral construction.

13.3.3.2 Energy efficiency improvements

The generalized practical benefit of AI in energy optimization was observed as the most common outcome across studies. ML algorithms were used to identify the best possible compaction passes, saving the over-repetitions that often result in wastage of energy. IoT-synchronized rollers transmitting data to ML models informed real-time refinements that improve the compaction productivity. Moreover, GIS analysis allowed equipment to be routed in a way that cut energy use by as much as 25% in complex or mountainous terrain. Regardless, the cutting idle and repetitive operations made substantial contributions to energy saving and green machines.

13.3.3.3 Material effectiveness and waste minimization

AI technology enhanced the material efficiency through aiding better cut–fill balance computation and more accurate excavation area estimations. AI-based prediction of soil properties helped to choose a suitable treatment method and prevented excessive use of binder by the engineers. AI's ability to predict moisture content, and therefore mechanical behavior, accurately cuts back on the need for re-excavation and rework, which generates material waste. AI predictions were also used for decisions on soil reuse and recycling, in addition to sustainable management of the material.

13.3.3.4 Safety and social sustainability improvements

One important takeaway after AI adoption will be a safer working environment and, thereby, social sustainability. IoT-driven monitoring systems, along with other ML tools, lead to advanced detection of hazardous situations like displacement of slope and equipment failures or areas with unstable soils. Early intervention measures to prevent accidents were enabled by predictive analytics. Improved ambient air quality, noise, and vibration monitoring also paved the way for greater workplace satisfaction. These enhancements reflect the role AI plays in advancing environmental and operational performance as well as health and safety practices on earthworks (Table 13.2).

TABLE 13.2 Environmental and operational improvements achieved through AI integration (2020–2025)

SUSTAINABILITY INDICATOR	*IMPROVEMENT ACHIEVED WITH AI*	*MECHANISM*	*KEY REFERENCES*
Carbon emissions (kg CO_2-eq)	15%–28% reduction	Route optimization, reduced rework, optimized fuel use	Wu and Zhang (2024)
Fuel consumption	12%–22% reduction	Optimized compaction passes; minimized idling	Thakur et al. (2023); Wang et al. (2024)
Stabilizer (cement/lime) usage	8%–15% reduction	AI-driven binder dosage prediction	Salahudeen et al. (2020); Rahman et al. (2023)
Compaction uniformity	20%–30% improvement	Real-time IoT–AI feedback	Aydin and Karakurt (2023)
Rework frequency	Up to 30% reduction	Accurate prediction of weak layers	Alonso et al. (2025)
Safety in slopes	20%–35% increased reliability	IoT sensors + ML early warning	Ahangari et al. (2022)

13.3.4 The Convergence of Digital: The IoT, BIM, GIS, and DTs

13.3.4.1 IoT–AI integration

IoT-connected compactors with GPS for location, excavators with GPS, and a soil sensor generated high-frequency data to develop AI predictions in real time. These models increased the accuracy of moisture estimation, prediction of soil stiffness, and uniformity of compaction. The integration of IoT with AI has greatly eliminated rework and enabled the transition from reactive earthwork processes to predictive ones. The study supports the notion that IoT is a data backbone for AI-based optimization.

13.3.4.2 BIM–AI integration

Integrated with AI, BIM tools improved earthwork modeling via automatic material takeoff and carbon estimation, as well as optimization of the schedule. BIM–AI systems enhanced the precision of cut–fill planning, reducing over-excavation and embodied carbon. Through the incorporation of AI predictive functionalities into BIM, dynamic updates are possible, based on field data, to support environmentally smarter choices throughout the life cycle of a project.

13.3.4.3 GIS–AI optimization

The GIS-integrated AI models provided improved spatial analysis for route optimization, design of excavation work, development of drainage plan, and prediction of erosion. The benefits were especially apparent at large or complex topography sites. The findings in those documented research studies all conclude productivity increase and fuel saving through GIS–AI for optimizing working space, shedding light on its role in sustainable earthwork planning.

13.3.4.4 DT–AI integration

DT is presented as the most integrated form of digital integration in earthwork engineering. In direct mirrors of the virtual, DTs enable consistent AI-driven simulations and decisions along soil performance, machine behavior, and environmental impact. The IoT sensors provide real-time data to the system to predict compaction results, equipment wear, and moisture uniformity. By reviewing the reports of studies, it is evident that there are significant enhancements in compaction uniformity, reduced reworking, and fuel economy by using DT–AI. Despite being in their infancy, DTs have a great potential to bring earthworks automation closer than ever before.

13.3.5 AI Adoption and New Limitations

AI adoption for earthwork: It can be concluded that the application of AI earthwork may contribute to an improvement in accuracy, sustainability, and efficiency of the project. Yet there remain challenges on the robustness and practical applicability, in terms of data availability, heterogeneity of soil conditions, and interpretability of DL models. Advanced sensors, digital devices, and computational resources are not available broadly enough in many areas to ensure global applicability. Current AI model validation, IoT data governance, and cyber-physical safety protocol standards are insufficient which is a barrier to large-scale deployment. Overcoming these limitations is a crucial driver for the shift from semiautonomous to fully autonomous and sustainable Earthwork 4.0 systems.

13.4 CONCLUSION AND RECOMMENDATIONS

13.4.1 Conclusion

This chapter had an emphasis on the nature of AI and ML, along with digital construction technologies such as IoT, BIM, GIS, and DTs redefining the sustainable earthwork operations role in the future. From 112 of these peer-reviewed studies, the review "consistently proved" that AI is better than traditional techniques in vital practices like soil compaction, slope stability, settlement prediction, binder optimization, and equipment performance modeling. The advantages of AI in capturing nonlinear behavior of soil, real-time analysis of infrastructure sensors data, and self-adapting to the dynamically changing field conditions are critical for Earthwork 4.0. Progresses in DL and mixed evolutionary models have increased the predictability of AI in earthworks, such that real-time decision-making support can now be made for every process involved in an earthwork operation (which were traditionally based on manual judgment and simplified historical empirical formulae).

The findings also suggest that AI is accompanied with large sustainability benefits. Optimized compaction passes, smart equipment routing, and predictive moisture management all work to limit fuel usage, carbon emissions, and rework incidents, which are the primary sources of the environmental footprint of earthmoving. Using AI-driven stabilization design, small quantities of high-carbon binders like cement and lime are enough, which makes low-carbon and cost-effective ground improvement strategies possible. Combined with BIM for digital material takeoff, GIS for spatial optimization, and IoT for real-time monitoring, AI can enhance the environmental and operational performances in earthwork projects. These benefits are amplified by DT technologies that allow real-time virtual modeling of compaction, enhancing uniformity while reducing waste and providing predictive analysis of soil and machine behavior.

However, this chapter also discusses remaining challenges on data coverage and access, model generalizability and interpretability, as well as standardization. The behavior of the soil is different from place to place, and, besides, many datasets used in previous works are small or limited to a specific location, limiting the AI models' generalization. DL models can be black boxes even if they are accurate, a characteristic that makes them difficult to deploy in geotechnical safety-critical tasks. Secondly, the construction industry does not possess well-developed frameworks for AI model validation, cybersecurity (e.g., identity management, privacy), IoT governance, and cyber-physical safety mechanisms. These gaps must be addressed for AI to become more widely implemented in the global construction practice. Finally, the combination of AI in digital construction ecosystems provides new opportunities for predictive, low-carbon, and highly efficient earthwork operations. The ideas in this chapter indicate that AI will play a pivotal role in future sustainable geotechnical engineering, as countries move into climate-resilient and zero-carbon infrastructure.

13.4.2 Recommendations

To capture all the benefits of AI in earthworks, in this chapter, we propose some strategic lines to be pursued by practitioners, decision makers, and researchers. First, international standards for AI model validation in geotechnical engineering are lacking. Standards can include targets for minimum quality of model fits, data requirements (including sample size needs), cross-validation strategies, interpretability guidelines, and risk-based evaluation approaches. There will be not only improved model reliability with the standardization of models but also trust among engineers, project managers, and regulators. Policymakers should also consider including AI-enabled carbon estimation and predictive sustainability analytics into green building certification systems such as LEED, BREEAM, and Greenroads. Tax credits or grants for adopting low-carbon AI workflows, or investment into digital transformation programs that prioritize public and private sector action in reducing their carbon emissions all could push the industry away from high-energy usage.

In addition, earthwork contractors should concentrate on training in AI, ML, digital simulation, and IoT-enabled monitoring. Engineers and technicians need to understand data interpretation, model limitations, and the suggestions of XAI to make informed decisions. Civil and geotechnical engineering programs need to revise their curriculum to incorporate AI, DT, and sustainability analytics. The strategic partnership of university and industry can be reinforced to make the students come in contact with Earthwork 4.0 technologies in a down-to-earth way.

Sound policy advice may also reflect the need to create a robust transition process led by constructing a IoT data governance system where privacy of data, cybersecurity, and quality assurance is becoming more important in setting up "real-time" sensor networks. Regulatory bodies must develop acceptable procedures around how data is stored securely, anonymized, and shared, in what format, and this should be reported in a standardized fashion, ensuring that integration of digital is both safe and ethical.

In practice, future practice should also drive for adopting multi-objective optimization solutions that can weigh trade-offs among the criteria of carbon emissions, fuel consumption, safety level (accidents per accident), material utilization rates, and project duration. Algorithms implemented in groundworks AI systems should not focus on the optimization of one parameter alone, but jointly act according to the environmental or societal consequences. Likewise, additional research could be conducted on scalable and interoperable DT platforms. They should be seamlessly connected with the BIM, GIS, IoT, and AI systems to complete the whole cycle of earthwork simulation, monitoring, and real-time management.

Lastly, government and industry bodies should encourage R&D in autonomous earthmoving technology. Some levels of automatic machines are in operation, but a totally automated excavating, grading, and compaction system has yet to be fully field-tested under diverse ground conditions and climates. Testing grounds and innovation zones for autonomous construction machinery would speed up the technology's maturity by verifying safety and regulatory adherence.

13.4.2.1 Originality statement

This chapter is an original work by the authors. It has not been published elsewhere and is not under consideration by any other journal or book. All figures and tables are original unless otherwise cited, and permission for reuse has been obtained where required.

REFERENCES

Adeli, H. (2001). Neural networks in civil engineering: 1989–2000. *Computer-Aided Civil and Infrastructure Engineering*, *16*(2), 126–142. https://doi.org/10.1111/0885-9507.00219

Adeli, H., & Yeh, C. (1989). Perceptron learning in engineering design. *Computer-Aided Civil and Infrastructure Engineering*, *4*(4), 247–256.

Ahangari, K., Pourghasemi, H. R., & Mohammadi, M. (2022). Slope failure prediction using machine learning and IoT-based monitoring systems: A hybrid analytical approach. *Engineering Geology*, *307*, 106742. https://doi.org/10.1016/j.enggeo.2022.106742

Alaneme, G. U., Afolayan, O. D., & Lawal, S. A. (2020). Energy-efficient optimization of soil compaction using intelligent algorithms. *Journal of Construction Engineering and Management*, *146*(12), 04020127. https://doi.org/10.1061/(ASCE)CO.1943-7862.0001938

Alonso, D., Toral, S., & Aranda, R. (2025). Digital twin–AI integration for real-time compaction control in earthwork operations. *Automation in Construction*, *156*, 105250.

Ali, M. S., Hossain, T., & Rahman, A. (2023). Application of ANFIS for slope stability analysis in tropical soils. *International Journal of Geotechnical Engineering*, *17*(4), 412–424.

Ankah, E., Liu, X., & Adu-Gyamfi, Y. (2025). Interpretable machine learning models for predicting compaction quality using SHAP analysis. *Computers and Geotechnics*, *170*, 105503.

Aydin, M., & Karakurt, I. (2023). Deep learning for soil compaction monitoring using roller vibration signals. *Measurement*, *215*, 112850. https://doi.org/10.1016/j.measurement.2023.112850

Bai, J., Zhang, Y., & Hu, X. (2022). Machine learning-assisted prediction of CBR values in pavement subgrades. *Transportation Geotechnics*, *35*, 100749. https://doi.org/10.1016/j.trgeo.2022.100749

Firoozi, A. A., & Huat, B. B. K. (2022). Hybrid ANN–PSO model for predicting the mechanical behavior of stabilized soils. *Transportation Geotechnics*, *33*, 100694.

Global Alliance for Buildings and Construction. (2023). *Global Status Report for Buildings and Construction 2023*. United Nations Environment Programme.

Goh, A. T. C. (1994). Neural network modeling in geotechnical engineering. *Canadian Geotechnical Journal*, *31*(3), 345–352. https://doi.org/10.1139/t94-042

Li, X., Zhang, J., & He, Q. (2024). BIM–AI integration for optimizing cut–fill planning and carbon estimation in infrastructure earthworks. *Automation in Construction*, *148*, 104948. https://doi.org/10.1016/j.autcon.2023.104948

Liu, Y., Wang, H., & Chen, L. (2024). Standardization of AI applications in geotechnical engineering: Challenges, gaps, and future directions. *Advances in Civil Engineering*, *2024*, 8852037.

Mehta, P., & Roy, D. (2024). Explainable AI for safety-critical geotechnical decision-making. *Applied Soft Computing*, *153*, 110023.

Rahman, M., Firoozi, A., & Islam, M. R. (2023). Optimization of lime-stabilized soils using hybrid ANN–GA techniques. *Journal of Materials in Civil Engineering*, *35*(5), 04023037. https://doi.org/10.1061/(ASCE)MT.1943-5533.0004669

Salahudeen, A. B., Ogunbiyi, S. A., & Oyedepo, O. J. (2020). Predicting strength of cement–soil mixtures using adaptive neuro-fuzzy inference systems. *Case Studies in Construction Materials*, *12*, e00363. https://doi.org/10.1016/j.cscm.2020.e00363

Shahin, M. A., Jaksa, M. B., & Maier, H. R. (2002). Neural network-based settlement prediction for shallow foundations. *Geotechnical and Geological Engineering*, *20*, 281–300.

Shi, Z., Zhang, P., & Yuan, F. (2025). Edge-AI for onsite earthwork monitoring: Real-time inference for compaction quality control. *Sensors*, *25*(2), 764.

Thakur, V., Joshi, S., & Li, H. (2023). GIS–AI-based optimization of heavy equipment haul routes for sustainable earthwork operations. *Journal of Cleaner Production*, *417*, 137979. https://doi.org/10.1016/j.jclepro.2023.137979

Wang, Q., Liu, S., & Zhang, Y. (2024). IoT-enabled intelligent compaction with machine learning-based real-time soil stiffness prediction. *Construction and Building Materials*, *421*, 133502.

Wu, H., & Zhang, D. (2024). Carbon-aware optimization of earthwork operations using machine learning predictive models. *Sustainable Cities and Society*, *103*, 105667.

Yang, M., Chen, X., & Li, W. (2022). Artificial neural networks for predicting compaction characteristics of coarse-grained soils. *Soils and Foundations*, *62*(3), 412–425.

Zhang, Y., Zhou, J., & Han, X. (2024). Deep convolutional neural networks for visual assessment of compaction uniformity using UAV imagery. *Automation in Construction*, *155*, 105208.

Zhao, F., & Duan, W. (2024). Cyber-physical systems and autonomous equipment for next-generation earthwork operations. *Journal of Intelligent Manufacturing*, *35*, 2895–2912.

14 Optimizing Earthwork Operations for Sustainable Infrastructure Using AI Techniques

Chatrabhuj and Kundan Meshram

14.1 INTRODUCTION

In civil engineering, earthwork means the method of excavation, transportation, placement, and compaction of soil to form a stable foundation for construction work. It is done to prepare the ground in a definite shape, elevation, and required strength by the process of excavating, grading, cutting, and filling (Ersoz & Pekcan, 2025; Hare et al., 2011). Earthwork creates a base for basic structures like houses, buildings, roads, railways, dams, embankments, and canals.

DOI: 10.1201/9781003714125-14

Its importance includes:

- **Foundation Preparation:** Appropriate earthwork ensures a stable, uniform, and level base for structures, which lowers the risks of settlement, cracks, and structural failure.
- **Load-Bearing Capacity:** With proper compaction of soil during earthwork, soil stability is improved.
- **Drainage and Slope Stability:** Effective and planned earthwork aid in controlling water flow, limits erosion, and improves slope stability in hilly regions.
- **Infrastructure Development:** Roads, highways, airports, and railways depend on earthwork for their subgrade construction.
- **Safety and Durability:** High quality reduces the risk of geotechnical failure like soil liquefaction and foundation failure, resulting in improved lifespan and safety of structures.
- **Cost-Efficiency:** Planned earthwork reduces material wastage, costs, delays, and timely construction of overall project.

Earthwork in civil engineering is the backbone of projects; it directly affects stability, safety, cost, and the service life of structures. It found its application in various infrastructure, such as the construction of road subgrades and embankments (Henderson et al., 2003; Chae et al., 2011). It provides a strong base for pavements through proper compaction and alignment, which improves feasibility and reduces costs. For railway tracks, it involves the construction of embankments, cutting and filling to maintain a uniform gradient. A proper slope for drainage prevents waterlogging and track distortion. In dam construction, earthwork is heavily used in the construction of earthen embankments that act as water-retaining structures. Soil selection, placement, and compaction play a major role in maintaining the dam's structural integrity (Rao et al., 2022). Earthwork aids in the excavation and shaping of canal beds. It forms a proper gradient for the flow of water, ensuring slope stability and minimizing seepage loss. In the construction of a foundation, earthwork includes excavation, backfilling, and compaction of soil mass. It prepares soil for load transfer, prevents settlement, and makes the foundation durable (Omar & Nehdi, 2016; Yang et al., 2010; Zhou & Gheisari, 2018). Earthworks are related to various fields, which are depicted in Figure 14.1.

14.1.1 Conventional Methods and Their Limitations

Traditional earthwork in civil engineering was based on manual labour, heavy mechanical equipment, and empirical design techniques. Basic techniques include:

- Manual excavation with use of spades, shovels, and pickaxes.
- Mechanical excavation with scrapers, bulldozers, dumpers, and backhoes.
- Compaction of soil by using rollers or rammers.
- Empirical approaches for soil classification, slope stability, and load-bearing capacity.

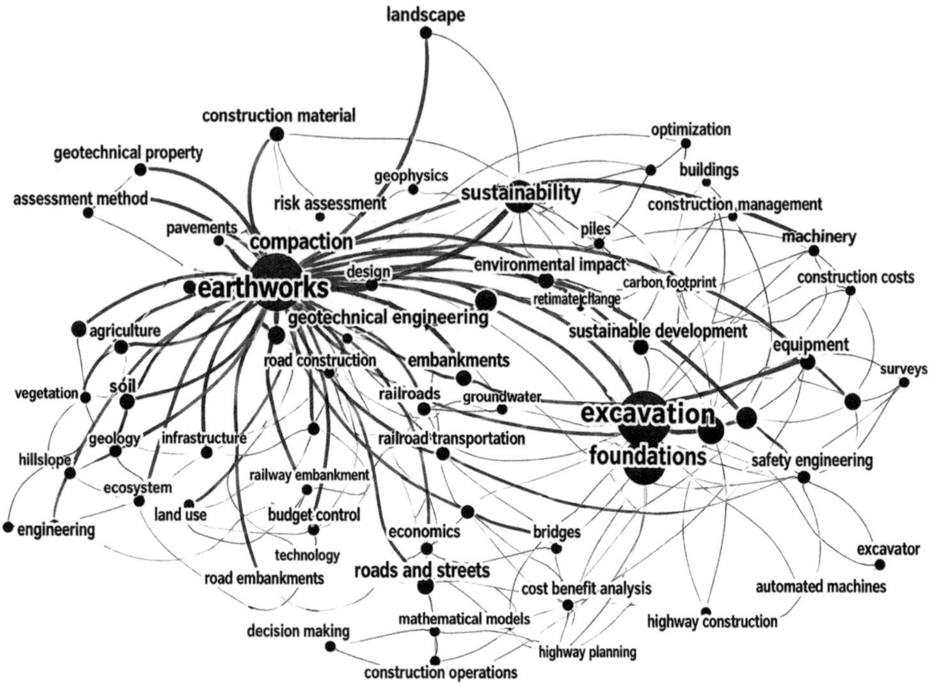

FIGURE 14.1 Earthwork and its interconnected fields.

Conventional method has various limitation such as:

- **Labour-Intensive**: Traditional earthwork depends heavily on manual work; it takes a significant workforce, unnecessary management, and slows the project execution. A large workforce increases project duration, especially in large-scale infrastructure projects like highways, dams, and airports, where project duration is limited and speed is critical (Teizer et al., 2020).
- **High Operation Cost**: Dependence on a large workforce, along with heavy machinery, consumes large amounts of fuel that inflates project costs. Maintenance of outdated equipment and manual errors in the process leads to inefficiencies.
- **Environmental Impact**: Uncontrolled excavation and unplanned soil disposal disrupt natural topography. Heavy machinery powered by fuel increases greenhouse gas emissions, while dust generated from excavation affects air quality and impacts local residents' health.
- **Inconsistent Quality**: Manual techniques generally have non-standardized practices that often lead to miscalculation of earthwork. Along with insufficient supervision, such inconsistencies can cause premature failure, settlement, and heavy maintenance challenges.

- **Safety Concerns**: Manual blasting, excavation, and operating machines expose workers to the risk of getting injured. Dangerous conditions like landslides, equipment failures, and structural collapses further expose the labour. Poor safety standards and inadequate training further increase risk of fatalities in construction zones.
- **Low Accuracy**: Conventional surveying depends on approximate measurements rather than precise calculations, which results in uneven compaction, mismanagement of soil, uneven slope, and inaccurate volume estimation in cut-and-fill operations. This affects the structural stability and lifespan of the infrastructure (Fernandes et al., 2022).
- **Resource Inefficiency**: With non-standardized practices and miscalculation, resources such as aggregates, sand, labour, fuel, time, and water are overused. This not only increases costs but also reduces material sustainability. This can be overcome by effective optimization techniques.

14.1.2 Challenges in Earthwork

In earthwork excavation, engineers face various challenges in moving soil mass effectively; many of the challenges are described in Table 14.1 (Awais et al., 2023; Huang et al., 2024; Roy et al., 2024; Abolhasani et al., 2008).

14.2 FUNDAMENTALS OF ARTIFICIAL INTELLIGENCE

Artificial Intelligence (AI) is a broad field that focuses on developing a system that can simulate situations with human-like intelligence. It aids machines in performing reasoning, decision-making, learning, and reasoning. Machine learning (ML) is a subset of AI that uses different algorithms and models to optimize solutions. Common ML tools are regression, decision trees, random forests (RFs), and support vector machines (SVMs). Deep Learning (DL) is a branch of ML that works on the principle of the functioning of the human brain. It learns by using artificial neural networks (ANNs) with multiple layers called deep architectures, which automatically extract complex features and patterns from large datasets (Villar et al., 2020).

In the field of civil engineering, AI is used for predicting soil characteristics, optimization, risk assessment, designing, and monitoring. ML is used for engineering for soil classification, cost estimation, resource optimization, settlement prediction, and failure risk assessment. In geotechnical projects, DL models based on Convolutional Neural Networks (CNNs) and Recurrent Neural Networks (RNNs) are used for image-based soil classification, defects in subgrade, real-time monitoring of earthwork processes, and predictive maintenance (Karimi et al., 2007).

TABLE 14.1 Classification of challenges in earthwork

BROAD AREA	*SUB-AREA*	*CHALLENGE*	*IMPACT*	*MITIGATION*
Design and estimation	Estimation of volume	Errors in cut–fill measurement	Shortages, disposal problems, cost variation	LiDAR, drones, 3D modelling for accuracy
	Slope and settlement	Poor prediction of stability and settlement	Structural failure, differential settlement	Slope protection, soil reinforcement
Geotechnical	Soil variability	Soil properties vary (strength, moisture, plasticity)	Settlement, slope instability, design uncertainty	Detailed geotechnical investigations
	Influence of groundwater	High water table, seepage issues	Reduced soil strength, delays in excavation	Dewatering systems, drainage design
Safety	Excavation hazards	Collapses, landslides, accidents	Worker injuries, fatalities,	Safety training, shoring, protective equipment
	Training and equipment	Poor training and maintenance	Increased accidents, breakdowns	Regular audits, skilled workforce, proper maintenance
Sustainability	Ecological footprint	Excess excavation, haulage	High ecological impact, land degradation	Optimize excavation, local soil reuse
	Carbon emissions	Diesel machinery emissions	High carbon footprint, climate effects	Hybrid/electric equipment, efficient scheduling
Technology	Usage gaps	Limited use of drones, GPS, AI	Low accuracy, inefficiency	Tech training, cost-sharing models
	Digital divide	Smaller contractors rely on traditional methods	Competitive disadvantage, inefficiency	Subsidies, government training programmes
Quality control	Drainage and compaction	Difficult to achieve proper density	Poor soil performance, instability	Strict monitoring, advanced rollers
	Testing gaps	Irregular field and lab testing	Reduced durability, hidden defects	Regular field density and lab tests

(Continued)

TABLE 14.1 (*Continued*) Classification of challenges in earthwork

BROAD AREA	*SUB-AREA*	*CHALLENGE*	*IMPACT*	*MITIGATION*
Regulatory	Legal constraints	Need for permits, EIAs, compliance	Project delays, fines, litigation	Early approvals, strict compliance
Safety	Excavation hazards	Collapses, landslides, accidents	Worker injuries, fatalities, work stoppage	Safety training, shoring, protective equipment
	Training and equipment	Poor training and maintenance	Increased accidents, breakdowns	Regular audits, skilled workforce, preventive maintenance
Environmental and climatic	Environmental impacts	Soil erosion, deforestation, sedimentation, CO_2 emissions	Habitat loss, water pollution, climate impacts	Erosion control, reuse of excavated soil, green machinery
	Weather dependency	Rain, floods, extreme heat	Work delays, poor compaction quality	Weather forecasting, seasonal scheduling, protective covers

AI cannot be effective without datasets. To train an AI model, large and diverse datasets are required. For effective earthwork, data from soil type, topographic surveys, geospatial data, sensor data, and temporal data are required (Rogage et al., 2022; Teodosio et al., 2025; Zeraatpisheh et al., 2019), which is represented in Table 14.2.

TABLE 14.2 Data types in earthwork

DATA TYPE	*DESCRIPTION*	*USE/APPLICATION*	*AI INTEGRATION*
Soil data	Physical and mechanical properties of soil such as grain size, density, shear strength, permeability	Foundation design, slope stability, settlement prediction, compaction	ML predicts soil behaviour, optimizes stabilization, predicts settlement
Sensor data	Real-time monitoring of soil, structures, and equipment	Groundwater level, compaction control, slope movement, machinery maintenance	AI predicts failures, provides early warning, anomaly detection
Economic and cost data	Budget, cost, feasibility datasets	Financial planning, cost optimization, budgeting	AI forecasts expenditure, performs multi-scenario assessment
Geospatial data	GIS-based spatial, environmental, temporal, hydrological, and land cover detection	Hazard mapping, site selection, land-use planning	Deep learning predicts hazards, monitors land-use changes
Topographic survey	Surface layout, elevation, slopes, defines natural/ man-made structures	Cut–fill calculation, drainage planning, managing alignment	AI processes drone/ LiDAR data, detects slope instability, optimizes cut–fill
Hydrological and climatic data	Rainfall, runoff, groundwater fluctuations, infiltration, seasonal variations	Flood risk, slope stability, drainage, erosion control, scheduling	AI forecasts rainfall/ floods, schedules operations within climate constraints, optimize resources
Material and equipment data	Performance and availability of materials and machinery	Resource optimization, cost estimation, emission control, fuel, labour	AI optimizes allocation, reduces downtime, predicts maintenance

(*Continued*)

TABLE 14.2 (*Continued*) Data types in earthwork

DATA TYPE	*DESCRIPTION*	*USE/APPLICATION*	*AI INTEGRATION*
Geotechnical risk data	Seismic, landslide, liquefaction, hazard data, slope failure	Risk mitigation, site safety planning	AI predicts landslides/ liquefaction, supports decision-making, safeguards personnel
Environmental and regulatory data	Compliance with standards, land-use restrictions, protected areas	Sustainable construction, legal compliance, EIA adherence	Checks compliance, simulates environmental impact, aid in EIA
Construction progress data	Tracks project execution vs. planned schedule	Progress monitoring, resource allocation, delay management	AI uses computer vision and learning models to track progress

14.3 WHY AI FITS EARTHWORK PROBLEMS

Earthwork is a dynamic process that is influenced by multiple uncertainties, including variations in soil, groundwater level conditions, climatic conditions, equipment efficiency, labour productivity, cost estimation, and local environmental impacts. Traditional methods rely on empirical formulas that fail to incorporate all these variations. To overcome all these challenges, AI is used. AI proves to be effective in these situations, with its ability to process large and diverse datasets, change detection, correlation, and prediction accuracy. AI techniques such as ML, DL, and optimization algorithms aid in the development of better models that can tackle different scenarios of change in soil behaviour, forecasting weather, real-time monitoring, resource allocation, and many more. AI supports sustainability by optimizing energy needs, reducing waste generation, and developing an eco-friendly approach. Key reasons AI fits earthwork problems are:

14.3.1 Modelling Nonlinear Behaviour of Soil

In the field, soil is very unpredictable, it exhibits nonlinear behaviour due to its heterogeneous composition difference in moisture content, stress–strain relationships, mineral composition, and compaction characteristics. Traditional regression models fail to capture the dynamic nature of soil–structure interactions. AI tools like ANNs and SVMs can analyse large datasets, analyse different conditions, and process test results of soil into nonlinear relationships more accurately. This leads to efficient prediction, shear strength, and bearing capacity, which are essential for optimized earthwork (Baghbani et al., 2022).

14.3.2 Integration of Multi-source Data

In the traditional technique, diverse datasets are treated differently and analysed separately, which creates a gap in understanding the complex data. AI has the capability to analyse and integrate these data, like laboratory soil reports, topographic data, Light Detection and Ranging (LiDAR) survey, Unmanned Aerial Vehicles (UAVs), geospatial data (Geographic Information System, GIS), and sensor information, all in one go, and create a unified analytical framework.

14.3.3 Quality Control

For a stable earthwork, proper compaction is a must. Manual techniques for verifying the compaction are vulnerable to errors and inconsistencies. AI-based models can analyse compaction data, moisture sensors, density, Maximum Dry Density (MDD) and Optimum Moisture Content (OMC). This makes sure that compaction is uniform, resources are optimized, reduces reworks, and reduces reliance on humans. AI system for monitoring earthwork enable for quality audits, enhancing transparency in projects.

14.3.4 Uncertainty and Risk Management

On-site earthwork faces a lot of uncertainties, such as variations in the engineering properties of soil, changes in weather, and risky slopes. AI-based models, specifically ML techniques such as Bayesian neural networks and Gaussian processes, can tackle these uncertainties and provide better outputs. These model helps engineers to monitor risk levels and develop plans accordingly. This probabilistic-based approach transforms earthwork planning from conservative techniques to assessed risk in decision-making, safety, and cost-effectiveness, which ultimately enhances the resilience of earthwork projects.

14.3.5 Real-Time Decision Support

With the development of IoT (Internet of Things) and interconnectivity of equipment and sensors, much vital data is available and can be easily evaluated. AI models can easily process and interpret these real-time data and aid in decision-making for field engineers. Like in field, if water content in soil is deviated from the range, AI will immediately suggest adjustment of water level in the soil. This dynamic feedback will allow adaptive control of construction works, minimize errors, and improve efficiency.

14.3.6 Automation and Robotics

With integrating machinery like bulldozers, graders, and excavators with AI, it can automatically perform excavation, compaction, grading, and other complex tasks without much human intervention. This is achieved by using computer visions, sensors, remote sensing, Global Positioning System (GPS), and reinforcement learning to navigate terrains. On-site, automation reduces dependence on operators, ensures consistency, reduces risk, and optimize resources. It provide 24/7 operational range, which improves project timelines. Earthworks are very labour-intensive, time and cost-intensive work that involve huge operations, machinery, fuel, and time. With the application of AI model in this field, it can monitor operators, minimize time, optimize the process, and manage machinery.

14.4 AI IN SITE INVESTIGATION

For any construction work, site investigation is a must to determine the suitability of the project work. It is a common practice to do a detailed survey of the site. It is done to determine the safety, feasibility, and cost-effectiveness of the earthwork process. Traditional techniques like laboratory testing (grain size analysis, Atterberg limits, shear strength tests) and field test (Standard Penetration Test (SPT), Cone Penetration Test (CPT), borehole drilling) are done to calculate the engineering properties of soil. This is very slow and costly; with AI in the role of site investigation, it gives an innovative solution to overcome these challenges for an accurate, rapid, and scalable solution (Nyokum & Tamut, 2025; Baghbani et al., 2022). These are given in Figure 14.2.

AI is a transformative tool in the field of site investigation by improving its prediction capability, integration, automation, and efficiency at every level. Popular models like ANNs, RFs, CNNs, and SVMs can predict soil parameters accurately even with smaller datasets. This reduces dependency for larger datasets from laboratory and field reports. With the integration of geospatial and remote sensing techniques, including UAV and satellite imagery, AI will correlate terrain, vegetation cover, and moisture distribution on soil maps. ML technique enables better and automated soil classification into clayey, sandy, silty, loam, and gravel types. AI can establish a relationship between laboratory test results (e.g., triaxial compression, consolidation) and in-situ test (SPT, CPT)

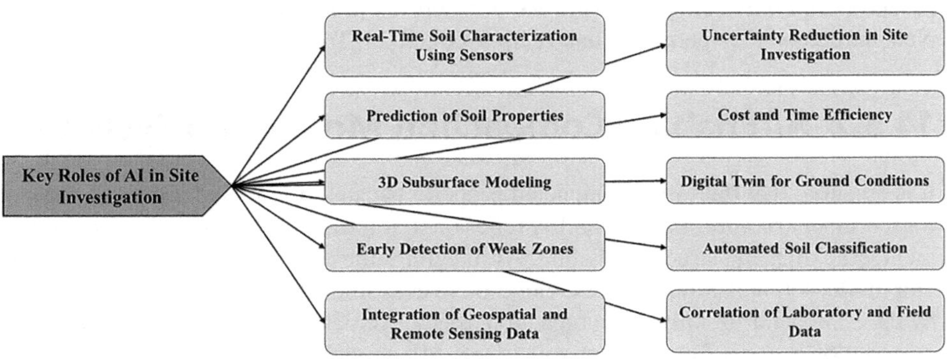

FIGURE 14.2 Roles of AI in site investigation.

results. Upon integrating AI with advanced geospatial techniques, it creates a 3D subsurface modelling, resulting in an accurate prediction of soil profiles. It can train models on geotechnical databases with the aim of detecting weak zones such as critical slopes, presence of swelling soil, and liquefiable sands. AI significantly reduces cost and time while improving efficiency, safety, and reliability (Wang et al., 2023).

14.5 AI IN EXECUTION AND MONITORING

Construction and execution of the earthwork project included the implementation of the designed plan on-site, by means of excavation, grading, embankment filling, compaction, hauling, and monitoring. In past, this process was very much dependent on manual labour, supervision, checks and measurements, which are vulnerable to errors, time-consuming, and had a limited range. AI is revolutionizing with automated, real-time, vast coverage, predictive analysis, and improved safety on the field of earthwork.

14.5.1 Smart Earthmoving Equipment

With development, modern construction machinery for Excavation and Digging – Excavators, Backhoe Loaders, Draglines, Trenchers, for Loading and Hauling Machines – Wheel Loaders, Bulldozers, Dump Trucks, Scrapers for Compaction Machines – Drum Rollers, Plate Compactors, Rammers for Grading and Levelling Machines – Motor Graders, Track-Type Tractors, and other machines like Hydraulic Hammers, Pile Drivers, and Scraper Pan Machines are equipped with AI-assisted capabilities. These heavy machines utilized various other tools such as GPS, Global Navigation Satellite System (GNSS), LiDAR sensors, accelerometers, and onboard AI algorithms for automated task completion. This helps in auto-grading and levelling of soil mass as per plan, and reduces deviation from the predefined path. It also prevents over-excavation and under-compaction due to different soil types, moisture content, and compaction requirements. Real-time data is provided to the operator for supervision, which increases the precision and uniformity of earthwork. It improves productivity by reducing the complete burden on the operator (Yuan & Moayedi, 2020; Tran et al., 2021).

14.5.2 AI-Enabled Compaction Monitoring Systems

It is essential that the soil is compacted properly because it will hold a solid foundation for the superstructure, and its stability relies on it. Improper compaction results in settlement, cracking, and structural failure. AI integrates the data from sensors to rollers, like amplitude, soil moisture, stiffness, vibration to determine the number of passes required by the compactor for uniform compaction, predict vehicle speed, pressure required, and compaction pattern for different soil types. ML models constantly learn from previous compaction data to improve prediction. It maintains uniform compaction, which improves long-term structural performance.

14.5.3 Computer Vision for Real-Time Earthwork Quality Monitoring

AI uses visions from devices such as cameras, sensors, and images to monitor and map earthwork operations. It is used to detect deviation from the original plan, identify uneven placement, loose soil, over-excavated areas, and monitor safety protocol. Modern DL algorithms like CNNs, RNNs, Long Short-Term Memory (LSTM), Gated Recurrent Units (GRU), Generative Adversarial Networks (GANs), and Graph Neural Networks (GNNs) can easily classify soil type, identify hazards, and assess compaction quality based on images. Main benefits include reduced manual inspection and surveys, giving immediate feedback, and aiding documentation.

14.5.4 UAV-Based Volume Estimation

Modern UAVs are equipped with high-resolution cameras (RGB Images), LiDAR sensors, and multispectral imaging that can capture vast spatial data across the construction site. AI processes this data to develop 3D terrain models and Digital Elevation Models (DEMs), where one can easily calculate cut-and-fill volume accurately, enabling a balance in earthwork. It can track the development of a project, compare actual and planned design, detect erosion, bottlenecks, and waterlogging on-site. UAVs improve resource allocation, tracking, mapping, large-scale data collection, and efficient risk management.

14.5.5 Integration of Devices

The actual power of AI lies in the integration of different equipment like UAVs, machinery, equipment, sensors, and software. Data from the smart machines and computer vision are fed into a central AI-based decision support system that predicts site condition, vulnerable zones, settlement condition, adjusts earthwork operations, equipment deployment, workforce allocation, and optimizes scheduling (Parente et al., 2015).

14.5.6 Overall Benefits

AI improves construction accuracy by precise classification of soil, compaction, and excavation, while improving effectiveness and productivity. It supports real-time decision, which will give actionable insight for better management and risk mitigation. It will reduce exposure to hazardous construction zones, minimize fuel, labour, rework, and material waste. AI limits the impact of earthwork on the environment by reducing soil movement and energy consumption. It will identify and predict equipment failures before they occur, which reduces their downtime. AI can easily monitor earthwork quality to ensure better quality of work. It can allocate machinery, materials, and workforce efficiently to reduce idle time (Villar et al., 2020).

14.6 CHALLENGES AND LIMITATIONS

In the current scenario, AI is being adopted in earthwork projects to enhance efficiency, accuracy, and decision-making. However, its implementation is still challenging. Factors such as data availability, soil complex nature, computational requirements, and resistance to adoption must be addressed. It is necessary to understand these challenges and identify practical methods to overcome them. Table 14.3 summarizes the key challenges of AI in earthwork, their impacts, fields of application, and strategies for mitigation (Parente et al., 2016; Correia et al., 2015).

TABLE 14.3 Challenges in integration of AI

CHALLENGE	*IMPACT*	*APPLICATION*	*SOLUTIONS*
Computational constraints	Slow predictions, inefficiency	Real-time monitoring, UAV data analysis	Optimize models, use cloud computing, deploy lightweight AI
Data quality and availability	Delays, inaccurate predictions	Soil testing, site investigation	Use of sensors, UAVs, preprocess and augment data, synthetic data, and transfer learning
Dependence on experts	Limited automation	AI-assisted design and monitoring	Human-in-the-loop systems, continuous feedback and learning
Environmental variability	Reduced prediction accuracy	Earthworks in changing climates	Update models with real-time data, adaptive AI algorithms
High cost	Budget issues, limited adoption	Construction monitoring, earthmoving	Pilot projects, open-source AI tools, staff training, stage implementation
Lack of standardization	Inconsistent results, errors and output	Site monitoring, project management	Follow best recommended practices, development of internal validation protocols
Limited interpretability	Difficulty in decision validation	Design optimization, soil prediction	Use of explainable AI techniques
Multi-source data integration	Inefficient decision-making	GIS, UAV imagery, sensor data	Data fusion, cloud computing, standardization of datasets
Regulatory and legal issues	Liability, compliance risk	Construction and infrastructure projects	Maintain proper documentation, combine AI predictions with regulatory standards

(*Continued*)

TABLE 14.3 (*Continued*) Challenges in integration of AI

CHALLENGE	*IMPACT*	*APPLICATION*	*SOLUTIONS*
Resistance to adoption	Low acceptance by engineers, labourers, government	General AI implementation	Training workshops, pilot demonstrations, involve end-users in development, awareness campaign
Sensor and equipment failures	Unreliable outputs, data gaps	IoT-based monitoring system, machinery tracking	Redundant sensors, regular maintenance, use of error detection algorithms
Soil complexity	Poor modelling of soil behaviour	Compaction, grading, slope stability	Hybrid AI-geotechnical models, deep learning, expert opinion

14.7 CASE STUDY ON THE IMPLEMENTATION OF AI IN EARTHWORKS

14.7.1 Implementation of AI-Powered Grading Optimization in a 65-Acre Land by Haskell Civil Engineers, USA

Haskell Civil Engineers is a U.S.-based construction and infrastructure development company. It undertook earthwork of 65 acres of land for earthwork management. The project aimed to deal with the task of cut-and-fill volumes, involving the handling and distribution of nearly 199,000 cubic metres of soil. Traditional workflows relied on manual trial-and-error mechanism limited to small underdeveloped software like CAD, leading to costly design, extension of budgets, and huge environmental impacts due to soil and air pollution.

To tackle these challenges, Haskell adopted an AI-powered grading optimization tool embedded in Autodesk Civil 3D, with the aim to integrate AI into the site design process. It was used for technical implementation, cost-efficiency, environmental performance, and regulatory implications of AI in earthwork.

14.7.1.1 Challenges

To grade 199,000 cubic metres of soil mass, the main challenge was to attain a balance between deadline and budget. Key difficulties were:

- Managing a large volume of earthwork and maintaining a balance between cut and fill.
- Time constraint: manual grading takes a lot of trials to achieve the desired results, which delays the timeline of construction activities.

- Risk of errors: minor deviation could lead to costly redesigning, improper drainage, and slope instability.
- Environmental imbalance: an imbalance in earthwork led to excess removal of soil, resulting in fuel usage and carbon emission.
- Regulatory compliance: the project has to meet the standards, rules, and guidelines of the federal body regarding issues of erosion control, water management, and sustainability as per the U.S. Environmental Protection Agency (EPA) and local municipal planning departments.

14.7.1.2 Implementation process

14.7.1.2.1 *Integration with Autodesk Civil 3D*
For effective grading and handling of soil mass, the firm used Autodesk Civil 3D's AI-powered grading optimization module. It involves importing various data, like topographic and geospatial, from survey and LiDAR datasets. It defines project parameters like building elevation, alignment, slope, and surface flatness. It aids in multiple simulations, which will lead to automatic adjustment of surface gradient to achieve minimal earth imbalance, utilizing ML algorithms to evaluate variations in terrain, optimize surface, and balance cut-and-fill volumes with high accuracy.

14.7.1.2.2 *Design based on data*
The system aids in real-time collaboration between civil engineers, planners, managers, and environmentalists. AI-based model formulates data dashboards which display effectiveness, cut-and-fill volumes, cost implications, and carbon footprint for individual design case scenario. This helps the team to easily compare different grading options. Make evidence-based decisions and develop an effective cost-benefit analysis for clients.

14.7.1.2.3 *Implementation deadline*
With the use of AI-based optimization software, it reduces the project construction time by 40%–50%, aiding in faster development and construction readiness. In practical case scenario, the project timeline to complete a project is 4–5 weeks; with AI it can be accomplished within 2 weeks.

14.7.1.3 Economic analysis and cost implications

14.7.1.3.1 *Reduced operational costs*
With the help of AI, it reduces the unwanted movement of materials, which gives significant financial benefits. It minimizes the hauling costs (fuel, equipment operation, and labour). It reduces import and export of materials which cuts external sourcing expenses. It also lowers the maintenance costs for machinery due to reduced operating hours. It is estimated that the AI-optimized grading saves nearly 8%–10% in total earthwork costs.

14.7.1.3.2 *High design efficiency*
Optimization based on data-driven and AI techniques helps in better design efficiency and helps in dealing with client negotiations. Engineers from Haskell demonstrated better return on investment using AI tools. They enhanced design-related issues, providing

real-time simulation modelling which supports cost-effective decisions during construction planning. They improved accuracy and reduced rework during construction.

14.7.1.4 Environmental and sustainability benefits

14.7.1.4.1 *Minimization of material transport*
With the application of AI in the role of earthwork movement, there are many key benefits related to environmental parameters. By precise cut-and-fill balance the need for importing or exporting earth was drastically reduced. This led to a reduction of 20%–22% of hauling distance, cutting off fuel use and emissions. This lowered the fuel consumption and carbon dioxide (CO_2) emissions. This technique also reduces the noise pollution and road wear caused by truck movement.

14.7.1.4.2 *Reduced carbon footprint*
Integration of AI techniques in optimization resulted in a decline of nearly 40–50 tons of CO_2 emissions compared to traditional methods. The sustainable outcome supported Haskell's internal corporate social responsibility (CSR) goals along with the U.S. Green Building Council's LEED certification framework for eco-friendly site design.

14.7.1.4.3 *Soil and water management*
AI aids in maintaining natural drainage patterns, reducing the need for additional rainwater infrastructure and erosion control mechanisms. Smoother terrain provides surface stability and low sedimentation risk in nearby water bodies, ensuring compliance with EPA stormwater regulations (NPDES Phase II).

14.7.1.5 Governmental and regulatory alignment

The developed project was formulated in such a way that it aligns with local and federal standards, like:

- **EPA Guidelines:** Optimized grading reduces the risk of erosion and surface water runoff, which helps in meeting environmental management requirements.
- **Municipal Codes:** AI-based software provides a robust and detailed planning map and procedure and record, simplifying submission for planning and zoning approvals.
- **Sustainability Incentives:** With reduced and low environmental impact from the project, it is eligible for state-level sustainability credits or green infrastructure incentives.
- **Occupational Safety and Health (OSHA):** With low earthmoving activity and reduced on-site heavy equipment exposure, it lowers the risk for workers.

With the alignment of these standards, Haskell showcased its potential in technology, streamlining both the design approval and construction permitting processes.

TABLE 14.4 Key outcomes of AI-optimized earthwork approaches

PARAMETER	*TRADITIONAL METHOD*	*AI-OPTIMIZED APPROACH*	*IMPROVEMENT*
Time duration	4–5 weeks	<2 weeks	40%–50% faster
Earthwork imbalance	5%–10%	<1%	High precision
Haul distance	100% baseline	75%–80% baseline	20%–25% reduction
Fuel consumption	100% baseline	70%–75% baseline	25%–30% savings
CO_2 emissions	100% baseline	60%–70% baseline	30%–40% lower
Cost	100% baseline	90%–92% baseline	8%–10% cost savings

14.7.1.6 Outcomes

The major outcomes of the earthwork by Haskell are given in Table 14.4.

14.7.1.7 Conclusion

Haskell Civil Engineers' success demonstrated the role of AI-driven tools in civil infrastructure planning as a more efficient, transparent, and sustainable process. The firm successfully balanced 260,000 cubic yards of earthwork across a 65-acre site while reducing environmental impacts, costs, and time. It demonstrates a software-based approach, like CAD platforms, that enhances precision without overhauls of projects. In many civil engineering projects sustainability and profitability are not mutually exclusive, intelligent grading reduces both environmental and economic costs. The developed mechanism can be scaled for other projects for different infrastructures, residential, and industrial developments to meet carbon neutrality and smart city goals.

14.7.2 Case Study: Earthwork Optimization for Sustainable Highway Construction in Portugal

Construction of highways is essential for the development of the nation; it is one of the most resource-intensive activities. It involves extensive excavation, transportation, and movement of a large volume of soil mass. In Portugal, an infrastructure project was formulated with the incorporation of intelligent earthwork optimization system to tackle these challenges. They integrated AI, Data Mining, and GIS technologies to develop effective, innovative, adaptive, and eco-friendly solutions for earthwork management. This case showcases a milestone in the application of digital and intelligent technology in cost management, operational productivity, and environmental sustainability in transportation engineering.

14.7.2.1 Challenges

Portugal's national highway expansion programme was proposed to improve transportation connectivity. With traditional methods, the management of earthwork in huge volumes led to several challenges like:

- **Manual Design Limitations:** Traditional designs were generally based on manual labours and engineers for all calculation such as cut-and-fill balance, haul routes, and material reallocation, often leading to inefficiencies and human error.
- **Resource Misallocation:** Due to human factors, material and equipment were not effectively distributed in construction sites. This leads to idle machines, unnecessary fuel consumption, improper time management, and ineffective workload distribution. Excavated soil and aggregates are generally reusable, which were transported off-site instead of being efficiently reused within the project area.
- **Environmental Impact:** Excessive movement of soil mass leads to air dust, consumption of fuel, emission of greenhouse gas, and land degradation.
- **Extended project duration and cost:** Unorganized logistics and management have led to unplanned earth movement, which results in prolonged construction time and inflated operational expenses.
- **Data Fragmentation:** Inconsistent, unmanaged, and isolated datasets for massive projects make it difficult to read, understand, and interpret the data, which leads to poor coordination between designers, contractors, and environmental regulators.
- **Lack of Predictive Planning:** It is observed that traditional models can only forecast static data, not dynamic data. The project timeline can be affected by dynamic variabilities like weather, difficult terrain, human factor, instrument failure, and machine performance in the field.

The challenges pave the way for the need for a more intelligent, efficient, integrated, and dynamic system, which is capable of optimizing earthwork.

14.7.2.2 Technological innovation

Portuguese highway project implemented a hybrid technology known as a multi-layered digital framework, which combines AI algorithms and GIS-based spatial analysis tools. The integrated system performed the following core functions:

- **AI-Powered Optimization:** ML-based optimization algorithm analysed terrain, soil characteristics, and design specifications to reduce hauling distance, fuel utilization, and equipment downtime. An AI model automatically detects the most efficient cut-and-fill balance and equipment deployment for more productivity.
- **Data Mining and Predictive Modelling:** For cut and fill, data engineers study historical data before planning infrastructure projects, in addition to that, data from real-time sensors was used to forecast equipment productivity, downtime, and fuel efficiency. Predictive models aid in scheduling, resource allocation, and completion of project within time.
- **GIS-Based Route Planning:** For efficient route planning, GIS technique was used for topography mapping, soil stratigraphy, and access roads to optimize material flow paths. The model optimizes and identifies shortest and conservative routes.

- **Data Integration:** Integration of GIS and AI tools in construction machinery equipment gives data regarding operational details (e.g., speed, location, load weight) in a centralized system which can be assessed easily. It can also be used to assess work schedules, machine deployment, and haul routes in response to terrain or weather disruptions.

14.7.2.3 Implementation process

- **Data Acquisition and Model Setup:** LiDAR and drones were used for high-resolution topographic data. Data of soil characteristics was used to integrate with GIS data to give geotechnical properties with spatial location. The developed model was trained on historical datasets including haul times, fuel usage, and machine efficiency. A multi-objective model was developed to establish a balance between cost reduction, environmental footprint, and operational efficiency.
- **Real-Time Operational Framework:** Within the duration of the project, data from the equipment and along with GPS was analysed continuously in real time. AI model monitored the development and progress of the earthwork, fuel and energy consumption, machine idle and downtime, and its route deviations. The data provided the best optimization and recalibration of best path for every loop which adjusted the scheduling to align with weather forecast and soil moisture conditions.
- **Stakeholder Collaboration:** AI model provides a digital view of the data in single platform. It aids in multi-disciplinary collaboration among civil and environmental design engineers, managers, and field operators. It helps in easy government inspections and audits. Each member can view all the data in the dashboard with customizable reports with improved transparency and traceability.

14.7.2.4 Key outcomes

The main improvements of the incorporation of AI in earthwork are summarized in Table 14.5.

The Portuguese case showcases a transformative potential of AI and GIS integration in earthwork management, which depicts data-driven automation for better decision-making and resource utilization. Equipped with AI-assisted system, it fosters policy innovation by acting as a bridge between engineering and biodiversity preservation goals.

14.7.2.5 Findings

The implementation of an intelligent earthwork management tool in Portugal's highway project led to an advancement in civil construction projects. By using AI data and GIS technology, the project achieved substantial gains in cost-efficiency, operational precision, and environmental responsibility. It reduces time by 50%, costs up to 40%, reduces

TABLE 14.5 Improvement with incorporation of AI model

PARAMETER	*TRADITIONAL METHOD*	*AI TECHNIQUE*	*IMPROVEMENT*
Construction cost	100%	50%–70%	30%–50% cost savings
Project duration	100%	40%–50%	50%–60% faster completion
Fuel consumption	100%	50%–60%	40%–50% reduction
CO_2 emissions	100%	40%–60%	40%–60% reduction
Material reuse	40%–50%	80%–90%	2× improvement
Equipment utilization	65%–70%	90%–95%	Enhanced operational efficiency
Design revision frequency	Frequent (manual recalculation)	Minimal (AI automated)	Over 70% reduction in design rework
Regulatory reporting time	100%	50%	50% faster environmental compliance reporting
Stakeholder communication delay	3–5 days average	Real time	Seamless coordination

fuel consumption by 45%, and improves equipment utilization. This case study confirms that intelligent, adaptive, and data-integrated systems will aid in development of green infrastructure which will bridge the gap between sustainable aspect of economic and environmental stewardship.

14.8 CONCLUSION

AI is rapidly transforming the domain of earthwork in civil engineering, giving unparalleled advantage in efficiency, safety, decision-making, monitoring, and sustainability. The chapter explores the advantages of the integration of AI in every aspect of earthwork, from planning, site investigation, design, construction, risk, and environmental impacts. Traditional methods have limitations such as human errors, dependency on labour, handling large and complex datasets, environmental mitigation, and climate-driven risks. AI addresses the problem very efficiently, by incorporating ML, DL, optimization algorithms, computer vision, digital twin technologies, and predictive analysis in earthworks. Application of AI in soil classification, slope investigation, soil stabilization, cut-and-fill balance, resource optimization, and equipment management has showcased high accuracy, reduced delays, and enhanced project safety. Upon integration of AI with other technologies like GIS, remote sensing, GPS, Building Information Modelling (BIM), IoT, and edge computing, it has potential in decisive planning and real-time solutions. These tools not only optimize efficiency but also aid in automated AI-driven earthwork. In the current scenario, AI in earthwork faces a lot of challenges like data quality, availability, high cost, training, skilled workers, computers, advanced sensors, safety consideration, data privacy, software utilization, and acceptability among labourers and ethical issues. By embracing these situations, civil engineers can achieve sustainable, efficient, and resilient earthwork practices.

REFERENCES

Abolhasani, S., Frey, H. C., Kim, K., Rasdorf, W., Lewis, P., & Pang, S. H. (2008). Real-world in-use activity, fuel use, and emissions for nonroad construction vehicles: A case study for excavators. *Journal of the Air & Waste Management Association*, *58*(8), 1033–1046. https://doi.org/10.3155/1047-3289.58.8.1033

Awais, M., Naqvi, S. M. Z. A., Zhang, H., Li, L., Zhang, W., Awwad, F. A., … & Hu, J. (2023). AI and machine learning for soil analysis: An assessment of sustainable agricultural practices. *Bioresources and Bioprocessing*, *10*(1), 90. https://doi.org/10.1186/s40643-023-00710-y

Baghbani, A., Choudhury, T., Costa, S., & Reiner, J. (2022). Application of artificial intelligence in geotechnical engineering: A state-of-the-art review. *Earth-Science Reviews*, *228*, 103991. https://doi.org/10.1016/j.earscirev.2022.103991

Chae, M. J., Lee, G. W., Kim, J. Y., Park, J. W., & Cho, M. Y. (2011). A 3D surface modeling system for intelligent excavation system. *Automation in Construction*, *20*(7), 808–817. https://doi.org/10.1016/j.autcon.2011.02.003

Correia, A. G., Parente, M., & Cortez, P. (2015). *Earthwork Optimization System for Sustainable Highway Construction*. Universidade do Minho. https://hdl.handle.net/1822/39061

Ersoz, A. B., & Pekcan, O. (2025). UAV-based automated earthwork progress monitoring using deep learning with image inpainting. *Automation in Construction*, *175*, 106211. https://doi.org/10.1016/j.autcon.2025.106211

Fernandes, P. G., Nobre Júnior, E. F., & Prata, B. D. A. (2022). Optimization of earthworks planning: A systematic mapping study. *Canadian Journal of Civil Engineering*, *49*(12), 1781–1795. https://doi.org/10.1139/cjce-2022-0185

Hare, W. L., Koch, V. R., & Lucet, Y. (2011). Models and algorithms to improve earthwork operations in road design using mixed integer linear programming. *European Journal of Operational Research*, *215*(2), 470–480. https://doi.org/10.1016/j.ejor.2011.06.011

Henderson, D., Vaughan, D. E., Jacobson, S. H., Wakefield, R. R., & Sewell, E. C. (2003). Solving the shortest route cut and fill problem using simulated annealing. *European Journal of Operational Research*, *145*(1), 72–84. https://doi.org/10.1016/S0377-2217(02)00206-0

Huang, L., Zuo, S., Yu, B., & Chen, S. (2024). Optimization of the earthwork excavation-filling balance and allocation for the upper reservoir of a pumped storage power station. *Journal of Energy Storage*, *83*, 110722. https://doi.org/10.1016/j.est.2024.110722

Karimi, S. M., Mousavi, S. J., Kaveh, A., & Afshar, A. (2007). Fuzzy optimization model for earthwork allocations with imprecise parameters. *Journal of Construction Engineering and Management*, *133*(2), 181–190. https://doi.org/10.1061/(ASCE)0733-9364(2007)133:2(181)

Nyokum, T., & Tamut, Y. (2025). Artificial intelligence in civil engineering: Emerging applications and opportunities. *Frontiers in Built Environment*, *11*, 1622873. https://doi.org/10.3389/fbuil.2025.1622873

Omar, T., & Nehdi, M. L. (2016). Data acquisition technologies for construction progress tracking. *Automation in Construction*, *70*, 143–155. https://doi.org/10.1016/j.autcon.2016.06.016

Parente, M., Cortez, P., & Correia, A. G. (2015). An evolutionary multi-objective optimization system for earthworks. *Expert Systems with Applications*, *42*(19), 6674–6685. https://doi.org/10.1016/j.eswa.2015.04.051

Parente, M., Correia, A. G., & Cortez, P. (2016). A novel integrated optimization system for earthwork tasks. *Transportation Research Procedia*, *14*, 3601–3610. https://doi.org/10.1016/j.trpro.2016.05.428

Rao, A. S., Radanovic, M., Liu, Y., Hu, S., Fang, Y., Khoshelham, K., … & Ngo, T. (2022). Real-time monitoring of construction sites: Sensors, methods, and applications. *Automation in Construction*, *136*, 104099. https://doi.org/10.1016/j.autcon.2021.104099

Rogage, K., Mahamedi, E., & Brilakis, I. (2022). Beyond digital shadows: A digital twin for monitoring earthwork operations in large infrastructure projects. *AI in Civil Engineering*, *1*, 7. https://doi.org/10.1007/s43503-022-00009-5

Roy, A., McCabe, B. Y., Saxe, S., & Posen, I. D. (2024). Review of factors affecting earthworks greenhouse gas emissions and fuel use. *Renewable and Sustainable Energy Reviews*, *194*, 114290. https://doi.org/10.1016/j.rser.2024.114290

Teizer, J., Neve, H., Li, H., Wandahl, S., König, J., Ochner, B., … & Lerche, J. (2020). Construction resource efficiency improvement by Long Range Wide Area Network tracking and monitoring. *Automation in Construction*, *116*, 103245. https://doi.org/10.1016/j.autcon.2020.103245

Teodosio, B., Wasantha, P. L. P., & Yaghoubi, E. (2025). Application of artificial intelligence in reactive soil research: A scientometric analysis. *Geotechnical and Geological Engineering*, *43*, 145. https://doi.org/10.1007/s10706-025-03097-z

Tran, K. M., Bui, H. H., & Nguyen, G. D. (2021). DEM modelling of unsaturated seepage flows through porous media. *Computational Particle Mechanics*, *8*(6), 1223–1238. https://doi.org/10.1007/s40571-021-00398-x

Villar, Y., Menéndez, M., Fernández-Muñiz, Z., & Bernardo, A. (2020). Sustainable earthworks: Optimization with the ICOM method. *Energy Reports*, *6*, 404–419. https://doi.org/10.1016/j.egyr.2020.08.060

Wang, L., Wu, C., Yang, Z., & Wang, L. (2023). Deep learning methods for time-dependent reliability analysis of reservoir slopes in spatially variable soils. *Computers and Geotechnics*, *159*, 105413. https://doi.org/10.1016/j.compgeo.2023.105413

Yang, J., Arif, O., Vela, P. A., Teizer, J., & Shi, Z. (2010). Tracking multiple workers on construction sites using video cameras. *Advanced Engineering Informatics*, *24*(4), 428–434. https://doi.org/10.1016/j.aei.2010.06.008

Yuan, C., & Moayedi, H. (2020). The performance of six neural-evolutionary classification techniques combined with multilayer perception in two-layered cohesive slope stability analysis and failure recognition. *Engineering with Computers*, *36*, 1705–1714. https://doi.org/10.1007/s00366-019-00791-4

Zeraatpisheh, M., Ayoubi, S., & Jafari, A. (2019). Digital mapping of soil properties using multiple machine learning methods in a semi-arid region, central Iran. *Geoderma*, *338*, 445–452. https://doi.org/10.1016/j.geoderma.2018.09.006

Zhou, S., & Gheisari, M. (2018). Unmanned aerial system applications in construction: A systematic review. *Construction Innovation*, *18*(4), 453–468. https://doi.org/10.1108/CI-02-2018-0010

15 Earthquakes in the Age of AI
Redefining Risk Resilience and Responsibility

Shima Al-Balushi, Anas Ansari, and Abdullah Ansari

15.1 INTRODUCTION

Earthquakes show the resilience of structures and human life. It is a constant threat to the residents and the infrastructure. Technology advances rapidly and this change allows artificial intelligence (AI) to provide important means for examining data relating to earthquakes and for reducing threats that occur from such events. AI provides methods that allow researchers to examine large datasets that show complex patterns across time and this approach reveals features in data that differ from patterns that traditional methods can identify. Using machine learning approaches and methods that use networks with multiple levels, researchers demonstrate success in identifying earthquakes and in detecting significant activity in seismic data. These methods also increase the rate of analysis and the level of performance in developing measures of seismic threats. The main contribution involves improving the capacity of cities to respond to events by developing plans for situations that require immediate action. AI also provides support for locating individuals, for conducting operations to provide assistance, and for organizing relief activities. This approach suggests direction for planning in cities that increases protection and that supports continued function over time. The current

DOI: 10.1201/9781003714125-15

work examines main uses in areas that include predicting events, following activity in real time, and estimating effects that result from earthquakes. The analysis also considers actual cases that show that these approaches can function in reducing threats from earthquakes and in providing protection for individuals and for resources.

15.2 TECHNOLOGICAL ADVANCEMENT

AI provides a new approach that supports work in the area of examining seismic activity and in improving the capacity of cities to respond to earthquakes. Traditional methods for analysis show increasing difficulty in providing the rate of processing, the level of precision, and the capacity to identify patterns that show relationships that are not direct in form. AI, using methods that involve learning from data and approaches that use structures with multiple levels of processing, addresses these issues by learning from datasets that contain large amounts of information, by identifying patterns that are not immediately visible, and by improving systems that follow seismic activity and that conduct analysis. This capacity appears in uses that range from detecting earthquakes and organizing seismic signals into categories and identifying different types of activity to estimating measures of movement in the ground and developing systems that provide warnings before events occur. Convolutional Neural Networks (CNNs) that use structures for processing spatial patterns can identify weak signals in seismic data that traditional approaches fail to detect. Recurrent Neural Networks (RNNs) that process information across time show high performance in examining data that vary over time to predict activity relating to seismic events.

Using this foundation, Atanasiu et al. (2008) and other researchers in the study demonstrated the importance of AI in reducing threats from earthquakes in settings in cities that show complex features. In these cities, traditional approaches show reduced effectiveness because of the presence of multiple factors that affect results from earthquakes, including conditions of structures that provide services, features of geological formation, and characteristics of populations that reside in these areas. By using decision tree and neural network algorithms, AI can integrate structural, spatial, and socioeconomic data to obtain accurate hazard maps and identify earthquake-prone areas to guide urban planning and mitigation strategies, such as retrofit of vulnerable buildings, enhancing emergency plans, and allocating resources for emergency response and rescue in a proactive way.

In addition to the above developments, Tang et al. (2022) developed an AI-oriented assessment framework that greatly reduces computational loads while maintaining high accuracy. The framework achieved 96% classification accuracy and extremely low regression errors. Their results demonstrated the practicality of AI in providing rapid and reliable assessments, particularly for steel structures.

Interestingly, they extended the model to other types of buildings as well as other hazard conditions to validate its usefulness for disaster preparedness and risk reduction actions. Similarly, Katsuumi et al. (2024) extended the scope of the application to an AI model that not only predicts seismic risks but also purifies the soil in the cities of urban and agricultural cities. Their model, which uses advanced machine learning and large databases, generates extremely accurate interpreted maps that identify earthquake-prone areas, as well as flood-prone areas, to ensure urban safety, water management, and agricultural city planning. In total, the above studies show that the application of AI in seismology and urban risk management is a powerful and data-driven method for disaster mitigation. Through improved hazard detection, risk mapping, rapid damage assessment, and soil risk management, AI can significantly contribute to the creation of safer, more resilient, and sustainable cities.

15.3 EARTHQUAKE PREDICTION AND MONITORING

Recent research highlights in the area of AI have reported new advances in the research of earthquake prediction and monitoring due to the limitations of traditional seismic methods. In more than 100 studies on earthquake prediction using AI, Al Banna et al. (2020) summarized the difficulty in prediction of earthquakes due to the complex and nonlinear nature of earthquakes which is limited by the traditional methods. These methods were based on data such as historical earthquake records, geophysical records, satellite images, and Global Positioning System (GPS) signals. These data were pre-processed using normalization, filtering, and feature selection. The main factors were magnitude, depth, timing, and location. Results showed that deep learning and hybrid models can detect these hidden patterns that traditional models might miss, but performance is strongly dependent on data quality and local seismic variations. The authors suggested that the combination of real-time data, regionally applicable models, and combinations of AI and geophysical methods should be used to improve early warning systems and reduce global risk. Building on the above studies, Ahamad and Singh (2025) focused on real-time seismic monitoring and risk assessment.

The review considered 20 recent studies with applications in high-rise buildings, underground transport, and energy infrastructure areas in fast-growing cities prone to disasters. The methods included CNN–Long Short-Term Memory (LSTM) hybrid models, fuzzy logic, ensemble learning, and expert-based probabilistic systems. One of the main contributions of this study was the combination of AI with Internet of Things (IoT) sensor networks and earthquake early warning systems to provide immediate awareness and rapid emergency response. The data used came from real seismic recordings, satellite images, microseismic data, geotechnical simulations, and historical records. These data were pre-processed using normalization, removal of noise, time-series transformation, and supervised learning. Results showed improvements in event prediction, estimation of structural damage, and mapping of hazards, but challenges still existed in terms of model interpretability, regional transferability, computational cost, and weak integration with national emergency systems. The study called for the adoption of explainable AI, edge computing, and standardized protocols. It also called for closer collaboration between data scientists, engineers, and policymakers to develop transparent and practical systems that enhance resilience.

15.4 DAMAGE ASSESSMENT

Earthquake damage assessment can be accomplished by developing site-specific algorithms. Morfidis et al. (2023) developed the Rapid Seismic Damage Assessment (RASDA) tool, an instrument to rapidly and efficiently assess damage of reinforced concrete buildings using artificial neural networks (ANNs). The RASDA tool allows for direct damage evaluation immediately after an earthquake without demanding immediate field inspections to make prompt safety decisions. Developed in a densely populated European urban environment, RASDA pre-inspected the building characteristics, the earthquake intensity, and the soil type, which were collected and processed by multilayer feedforward perceptron neural networks to classify the damage in minor, moderate, and severe damage categories within seconds, showing results comparable to conventional engineering assessments. The above results corroborate the recent global earthquakes' experiences. Apparently, the existing constructions are highly vulnerable and need to be retrofitted.

Moreover, the data and the field evaluations have shown that pre-inspection of the buildings and the prioritization of the most dangerous structures may minimize the potential damage and enhance the post-disaster response planning. Rapid visual inspections and the modern assessment techniques complement RASDA by identifying the vulnerable buildings, enabling more effective preventive measures, and resource concentration. In summary, the studies presented show that rapid assessment tools such as RASDA provide accurate base data, preventive inspections minimize the immediate risks, and comprehensive AI-based approaches improve predictions, planning, and response, altogether contributing to the global infrastructure resilience.

15.5 CASE STUDIES

15.5.1 Indonesia

Jena et al. (2020) studied to assess earthquake risk in Palu, Indonesia, which is one of the areas that are more prone to seismic activity, especially after the destructive earthquake and tsunami in 2018. This study aimed to identify the most vulnerable areas by constructing predictive models that incorporate geological and human factors. This way it is hoped that a clearer picture of areas that may be hit by earthquakes in the future and the degree of impact on population and infrastructure can be obtained. Random forest (RF), support vector machine (SVM) and logistic regression (LR) models were used in this study. All types of data mentioned above have added an extra dimension that helped the models to judge physically the possibility of earthquakes alongside the possible social and economic effects. So, different spatial data processing techniques and normalization processes were applied in order to improve the models' performance. The results revealed that the RF model achieved better prediction than the other two models in terms of prediction and identification of high-risk areas. The results revealed that the central and western regions of Palu are in greatest danger because these areas are close to active faults and soft soil that enhance ground shaking.

The team used the population density and infrastructure information and hazard maps to produce fine-grained maps with the information on areas most at risk in terms of physical threat and danger to life. The current research demonstrates that AI can provide a significant role in developing plans for events that occur in nature in areas where traditional methods for evaluation show insufficient performance. The results that models produce and the visual representations that show spatial distributions can provide important resources for individuals in positions in government, for individuals who develop plans for cities, and for groups that respond to situations that require immediate action to improve preparation and response that occur rapidly and to increase understanding of areas where developing preparation and increasing capacity to respond show particular importance.

15.5.2 Barcelona, Spain

The SERGISAI approach, which stands for SEismic Risk evaluation through integrated use of Geographical Information Systems and Artificial Intelligence techniques, provides a computer-based procedure for seismic risk assessment in urban areas by integrating seismic hazard information, vulnerability data, Geographic Information Systems (GIS), and Artificial Intelligence (AI) techniques (Zonno et al. 2003). Approaches used in previous work to examine risk or vulnerability often examine these factors in separate analysis, and this means that relationships between factors that determine overall risk levels are not considered. SERGISAI combines these two components in a single procedure. The approach uses both data that measure features and data that describe features. Information on conditions that produce events includes data from records of previous events, studies of structures in the region, and measures of motion in the ground, and information on vulnerability includes data on types of structures, condition of structures, patterns of occupancy, and features of infrastructure in urban areas. This information is processed using systems that provide values and procedures that combine values to produce measures of risk for different urban areas. Individuals with experience play an important role in confirming and improving the results to ensure that the findings reflect understanding from local contexts and experience from practice. The approach shows a main feature in that it contains separate components, and this allows it to be adapted to different urban areas even when data differ in availability, and this makes it useful for the various decisions that individuals responsible for urban areas need to make.

The procedure also includes methods for presenting results visually, and these include tables and maps, and these methods provide a means to communicate findings to different groups that include individuals who develop urban plans, teams that respond to events, and individuals who develop policy. Results indicate that SERGISAI provides an effective method for identifying areas with high risk and allows individuals in authority to identify where they need to provide treatment and allocate resources. The procedure that combines data on conditions that produce events and vulnerability in a systematic manner produces a representation of urban risk from events that shows more accuracy and reflects conditions more clearly than approaches used in previous work allow. The research also highlights the need to continuously update and incorporate new expertise to ensure assessments are accurate and relevant.

15.5.3 United States

Plevris (2024) stated the possible AI capacity to transform risk planning in six basic ways. Compared to other conventional ways, it considers the different possibilities of how AI-based innovations can make improvements in early warning methods, such as real-time monitoring of the structure's condition and dynamic approaches, which consider data streams with other potential hazards, together with landslides and tsunamis. Their work demonstrated the first application of AI in earthquake-resistant design in which generative design algorithms and mathematical analytics create structures that balance on the edge of safe, costly, and sustainable. The work also presented a new opinion on the ethical aspects of the use of AI in the event of failure; prompts the possibility of AI to draw the outline of a safer and more earthquake-resistant future; and considers the necessity of continuous technical knowledge. This study demonstrates some significant areas where AI-based innovations are making a significant impact. AI can be used in early fieldwork systems, which are faster and more accurate in predicting the coordinates of the earthquake, so that we can take precautions. AI can be used to work on the data collected from the sensors placed in the structures and to make faster and more accurate damage assessments.

There are some difficulties in using AI in this field, and to ensure that the AI tools are used in a sensible and equitable way, the ethical issues, such as accountability, transparency, and the possibility of bias, should be controlled sufficiently. This study attempts to develop a set of regulatory frameworks and guidelines that govern the regulation of the use of AI in earthquake risk mitigation. With these, we can use the benefits of AI in a way that minimizes the risks, and as a result, more effective and equitable disaster preparedness and response will be achieved as quickly as possible.

15.5.4 Jammu and Kashmir, India

The study by Ansari et al. (2023) develops a Multi-Hazard Damage Prediction (MhDP) model for tunnels exposed to earthquakes and landslides. Unlike single-hazard approaches, the model presented in this paper predicts combined damage from earthquake and landslides using ANNs more accurately in geologically and seismically active areas. The ANN was calibrated on geotechnical, geological, seismic, and landslide datasets and captured interactions between them. The input data for training the ANN model were field surveys, soil and rock properties, seismic intensity, and landslide susceptibility data. The results revealed that the MhDP model is more accurate than traditional methods and gives accurate damage forecasts. It also identifies the most important hazards. This tool helps the engineers to focus on the design, make, and prioritize the engineering structures safely while integrating AI into the geotechnical risk assessment for future infrastructure.

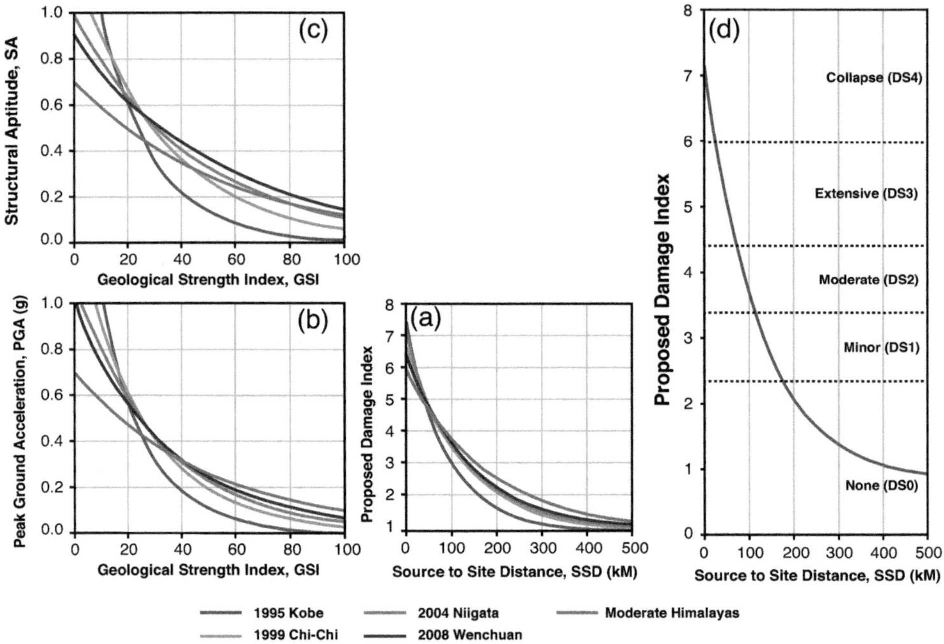

FIGURE 15.1 Comparative seismic response and tunnel damage trends for selected earthquake scenarios: (*a*) attenuation of the Proposed damage index with source-to-site distance (SSD), (*b*) variation of peak ground acceleration (PGA) with geological strength index (GSI), (*c*) variation of structural amplitude (SA) with GSI, and (*d*) damage-state classification based on Proposed Damage Index thresholds, ranging from no damage to collapse.

In addition to this, a deep learning Feedforward Neural Network model has been developed to predict tunnel damage and functionality after earthquakes (Ansari et al. 2023). The city of focus was Jammu and Kashmir, which is a high seismic region having several important tunnels. The model was trained on data from major earthquakes such as Chi-Chi, Mid-Niigata, Wenchuan, and some tunnel-specific parameters such as size, depth, rock strength, and design. It enhances the prediction accuracy, identifies the most effective damage-causing parameters, and also has visual outputs for rapid visual inspection after earthquakes (Figure 15.1). The applicability of the model has been shown worldwide and the application of AI to ensure tunnel safety and earthquake resilience has been demonstrated.

15.6 SUMMARY

These studies in combination illustrate how MhDP provides an integrated approach to addressing multi-hazard interactions, whereas the deep learning model discussed above is focused on earthquake-specific tunnel damage. Hence, it can be seen that an integrated strategy for tunnel safety in high-seismic-risk regions has been developed using these studies. The application of AI in earthquake engineering has shown potential as a revolutionary approach for reducing seismic risk and managing earthquake disasters. New studies have shown how AI and machine learning can be used effectively for seismic hazard prediction, structural damage assessment, and early warning systems. These studies have explored various AI techniques such as deep neural networks, convolutional networks, and feedforward neural networks for analyzing real-time earthquake data, predicting the damage to buildings, bridges, tunnels, and other structures, enhancing rapid visual inspection, and more.

Additionally, AI has also been combined with geological, geotechnical, and structural data to generate high-resolution hazard maps and assessments for aiding urban planning, earthquake early warning systems, emergency response, and infrastructure resilience.

Although AI models provide greater accuracy and speed than conventional methods, challenges still exist in terms of data availability, interpretability, regulatory standards, and ethical considerations. In summary, these advancements represent a significant step forward toward an intelligent, data-driven, and proactive approach to earthquake risk management—making cities smarter and safer.

REFERENCES

Ahamad, F., & Singh, V. K. (2025). AI-driven predictive modeling for real-time seismic activity monitoring and earthquake risk assessment. *International Journal of Scientific Research in Computer Science, Engineering and Information Technology,* 11(2), 3913–3921. https://doi.org/10.32628/CSEIT25112769

Al Banna, M. H., Taher, K. A., Kaiser, M. S., Mahmud, M., Rahman, M. S., Hosen, A. S., & Cho, G. H. (2020). Application of artificial intelligence in predicting earthquakes: state-of-the-art and future challenges. *IEEE Access*, 8, 192880–192923.

Ansari, A., Rao, K. S., Jain, A. K., & Ansari, A. (2023). Deep learning model for predicting tunnel damages and track serviceability under seismic environment. *Modeling Earth Systems and Environment*, 9(1), 1349–1368.

Atanasiu, G. M., Leon, F., & Popa, B. F. (2008). Seismic risk mitigation in urban areas based on artificial intelligence methods. In *Proceedings of the 14th World Conference on Earthquake Engineering,* Beijing, China, October 12–17, 2008.

Jena, R., Pradhan, B., Beydoun, G., Alamri, A. M., & Sofyan, H. (2020). Earthquake hazard and risk assessment using machine learning approaches at Palu, Indonesia. *Science of the Total Environment*, 749, 141582.

Katsuumi, A., Cong, Y., & Inazumi, S. (2024). AI-driven prediction and mapping of soil liquefaction risks for enhancing earthquake resilience in smart cities. *Smart Cities*, 7(4), 1836–1856.

Morfidis, K., Stefanidou, S., & Markogiannaki, O. (2023). A rapid seismic damage assessment (RASDA) tool for RC buildings based on an artificial intelligence algorithm. *Applied Sciences*, 13(8), 5100.

Plevris, V. (2024). AI-driven innovations in earthquake risk mitigation: a future-focused perspective. *Geosciences*, 14(9), 244.

Tang, Q., Dang, J., Cui, Y., Wang, X., & Jia, J. (2022). Machine learning-based fast seismic risk assessment of building structures. *Journal of Earthquake Engineering,* 26(15), 8041–8062. https://doi.org/10.1080/13632469.2021.1987354

Zonno, G., Garcia-Fernandez, M., Jimenez, M. J., Menoni, S., Meroni, F., & Petrini, V. (2003). The SERGISAI procedure for seismic risk assessment. *Journal of Seismology*, 7(2), 255–277.

16 Seismic Data Analysis for Tsunamigenic Earthquake Detection Using Soft-Computing Techniques

Tammineni Gnananandarao, C. H. Ajay, B. A. V. Ram Kumar, Hemanth Kumar Yerrabolu, and J. Y. V. Shiva Bhushan

16.1 INTRODUCTION

Coastal areas across the world are faced with various natural disasters, and tsunamis are some of the worst disasters. Tsunamis usually have minimal lead time; upon the arrival of the waves, its high kinetic energy may destroy whole coastal settlement in a few minutes. The Indian Ocean tsunami of 2004 continues to be a historic occurrence

DOI: 10.1201/9781003714125-16

because of its devastating level. The disaster that struck as a result of a giant megathrust earthquake just off the coast of Sumatra resulted in the deaths of about 2.3 million people across up to 14 different countries. In India, more than 18,000 people were killed, with the majority of deaths taking place in the states of Tamil Nadu and Andhra Pradesh as well as the Andaman-Nicobar Islands. More than the human cost, the economic costs are also broad; it was estimated that an average tsunami can cause damage in the range of US19 million, but larger ones are bound to cost significantly higher. Therefore, strict scientific examination of tsunami and the improvement of early warning systems is inevitable. The creation of a tsunami mostly occurs after a large amount of seawater is moved in a sudden manner. This displacement can be due to submarine earthquake, submarine volcano eruptions, submarine landslides, or, at extremely rare rates, meteor impacts in the ocean. Of them, most tsunamis are caused by submarine earthquakes, particularly the shallow-focus ones. In the case of an earthquake taking place beneath the sea floor, the vertical displacement so generated may suddenly raise or lower the sea bottom, forming the water column above it and causing long-period waves which propagate away. Although these waves seem small in the pelagic region since they have long wavelengths, they intensify to enormous magnitudes when they roll into shallow coastal bathymetry which leads to dangerous surges.

The area of the Sumatra forms one of the major seismic regions of the world as it is where the plates of Indo-Australian and the European tectonic systems meet. The related subduction interface is characterized by increased seismicity and there are several major earthquakes formed by the same interface, such as the 2004 mega-earthquake that had caused the tsunami. The seismicity in this zone affects several nations around the Indian Ocean, that is, Indonesia, Sri Lanka, India, Thailand, as well as Maldives, Malaysia, and even Singapore. The eastern seaboard of India and the southern part of Sri Lanka are directly opposite to Sumatra on the other side of the ocean, making them very vulnerable. This means that any considerable tsunamigenic earthquake that occurs in the Sumatra sector is bound to be immediately dangerous to these areas depending on the wave superficiality route and oceanographics. The scientific community has made significant strides in developing the knowledge regarding the cause and spread of tsunami over the past decades. However, there is one issue that always appears to be a challenge and that is to establish whether a particular submarine earthquake will produce a tsunami. Even the tsunamis created by seismic events that take place beneath the ocean floor are not always accompanied by a tsunami. The magnitude, fault orientation, directionality of the rupture, and vertical displacement of the fault are parameters that affect the potential of creating a tsunami. Generally, the most common earthquakes that can cause tsunamis are shallow (less than 70 km) earthquakes that have either thrust or reverse fault kinematics; however, there are notable exceptions and this requires the early warning systems to rapidly analyse the seismic information to determine the potential of risk. To alleviate the effects of tsunamis, some countries have set up special early warning stations. The Indian Tsunami Early Warning Centre (ITEWC) (at the Indian National Centre of Ocean Information Services (INCOIS)) at Hyderabad always remains on watch in India. The centre captures information provided by a network of sensors which comprise seismic stations, coastal tide gauges, deep ocean pressure transducers, and Global Positioning System receivers that measure the ground deformation and consequently processes the data in near-real time. When an undersea earthquake with considerable magnitude is detected, immediately the system starts to run numerical models to assess the possibility of tsunami formation and to predict further propagation of waves.

The tide-gauge systems along with deep-sea pressure gauges are used to confirm the formation of a wave of the tsunami. The rate of their data transmission is high and hence they can be easily cross-verified by the authorities and a clearer and more timely warning can be issued without waiting too long. However, the most serious issue remains in the first minutes after an earthquake: the timely ability to suspect an event in a tsunami is extremely important because during this period there is a limited amount of time to contain and mitigate the situation.

The use of conventional seismological methods, although effective, is at times not quite sufficient in its ability to capture the complexity of the seismic patterns of signals. Due to such limitations, an increasing group of scientists has turned to the approach of machine learning and artificial intelligence. These strategies help in speeding up seismic data interpretation, synthesis of experience of past events, and consequently increasing the identification of trends related to earth tremors that trigger the tsunami effects. Also machine learning algorithms are capable of dealing with nonlinear processes and nonlinear phenomena and much more efficiently than conventional analytical algorithms with large amounts of data.

The Sumatra region being prone to shallow-focus seismicity and known to have generated large tsunamis makes it the perfect testbed in the development of machine learning models. The data produced in seismic stations like PALK that capture high-quality broadband waves originating at the Sumatra margin is useful. Three-component broadband seismograms can be analysed to derive those salient parameters of the earthquake that characterize the movement of the ground and the source mechanics. The current research builds on this body of work by applying several machine learning methods to predict the tsunami-generating capability of the seismicity of the Sumatra island, using the measurements recorded at the PALK station. This study will enhance the work of the early warning systems and help tackle the global efforts to decrease the risks of tsunami by focusing strategically on superficial events and introducing the comparative analysis through multiple machine learning models.

16.2 BACKGROUND

The modern methodologies used include tsunami creation caused by earthquakes, the spreading of the oceanic tsunamis, and the consequent inundation distances. These are complemented by the modelling of the run-up heights in prone areas of the coastline (Titov and Gonzalez 1997). One of the fundamental processes is the solution of the earthquake source by the hydro-dynamic equations through which the components of the static displacement of the earthquake source under the influence of the boundary conditions at the ocean floor are solved (Comer 1984; Okal 1982). Boundary conditions include fault length and fault width; the mean static displacement can be determined by the magnitude that is set and the parameters of the focal source of an earthquake. One of the prospective approaches involves the development of a model which is founded on the sea surface deformations caused by shear and tensile faults in the half space (Okada 1985). However, parameters like magnitude and hypocentre only come in minutes of the earthquake, whereas focal parameters and seismic information are usually achieved by a combined, cooperative networks of various stations some thirty minutes or so after the earthquake. This latency can be prohibitive in areas that are close to the source of tsunami, thereby increasing the risk of evacuation. Fault parameters are frequently therefore taken to allow a worst-case scenario of 45° dip, 90° slip. Nevertheless, given a lack of time, an inventory of historical data on worst-case scenarios occurring in a particular region is prepared on the basis of hypocentral location and magnitude. Magnitude has been correlated to fault parameters to form some empirical relationships that are specific to the region (Papazachos et al. 2004). Thus, the first tsunami warning is often based on the pre-computed analysis of data and simulation results; the further focal parameter information is used to approve or reject the tsunami production, as confirmed by the Bottom Pressure Recorder (BPR) and tidal gauge records. Experiments in the current research revealed instances of Artificial Neural Network (ANN) and the Random Forest Regression (RFR) method successfully identifying early outliers of shallow-focus earthquakes at a distance of 2000 km given an earthquake report on the seismogram and a single three-component broadband station (Gnananandarao et al. 2021). This technique has a significant potential, and a pre-potential tsunamigenic earthquake (PTE) takes place under the sea surface and may cause a tsunami under some supplementary conditions. These are also the soft-computing methods used in geotechnical engineering studies as of late (Onyelowe, Gnananandarao, Jagan, Ahmad and Ebid 2022; Gnananandarao, Dutta, and Kumar 2022). This research aims at discovering PTEs as tsunamis when only very limited information is available about earthquake source, thereby reducing the number of casualties in the case of locally caused tsunamis of earthquakes in the earth everywhere especially in areas that are not well monitored. The first cause of tsunamis prediction in this study is entirely based on a single three-component seismogram; though it might not be the data required to accurately predict all of the numbers of earthquake sources, it is sufficient information to determine whether an earthquake is tsunamigenic or not.

16.3 METHODOLOGY

16.3.1 Gathering Data

The research gathered information on earthquakes in the area of the Sumatra and ransacked seismic records of the three-component broadband seismograms of the PALK station. These seismograms provided necessary ground movement information which were necessary in drawing parameters that are used in forecasting the potential tsunamis. Since PALK is sensitive to very clear signals offered by Sumatra, the records are appropriate in obtaining the necessary inputs.

16.3.2 Choosing the Event

Only shallow-focus earthquakes, that is, those with depth that does not exceed 70 km, were chosen to be analysed. These occurrences have a greater probability of producing tsunamis as compared to earthquakes that are deeper. In a bid to ensure consistency and make sure that dataset was relevant in tsunamigenic behaviour, events beyond this depth threshold were not considered at all.

16.3.3 Getting the Parameters

Key information was then determined based on every three-component seismogram, mostly, the characteristics of the waveforms and a few attributes pertaining to the sources. The results of these extracted values comprised the input parameters of all machine learning models used. These descriptive characteristics were assumed to represent better the behaviour of the earthquake and would allow the models to more accurately identify the occurrence of tsunamis.

16.3.4 Machine Learning Models

In order to assess and rank the potential propensity of tsunamis' generation, five discrete machine learning algorithms were utilized: (i) ANNs, (ii) M5P Model Tree, (iii) Support Vector Machine with Polynomial Kernel (SVMPoly), (iv) Support Vector Machine with Radial Basis Kernel (SVMRBK), and (v) RFR. These models have been selected mainly due to the work done previously which has shown that they outperform when working with nonlinear problems particularly in managing complicated patterns of seismic data.

16.3.5 Training and Testing the Model

The entire dataset was divided to obtain training and testing subsets in order to conduct a rigorous evaluation of each model. Training of models was done to categorize the earthquakes into tsunami and non-tsunami. Hyperparameters were optimized during training because there was an iterative approach to train our models, ensuring our predictive performance was improved to reduce problems like over-fitting or under-fitting.

16.3.6 Valuation of Performance

The standard statistical measures were used to determine model efficacy: accuracy, correlation coefficient, error measures: Mean Absolute Error (MAE), Mean Absolute Percentage Error (MAPE), and Root Mean Squared Error (RMSE). The analyses of these metrics enabled the models to be compared to determine which was the most reliable predictor to determine the likelihood of having a potential tsunamigenic earthquake. The model that offered the smallest errors and the best correlation was considered as being the most reliable.

This chapter takes a composite measure to incorporate the above methodologies, ANN, M5P model tree, SVM using the polynomial and radical basis kernels as well as the RFR to examine the tsunamigenicity of earthquakes that have been registered in the region of Sumatra. Only three-component seismograms, taken at the PALK station, are used to make the predictions, and the primary focus is laid on shallow-focus events (depth below 70 km) as events, which are considered most applicable to regional hazards of a tsunami.

16.4 SOFT-COMPUTING TECHNIQUES

16.4.1 Artificial Neural Networks

An ANN is a computational model; it is analogous to conceptually simple forms of pattern learning in the human brain. They are especially applicable in areas where non-linear relations are taken as inputs and outputs which could not be described using a simple analytical approach, e.g. in geophysical or seismic data. An ANN consists of many processing units, also known as nodes or neurons, which are linked with massively weighted links. The effect that one neuron has on another is changed by these weights. First, the values of weights are usually assigned randomly to the point that the network will lack any concrete knowledge about the data. Learning, therefore, only comes after repeated exposure to models.

The learning in an ANN occurs in the training phase. In training the network makes a prediction, compares it to the ground-truth value, and then tries to minimize the difference between the prediction and ground-truth value by updating the weights during the successive training stages. Backpropagation is one of the common ways of implementing such changes. Essentially, error signal starts backwards travelling through the network layer after layer, explaining the role of each connection in the final misprediction. The weights are then adjusted in an incremental fashion by the algorithm, to reduce the error, which is repeated over several epochs until an acceptable predictive performance of the network is achieved (Caesar, Gloger, Mandler 1993).

The standard backpropagation is structured into different layers. The input layer gets external data and passes this data onto the next layer without having to do any calculations. It is then followed by one or more hidden layers in which computational functions are carried out. Every neuron calculates a weighted average of its inputs and applies a nonlinear activation function, usually a smooth, differentiable one such as the sigmoid or the rectified linear unit (ReLU), to enable it to adapt its weights correctly during training. The final prediction of the network is presented in the output.

Common diagrammatic representations of such networks usually use coloured circles to represent types of neurons. In the illustrative schema, the circles with heavy espigot on top of them represent input neurons, which are just a means of relaying values into the system. Green circles represent the neurons that are hidden and in which learning of the primary pattern takes place. The circle in red is the output neuron which produces the final prediction of the model. The inconsistency between this prediction and the actual value causes the error signal to be computed and this section of the backpropagation then propagates the error signal backwards, in turn, causing dithering changes to weight updates to optimize performance.

Learning paths are not necessarily moving in a monotonic order. Networks selected may slow convergence, be stuck in local minima, or fall prey to over-fitting, i.e. memorizing training instances rather than executing generalizable patterns. The mitigation strategies involve modification of the number of hidden layers, experimentation of various activation functions, learning rate, among others. This empirical, iterative tuning is, to a great extent, necessary in applied settings, e.g. seismic or earthquake related attempts, no single architecture can be said to optimally perform on heterogeneous data.

It is important to note as well that ANNs usually need a considerable amount of data to attain generalization. Noisy or little data could interfere with the ability of the model to observe underlying trends. However, ANNs are viewed as powerful methods in that as soon as the concerned associations are learned, they can capture multivariate interactions, which are complex and are inaccessible to traditional methods of regression or simple statistical models. This flexibility is extremely useful in forecasting earthquakes and tsunamis, where the indicators carry less noticeable, behind-the-scenes details.

In a nutshell, regardless of how complicated ANNs may first seem, the principles of their work are relatively simple in concept: over time, the network runs and updates the internal weights in response to the examples being provided until the resulting predictions come close enough to the target values. This mechanism is similar to the learning mechanism of human beings in terms of the mathematical equipment used, that is, through repetition and feedback. The ability of ANNs to adjust to nonlinear associations makes them a powerful modelling tool to a broad range of application in science and technology in general, such as the one under investigation (Dutta et al. 2018, 2019; Gnananandarao et al. 2019). The architecture of the neural networks is shown in Figure 16.1.

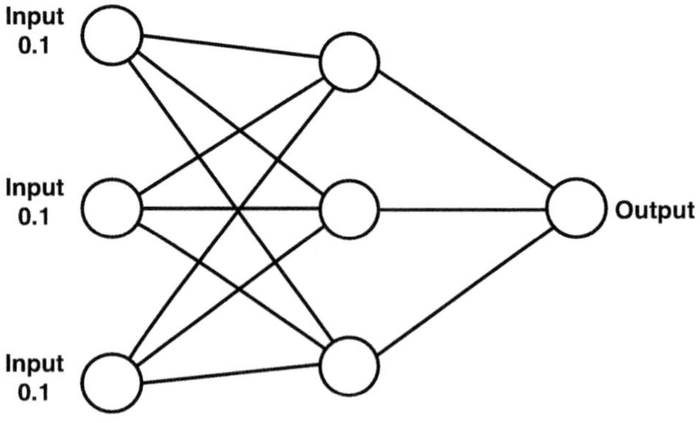

FIGURE 16.1 Architecture of a neural network.

16.4.2 M5P Model

The M5P model tree is a decision tree form which combines both traditional binary branching and linear regression models at the end nodes. M5P model, unlike classical decision tree which produces categorical labels, is provided to produce continuous values, and can be applied when the variable of interest is a number. Using a divide-and-conquer paradigm a tree construction repeatedly chooses, at each node, a feature that achieves the largest possible reduction in prediction error, often in the form of the largest possible decrease in the standard deviation. The lower the standard deviation, the homogeneous are the observations within the partition that was obtained as a result, and this allows a more precise linear relationship to be applied at the leaf.

The process of splitting can result in the tree becoming quite complex and deep. It is a common behaviour of tree-based methods; they often tend to over-fit by breaking the data down into smaller and smaller groups, when that breaking down does not have any substantive explanatory strength itself. This makes the resulting model too selective and far too noisy, so as to reflect the underlying pattern. To overcome this, M5P methodology has a pruning step. When the exhaustive tree is built, the algorithm will go backwards, trimming down branches that cannot produce significant changes in predictive power. The final tree is smaller in size, simpler to interpret, and has a better capacity to generalize as complex branches are replaced with simple linear regression models (Onyelowe et al. 2022, 2025).

This pruning and replacing plan is one of the major strengths of the M5P model tree because it combines the explainability of the linear regression and the flexibility provided by decision tree splits. In the ordinary regression leaves the relationships are smoother and continuous, but the above divisions define the data in different regions in which independent patterns might be seen. As such, the model first examines all permitted structures and then reduces them to a more stable form. The two-step process (first find the entire tree and then cut parts out) reduces over-fitting and produces a model that in most cases can work well on novel and previously unseen data.

Though such an approach might seem to be a completely mechanical process, in the practical sense it often has to be executed through some sort of trial and error. The tree can be too large or too small; in these cases, the hyperparameters may usually need some minor tweaking or the model may have to be re-trained several times to get a sufficient structure. To find very extensive technical details, one may consult the foundational papers which give the equations, as well as the splitting rules, in a more formal way (Quinlan 1992).

16.4.3 Support Vector Machines

SVMs are a group of procedures, based upon statistical learning theory, of classification and regression. The main goal is to reach the point that is ideal to divide two sets of observations. Instead of fitting any arbitrary line or curve, the SVM attempts to fit the one that maximizes the margin, i.e. the distance between the separating plane and the nearest points of each category. An increased margin is usually linked to high generalization ability in data that has not been seen before. The SVM also adheres to the principle of structural risk minimization which only minimizes the expected generalization error and not just the ability to fit the training data precisely. Therefore, the objective is that of a hyperplane which optimizes the margin while concurrently punishing misclassifications.

In practice, the data can hardly be easily separable. The observations made in different classes tend to overlap or may be irregular. In this case, SVM enables a boundary amount of errors in classification but with a penalty term. This penalty is controlled by a regularization constant which is denoted by C. Having a big value of C will force the model to reduce the classification errors that may result in over-fitting and having a smaller value of C will allow many more misclassifications which may also increase the performance of the generalization of the model.

The other higher note of SVMs is their ability to work with nonlinear data using the so-called kernel trick. The data can be transformed in a feature space in a higher dimension in terms of which the original input space is mapped, and this may make it more receptive to partitioning by a linear hyperplane. Importantly, such mapping is not always entered, but rather the kernel function (polynomial or radial basis function kernels) does the inner products in the higher dimensional space implicitly. As a result, SVMs have a strong mathematical basis and can be applied in both practical and nonlinear patterns of nonlinear separation (which are not simple) of data (Vapnik 1995).

16.4.4 Random Forest Regression

RFR is a machine learning technique that could be used in classification as well as regression. Its key concept is the combination of many single decision trees. An individual tree can be predisposed to over-fitting or biasness, so by building very many trees and averaging their outputs, Random Forest (RF) results in a more consistent, more predictive output. The trees in the forest are constructed randomly with a normally random element, that is, each tree is usually provided with a different but close subset of the training observations or input variables. The trees cannot take the same decisions due to this diversity of the points of view. In regression tasks, as was shown in this chapter, every tree provides a scalar estimate and the overall prediction is typically the arithmetic mean of all the tree values. The algorithm at every internal node randomly chooses a subset of potential variables, and tests a splitting criterion on this subset, which increases the resistance of the ensemble to over-reliance on any of the features.

RF also takes advantage of a process called a bagging, or bootstrap aggregation. In bagging, the training set is made by selection with replacement, whereby N observations are obtained in the actual dataset. The bootstrap samples are the same size as the original set, but about 67% of the single observations are usually also found in each bootstrap sample while the rest 33% are out-of-bag (OOB) data. These OOB samples give an internal way of estimating prediction error without additional validation set.

To construct every tree, there are two parameters that are required to be specified by the user: (i) the number of variables, to be randomly chosen at every split and denoted by m; (ii) the number of trees to grow, in total and denoted by k. Only the selected *m* variables are investigated (i.e. only the chosen ones are considered) to determine which one of them has the maximum effect in reducing error in each split. All sorts of quality metrics may inform this choice; the Information Gain Ratio is one of the most commonly used criteria. This method is typically used even though RF does not strictly conform to one set of criteria of all trees.

Upon all the trees being constructed, the RFR provides a prediction by averaging the results when all trees it consists of are taken into account. Since the output is a value, it is possible to calculate statistics like the mean-squared generalization error which gives information on what a model is likely to achieve on the unknown data. The total RF predictor is effectively a combination effect of all the trees, each of which has a small effect. It is also due to the fact that each and every tree is being trained on slightly different data and feature subsets that the resulting ensemble is normally more precise and less prone to over- or underspending than any single tree.

16.5 DATA

The literature was used to compile a complete dataset of 66 seismic events (Kundu et al. 2016). The given document is a systematic account of earthquakes that took place in the Sumatra region during the period between 2001 and 2016, and restricted by the earthquakes of a moment magnitude of 6.0 or more. Figure 16.2 shows the observation. It is shown that the events are shaped in yellow. These variables include depth (*H*), epicentral distance (*D*), azimuth (*A*), moment magnitude (*Mw*), and fault parameters (∅, δ, λ). The analysed results in Figure 16.3 present the distribution of the measured data in terms of histograms. In the classification case, a binary choice was used, which was marked as either 0 (for И) or 1 (for ¥). Correlation coefficient between Pearson was then used to measure the relationship between pairs of variables. The obtained coefficients are indicated in Figure 16.4. Figure 16.4 also clearly shows the relationship that exists between the epicentral distance and the longitude. Besides, long-distance reliance on the longitude in relation to epicentral distance is persuasively presented in Figure 16.4.

The histograms of the input parameters used in this chapter demonstrate how the input parameters intrinsically behaved, the physical properties of Sumatra earthquakes, and why nonlinear machine learning algorithms including ANNs had better performance in relation to rules-based models including M5P. The histogram of moment magnitude (*Mw*) depicts that 30% of events occur in this range 6.0–7.5, revealing that events less than 8.0 are isolated. As a result, moderate and strong earthquakes significantly overshadow the moderate size of events, and very large events are relatively small. Mathematically, this distribution offers an effective repertoire of patterns of energy release that are useful to classification exercises, as well as it creates natural nonlinearity to seismic wave propagation. Table 16.2 shows that since *Mw* does not rise with wave amplitude linearly, ANNs can learn to describe such nonlinearity more effectively than equation-based models like M5P do.

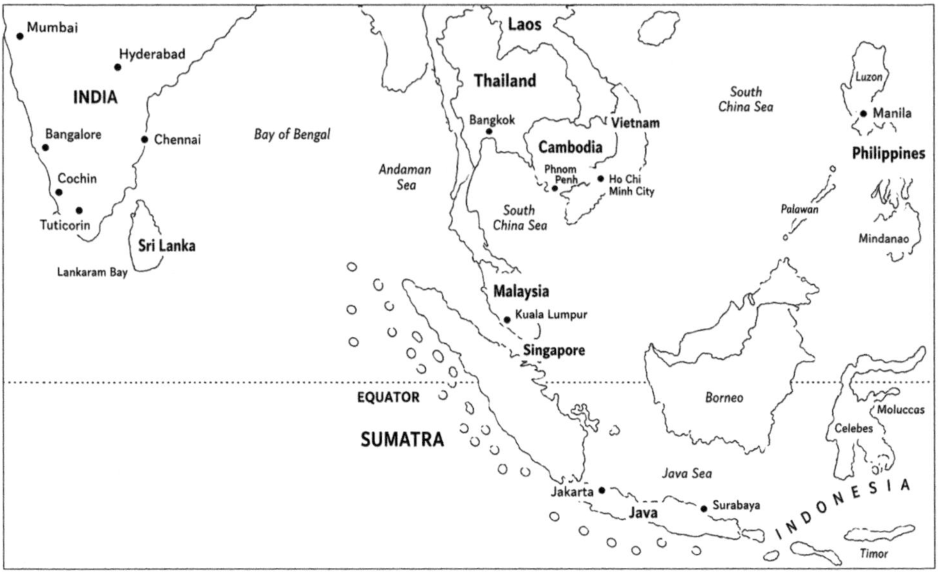

FIGURE 16.2 Potentially tsunamigenic earthquake events as yellow dots at Sumatra region and the location of a seismogram station in Sri Lanka (marked with a purple colour). (Courtesy Google maps.)

The depth (*H*) histogram shows that most of the events happen at shallow depths, as most events tend to fall in between 0 and 70 km, with hardly any of the intermediate to deep events. This pattern corresponds to tectonic environment of the Sumatra subduction zone, with the vast majority of megathrust earthquakes which produce tsunamis occurring at shallow depths. The depth has a direct effect on the binary model output and the tsunami risk, but this is a highly nonlinear effect. An example of this is that the potential of a tsunami can be significantly lowered by raising the depth by 10–30 km but increments by 80–100 km do not significantly alter the potential of a tsunami. ANNs automatically represent this nonlinearity, whereas M5P, which uses piecewise linear partitions, has difficulty doing so.

The histogram of the epicentral distance (*D*) shows that most of the events occur between 300 and 900 km from the seismic station; those events that are more than 1500 km are relatively few. As a result, earthquakes that are close to the station dominate the dataset. Distance has a significant effect on the amplitude of the waves, the spectral content, and the arrival time, but the interaction between distance and amplitude is nonlinear. The attenuation of seismic waves is determined by path geometry, crustal heterogeneity, and scattering of waves. ANNs are capable of learning such complex patterns, and M5P, with its linear segments, is not competent to reflect them.

The azimuth (*A*) histogram is generated with an equal distribution between 0° and 360°, which shows that earthquakes may have their origin in any direction to the station. Directional dependence also constitutes the controls over seismic wave polarity and amplitude in complex, nonlinear ways dependent upon the orientation of instruments

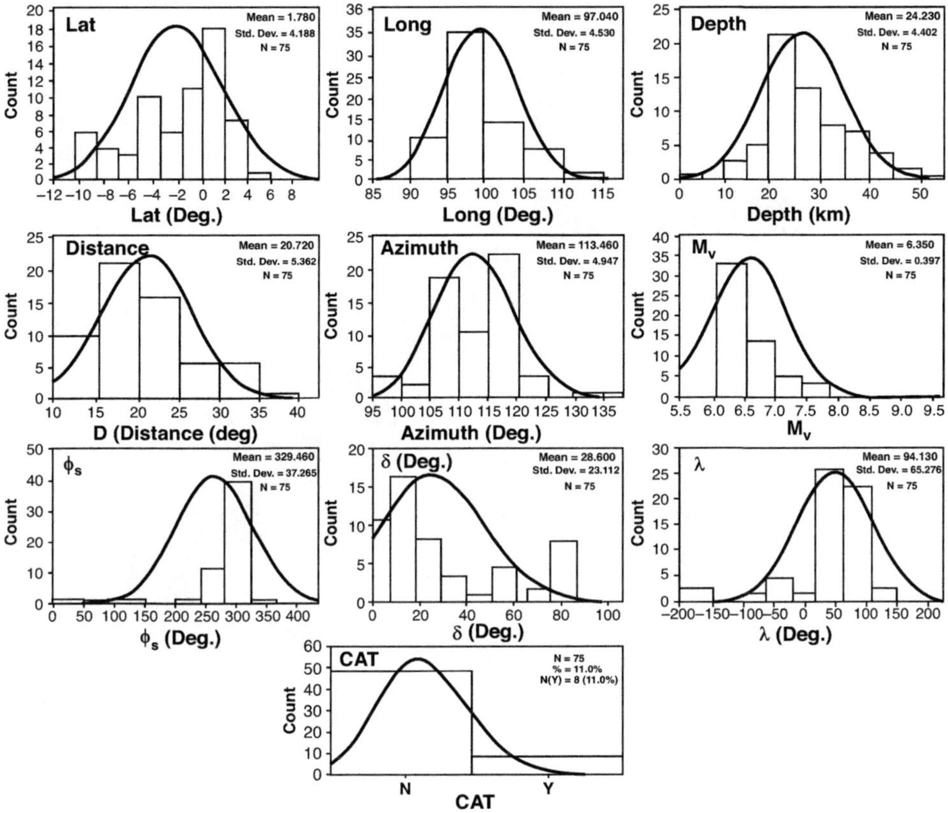

FIGURE 16.3 Histograms of the input and output variables used in this study.

and station geometry. ANNs represent these subtle directional interactions with their weighted inter-node relationships, but M5P that simplifies the problem space to a sequence of binary splits cannot represent minor directional differences.

The angles of strike (∅) range the entire 0–3600 range but are concentrated and tend to follow about the same direction as the orientation of the Sumatra trench. Although the effect of strike on rupture propagation is weaker and nonlinear, it still is a significant tectonic variable. Dip angles (*d*) are typically placed between 100 and 400, which is in line with the shallow-dip thrust faults in subduction zones. Depending variations in dip modify rupture geometry, slip direction, and tsunami likelihood in a nonlinear combination depending on strike and rake. The rake angle (*l*) histogram shows that the greatest number of events occurs between 800 and 1200, and this fact suggests that the opposite event, which is thrust faulting showing close relationship with the tsunami-producing mechanisms, is dominant. The nonlinear (and triaxial) interaction between strike, dip, and rake indicates that ANNs, capable of achieving multiple interdependent parameters, are stronger than M5P, based on a collection of simplified, rule-based decompositions.

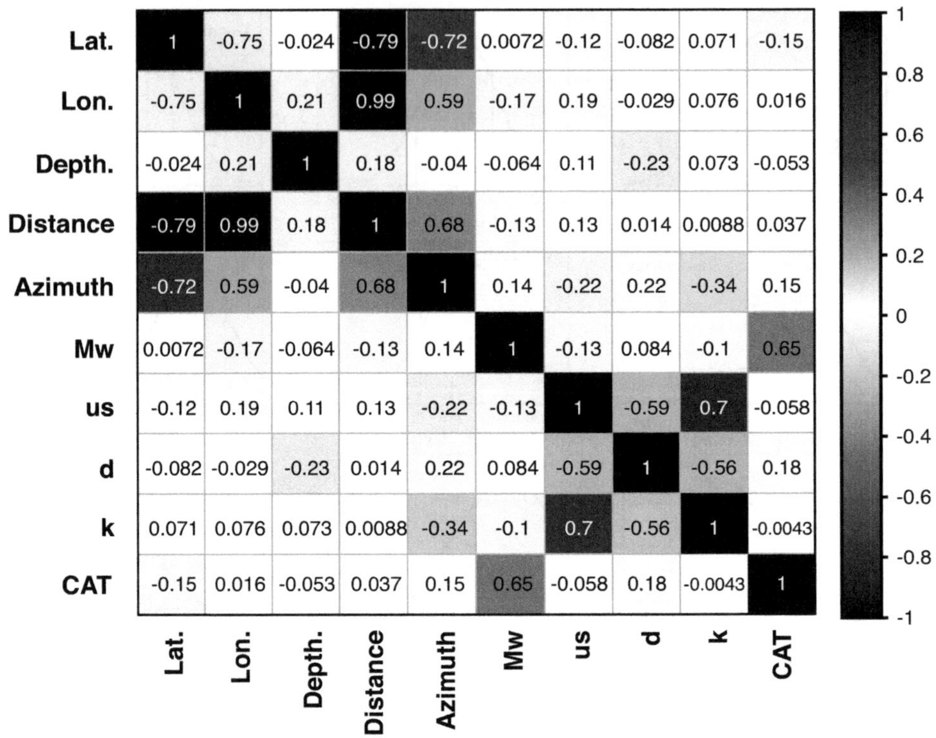

FIGURE 16.4 Pearson's correlation analysis.

The binary output histogram shows that there is an equal distribution of classes, as both class 0 and class 1 events occur almost with the same number, and thus does not tend to create bias that is associated with class imbalance during the training process. Indicators of nonlinearity and uneven distribution of the data include kernel density estimates that show that none of the predictor variables is normally distributed. As a result, ANNs are better in predictive accuracy and reduced rate of errors compared to the M5P model as the latter is limited to piecewise linear decision rule.

Through the heatmap of correlation, it was shown that the seismic data is nonlinear and complex in interrelationship. The output variable has low correlations with each of the inputs, meaning that there is dependency on multifactorial interactions to facilitate classification. The methodology of M5P suffers from strong multicollinearity – e.g. between distance and azimuth or depth and magnitude – because of the use of simple linear splits. ANNs, on the other hand, are very skilful at identifying nonlinear tendencies, dealing with multicollinearity, and modelling the relationship between two or more variables as a result of the layered architecture. The ability helps ANNs to discover more complex and deeper patterns equating to high accuracy, less error, and reliable predictions, hence the reason why ANNs are better than M5P in this relative comparison analysis.

16.6 RESULTS AND DISCUSSION

This chapter has constructed and tested five soft-computing models, including ANNs, M5P model tree, RFR, Support Vector Machine with Polynomial Kernel (SVMPoly), and Support Vector Machine with Radial Basis Kernel (SVMRBK) to categorize earthquakes in the Sumatra area that can produce tsunamis. About 70% of the data, which was obtained as single three-component broadband seismograms at the PALK station, was set aside as the data to be used in training the models; the other 30% set aside was to be used as independent test data and verification.

In the model development process, all the classifiers needed to be carefully tuned in terms of user-defined parameters in their quest to contribute to optimal performance. A very rigorous trial-and-error procedure was used to arrive at the final hyperparameter values. Table 16.1 gives the optimum combination of each classifier. In the case of ANN, there was only one hidden layer with five neurons ($M = 5$) that gave the best results. M5P model was reliable at the learning rate of 0.6, momentum of 0.1, 1000 iterations, and six hidden layers. The RFR model gave optimum results in case there were two features at a time ($K = 2$), the maximum depth was set to two ($m = 2$), and 100 decision trees ($I = 100$) were used. The optimal parameters of SVMPoly were a penalty coefficient C =2 and a degree of the polynomial of 1d and those that SVMRBK achieved were $C = 1$ and gamma = 0.5.

Table 16.2 shows the performance of each classifier in the course of training. The scores state that all five models correctly recognized most of the training samples and that the classifiers had a good deal of agreement in the vast majority of cases. Misclassification of the few events during the training process was established because few events were misclassified as indicated by gaffes in the frequent predictions (denoted by ¥). The generally accurate classifications in the training dataset confirm that all the soft-computing techniques were acquiring the inherent form in the information. Nevertheless, the fact that a few misclassifications took place highlights the nonlinear and complex interactions between earthquakes and tsunamis that cannot be completely explained by simple rule-based or linear models. After the training, predictive data was estimated based on the remaining 30% of the data.

TABLE 16.1 Best values for user-defined parameters for different machine learning algorithms

CLASSIFIERS USED	*USER-DEFINED PARAMETERS*
ANN	$M = 5$
M5P	Learning rate = 0.6, Momentum = 0.1, Iterations = 1000, Hidden Layers = 6
RFR	$K = 2$, $m = 2$, $I = 100$
SVMPoly	$C = 2$, $d = 1$
SVMRBK	$C = 1$, $g = 0.5$

TABLE 16.2 Comparison of the training data among the soft-computing models

		SOFT-COMPUTING MODELS				
S. NO	*CLASSIFICATION*	*ANN*	*M5P*	*RFR*	*SVMPOLY*	*SVMRBK*
1	И	И	И	И	И	И
2	И	И	И	И	И	И
3	И	И	И	И	И	И
4	И	И	И	И	И	И
5	И	И	И	И	И	И
6	И	И	И	И	И	И
7	И	И	И	И	И	И
8	И	И	И	И	И	И
9	И	И	И	И	И	И
10	И	И	И	И	И	И
11	И	И	¥	И	И	И
12	И	И	И	И	И	И
13	И	И	И	И	И	И
14	И	И	И	И	И	И
15	И	И	И	И	И	И
16	И	И	И	И	И	И
17	И	И	И	И	И	И
18	И	И	И	И	И	И
19	И	И	И	И	И	И
20	И	И	И	И	И	И
21	¥	¥	¥	¥	¥	¥
22	И	И	И	И	И	И
23	И	И	И	И	И	И
24	¥	¥	¥	¥	¥	¥
25	И	И	И	И	И	И
26	И	И	И	И	И	И
27	И	И	И	И	И	И
28	И	И	И	И	И	И
29	И	И	И	И	И	И
30	И	И	И	И	И	И
31	¥	¥	¥	¥	¥	¥
32	¥	¥	¥	¥	¥	¥
33	¥	¥	И	¥	¥	¥
34	И	И	И	И	И	И
35	И	И	И	И	И	И
36	И	И	И	И	И	И
37	И	И	И	И	И	И
38	И	И	И	И	И	И
39	И	И	И	И	И	И

(Continued)

TABLE 16.2 (*Continued*) Comparison of the training data among the soft-computing models

		SOFT-COMPUTING MODELS				
S. NO	*CLASSIFICATION*	*ANN*	*M5P*	*RFR*	*SVMPOLY*	*SVMRBK*
40	И	И	И	И	И	И
41	И	И	И	И	И	И
42	И	И	И	И	И	И
43	И	И	И	И	И	И
44	И	И	И	И	И	И
45	И	И	И	И	И	И
46	И	И	И	И	И	И

TABLE 16.3 Comparison of the testing data among the soft-computing models

		SOFT-COMPUTING MODELS				
S. NO	*CLASSIFICATION*	*ANN*	*M5P*	*RFR*	*SVMPOLY*	*SVMRBK*
1	¥	¥	И	И	¥	¥
2	И	И	И	И	И	И
3	И	И	И	И	И	И
4	И	И	И	И	И	И
5	И	И	¥	И	¥	И
6	И	И	И	И	И	И
7	И	¥	И	¥	¥	¥
8	И	И	И	И	И	И
9	И	И	И	И	И	И
10	И	И	И	И	И	И
11	И	И	И	И	И	И
12	И	И	И	И	И	И
13	И	И	И	И	И	И
14	И	И	И	И	И	И
15	И	И	И	И	И	И
16	И	И	И	И	И	И
17	И	И	И	И	¥	И
18	И	¥	¥	¥	¥	И
19	¥	¥	¥	И	¥	¥
20	¥	¥	¥	¥	¥	¥

Table 16.3 presents the results. ANN and SVMRBK obtained the best accuracy, having predictive accuracy of tsunami events in the range of 90%; M5P, SVMPoly and RFR also reached the levels of roughly 80%, 80% and 75%, respectively.

These results communicate a number of significant points. To start with, the fact that training to test performance dropped by a factor of 10% to 25% suggests that there is a slight over-fitting effect, which is also a common effect in small datasets that have complicated seismic relationships. Second, ANN and SVMRBK have a better ability to generalize. The architecture of ANN gives it the ability to model complex, nonlinear relationships among factors like the depth, strike, dip, rake, and the waveform characteristics of an earthquake unlike the radial basis kernel used by SVMRBK which provides the ability to model nonlinear edges in high-dimensional spaces.

Even though M5P, RFR, and SVMPoly had slightly lesser accuracies, they can still be used as supplementary instruments in classifying tsunami. RFR specifically indicates that ensemble-based decision trees may capture to some extent the diversity that exists in seismic data, together with ANN and SVMRBK, though they are less efficient in being able to capture highly nonlinear trends. The findings further indicate that it is important that further research should be provided with more information; the inclusion of more earthquakes in different parts of the globe through subduction would probably enhance the ANN model, making it more resilient and general.

Overall, the chapter shows that the methods of soft computing can be characterized as a simple, yet very effective approach in detecting tsunamigenic earthquakes. ANN and SVMRBK were found to be the most predictive according to the assessment of their predictive power, which suggests that the two might be useful in modelling in predicting tsunami in real time. The rest of the 30% of the dataset was used to test the models developed upon the completion of training. The results of the tests show that ANN, M5P, RFR, SVMPoly, and SVMRBK have 90%, 80%, 75%, 80%, and 90% rates of successful prediction of a tsunami, respectively, as can be seen in Table 16.3. Comparison of Tables 16.2 and 16.3 indicates that forecasting by ANN and SVMRBK is more accurate as opposed to M5P, RFR, and SVMPoly. However, RFR and SVMPoly still have the capability of forecasting tsunamis in the Sumatra. The predictions are extraordinary when it comes to the training phase, though 10% to 25% fall has been noticed at the testing stage. In this respect, therefore, future studies would recommend that the ANN model is further developed by employing a larger dataset across different regions to enhance the forecast on tsunami-causing earthquakes. The authors note that the use of soft-computing methodology is a cost-effective method of predicting tsunamis causing earthquakes.

16.7 CONCLUSIONS

Five soft-computing techniques are M5P, RFR, ANNs, SVMPoly, and SVMRBK. This chapter presents a method for the rapid classification of tsunamigenic earthquakes using a single three-component recording station located at a regional distance from the tsunami source. The investigation yields the following conclusions:

1. In this study almost all the five models are effectively approximating the tsunamigenic earthquake.
2. On making a detailed comparison, RFR provides the least approximation as compared to the other proposed models such as ANN, SVMPoly, and SVMRBK.

3. The estimation of potentially tsunamigenic earthquakes using ANN and SVMRBK achieved 100% accuracy in training and 90% in testing. In contrast, M5P recorded 97.82% accuracy in training and 75% in testing.

Overall, the chapter aims to give insights into using soft-computing techniques to predict earthquakes that can cause tsunamis. However, the data used is limited to the Sumatra region. If data from around the world were available, the models could be applied to a wider, global context, leading to better predictions of tsunamigenic earthquakes.

REFERENCES

Caesar, T., J. Gloger, and E. Mandler. 1993. "Pre-processing and feature extraction for handwriting recognition system." In *Neural Networks for Pattern Recognition*, 295–331. IEEE Xplore.

Comer, R. P. 1984. "The tsunami mode of a flat earth and its excitation by earthquake sources." *Geophysical Journal International*, 77 (1): 1–27. Oxford Academic. https://doi.org/10.1111/J.1365-246X.1984.TB01923.X.

Dutta, R. K., R. Rani, and T. Gnananandarao. 2018. "Prediction of ultimate bearing capacity of skirted footing resting on sand using artificial neural networks." *Journal of Soft Computing in Civil Engineering*, 2 (4): 34–46. https://doi.org/10.22115/SCCE.2018.133742.1066.

Dutta, R. K., A. Singh, and T. Gnananandarao. 2019. "Prediction of free swell index for the expansive soil using artificial neural networks." *Journal of Soft Computing in Civil Engineering*, 3 (1): 47–62. https://doi.org/10.22115/SCCE.2018.135575.1071.

Gnananandarao, T., R. K. Dutta, and V. N. Khatri. 2019. "Application of artificial neural network to predict the settlement of shallow foundations on cohesionless soils." In Anirudhan I.V., V.B. Maji (Eds.), *Lecture Notes in Civil* Engineering, 51–58. Singapore: Springer.

Gnananandarao, T., R. K. Dutta, and V. N. Khatri. 2021. "Tsunamigenic seismic activity (earthquakes) prediction from III-component seismic data." In Sitharam, T. G., S.Kolathayar, M.L. Sharma (Eds.), *Seismic Hazards and Risk. Lecture Notes in Civil Engineering*, 116 LNCE: 351–360. Springer. https://doi.org/10.1007/978-981-15-9976-7_31.

Gnananandarao, T., R. K. Dutta, V. N. Khatri, and M. S. Kumar. 2022. "Soft computing based prediction of unconfined compressive strength of fly ash stabilised organic clay." *Journal of Soft Computing in Civil Engineering*, 6 (4): 43–58.

Kundu, A., Y. S. Bhadauria, S. Basu, and S. Mukhopadhyay. 2016. "Prompt identification of tsunamigenic earthquakes from 3-component seismic data." *Physics of the Earth and Planetary Interiors*, 259: 10–17. Elsevier. https://doi.org/10.1016/J.PEPI.2016.08.003.

Okada, Y. 1985. "Surface deformation due to shear and tensile faults in a half-space." *Bulletin of the Seismological Society of America*, 75 (4): 1135–1154. GeoScienceWorld. https://doi.org/10.1785/BSSA0750041135.

Okal, E. A. 1982. "Mode-wave equivalence and other asymptotic problems in tsunami theory." *Physics of the Earth and Planetary Interiors*, 30 (1): 1–11. Elsevier. https://doi.org/10.1016/0031-9201(82)90123-6.

Onyelowe, K. C., T. Gnananandarao, and A. M. Ebid. 2022. "Estimation of the erodibility of treated unsaturated lateritic soil using support vector machine-polynomial and -radial basis function and random forest regression techniques." *Cleaner Materials*, 3. Elsevier Ltd. https://doi.org/10.1016/j.clema.2021.100039.

Onyelowe, K. C., T. Gnananandarao, J. Jagan, J. Ahmad, and A. Ebid. 2022. "Innovative predictive model for flexural strength, fck of recycled aggregate concrete from multiple datasets." *Asian Journal of Civil Engineering*, 8 (8): 1679–1693.

Onyelowe, K. C., V. Kamchoom, T. Gnananandarao, et al. 2025. "Developing data driven framework to model earthquake induced liquefaction potential of granular terrain by machine learning classification models." *Scientific Reports*, 15: 21509. https://doi.org/10.1038/s41598-025-07494-5.

Papazachos, B. C., E. M. Scordilis, D. G. Panagiotopoulos, C. B. Papazachos, and G. F. Karakaisis. 2004. "Global relations between seismic fault parameters and moment magnitude of earthquakes." *Bulletin of the Geological Society of Greece*, 36 (3): 1482–1489. National Documentation Centre. https://doi.org/10.12681/bgsg.16538.

Quinlan, J. R. 1992. "Learning with continuous classes." In *Australian Joint Conference on Artificial Intelligence*, 343–348. Singapore.

Titov, V. V., and Gonzalez, F. I. 1997. *Implementation and testing of the Method of Splitting Tsunami (MOST) model.*

Vapnik, V. N. 1995. "The nature of statistical learning theory." In *The Nature of Statistical Learning Theory*. Springer, New York. https://doi.org/10.1007/978-1-4757-2440-0.

17 Application of Non-Parametric Machine Learning Models in Estimating Penetration Rate of Tunnel Boring Machines

S. K. Shreyas and Arindam Dey

17.1 INTRODUCTION

Tunnels driven through soils or rocks serve as an indispensable component of urban transportation and underground mining operations. Such tunnels are commonly constructed adopting cut-and-cover, drill-and-blast or boring technologies (Saxena, 2018; Goodman, 1989). The cut-and-cover technique is the traditionally adopted method used to build shallow tunnels, where a trench is dug in the soil and is later covered by an

DOI: 10.1201/9781003714125-17

overhead support system that is capable of carrying the future surcharge on the tunnel. The drill-and-blast technique involves the use of explosives involving higher vibration levels of shorter duration burst. Although this is the classical method of tunnel excavation, it serves to be expensive as well as riskier to the lives of workers. In the boring technique, an easier and automatic excavation process is rendered using Tunnel Boring Machines (TBMs) that are equipped with circular full-face cutter head and disc cutters (Mahdevari et al., 2014). Mechanized tunnelling techniques are extensively applied in tunnel constructions through hard or soft rocks, involving process of excavation and support that are performed with the aid of ripper teeth, hydraulic rams and discs with an aim to provide. Adverse geotechnical conditions, namely spalling, bursting, squeezing and swelling of rocks as well as high water inflows, prove to be substantially governing the performance of TBM (Delisio et al., 2013). Furthermore, along the tunnel alignment, the in-situ variation in geological and geotechnical characteristics of the material often leads to the difficulty in predicting the rock mass behaviour through routine site investigations.

Several researchers from decades past have reported about the performance analysis and worked for the development of prediction models of TBMs (Tarkoy, 1973; Graham, 1976; Farmer and Glossop, 1980), the performance of which is commonly represented through their penetration rate (PR) and advance rate (AR). With the aid of empirical and/or experimental approaches to determine rock mass and machine parameters, several researchers correlated the TBM performance with PR and AR (Bamford, 1984; Rostami and Ozdemir, 1993; Sundaram et al., 1998; Sapigni et al., 2002), as well as between TBM penetrability and rock parameters (Graham, 1976; Farmer and Glossop, 1980; Cassinelli, 1982; Yagiz, 2008; Mikaeil et al., 2009). Since the estimation of PR is affected by several factors (geological and geomechanical factors of rock, as well as machine design parameters) that can vary significantly along the tunnel alignment, the development of logical correlations among such parameters remains critically important. In several cases, the existing empirical and analytical approaches suffer from the limitation in addressing the simultaneous effects of contributing factors concurrently influencing the PR. Following these limitations, instead of using conventional analytical solutions, attempts are made towards the usage of machine learning (ML) techniques in developing such complex correlations.

In the context of assessing the PR of TBM, various Artificial Intelligence (AI) techniques such as artificial neural networks (ANNs), neuro-fuzzy methods and other hybrid intelligent models have been used by various researchers (Grima et al., 2000; Okubo et al., 2003; Benardos and Kaliampakos, 2004; Javad and Narges, 2010; Eftekhari et al., 2010; Armaghani et al., 2017). The drawbacks suffered by some of these techniques (such as neural networks and support vector machines) related to their sensitivity towards unbalanced datasets, presence of outliers and issues with data units are well established (Krawczyk, 2016). Ensemble methods such as Random Forest (RF) and Gradient Boosting (GB) are insensitive to the aforementioned drawbacks, which makes the designing process of these algorithms much simpler than neural network models (Kohestani et al., 2017). These ensemble methods use decision trees as the base predictors (Geron, 2017; Goel and Abhilasha, 2017), which acts as a decision support tool that uses a tree-like hierarchical branching for assisting decisions (Friedman et al., 2000).

RF method is successfully implemented in various nonlinear problems of geotechnical and rock engineering (Ham et al., 2005), such as rock burst classification (Zhou et al., 2016a), pillar stability in hard rock (Zhou et al., 2016b), spatial distribution of organic carbon data of soil profile (Grimm et al., 2008) and cone penetration test (CPT)-based liquefaction potential prediction (Kohestani et al., 2015). On the other hand, GB method was successfully applied for predictions of stope stability (Qi et al., 2018c), strength of cemented paste backfills (Qi et al., 2018a, 2018d; Qi and Fourie, 2019), traffic behaviour (Dabiri and Abbas, 2018) and determination of rock mass properties (Qi et al., 2018b). The feasibility of using ML to predict the TBM performance is always primarily affected by the limitations in the quantity of datasets. In order to examine and evaluate the performance of ensemble methods to predict the PR of TBM, in this study, non-parametric ML models such as RF and GBRT are employed for a considerably large dataset comprising strength characteristics of rock through which the tunnel boring is in action.

17.2 DATASETS FOR PRESENT ML-BASED STUDY

The central importance of this study adheres to the PR of a shield TBM based on the specification of strength parameters of the rock and the TBM cutter. Based on the data collected from previous research for three different TBM projects, namely the Queens water tunnel in the USA, Karaj–Tehran water surface tunnel in Iran and Gilgel Gibe II hydroelectric project in Ethiopia (Javad and Narges, 2010), a large dataset with 185 samples related to PR of TBMs through the rocks is collated and the same is used to assess the feasibility of the RF and GBRT approaches on the chosen problem. Based on the site inputs, the PR of shield TBM through the rock deposits is assessed based on the strength parameters of rock, namely unconfined compressive strength (UCS), rock quality designation (RQD) and the distance between planes of weakness (DPW), for which the statistical inference is provided in Table 17.1.

TABLE 17.1 Statistical characteristic of rock characteristic data used in the present study

PARAMETERS	*MINIMUM*	*MAXIMUM*	*MEAN*	*STANDARD DEVIATION*
Uniaxial compressive strength (UCS, in MPa)	13.4	199.7	142.29	29.43
Rock quality designation (RQD)	40.6	99.88	91.04	15.34
Distance between planes of weakness (DPW, in m)	0.05	2	0.87	0.67
Penetration rate (PR, in m/hr)	1.27	14.43	2.33	1.31

17.3 EXPLORATORY DATA ANALYSIS

Exploratory Data Analysis (EDA) looks forward to initial investigations over collated dataset for trend and pattern recognition, anomaly detection and hypothesis checks through representative statistical and graphical portrayals. In the present study, 'Pandas', an open-source Python library for data analysis, with aid from 'describe()' function, is used to generate the summary statistics on the data collected, and the same is presented in Table 17.2. To visualize the frequency distribution of the features, corresponding probability density function (pdf) is plotted for each feature based on the Kernel Density Estimate (KDE). For the one-dimensional data, as used in this study, histogram plot is provided as a simple density estimator that segregates the data into discrete bins and, thereby, counts the number of points that satisfies the limit of each bin. This entire process is achieved using 'Seaborn' visualization library in Python that provides statistical graphs in order to perform both univariate and multivariate analysis. To make density plots in 'seaborn', either the 'distplot' or 'kdeplot' function can be used. The 'distplot' function makes multiple distributions with one function call as shown in Figure 17.1. It can be observed that UCS features show more concentration of data points near its mean value, RQD is negatively skewed with a tail on the left side of distribution, DPW shows a bell-shaped curve with two distinct peaks and PR is positively skewed with a tail on the right side of distribution. Furthermore, the correlation among individual features is assessed using the 'corr()' function, and the same is represented through a correlation matrix heatmap, as shown in Figure 17.2. It can be observed that RQD is largely and positively correlated to DPW, whereas UCS and PR exhibit a high negative correlation. Subsequently, PR exhibits a negative correlation with all other input features.

Figure 17.3 shows a box-and-whisker plot for both the input and output features, thereby exhibiting the distribution of quantitative data and aiding a comparison of the variables. Apart from the maximum and minimum magnitudes, the plots also depict the median, first and third quartile of the distribution. The box spans between first and third quartiles. The data points that fall out of box, as well as whiskers, are considered extreme points from the mean value of the feature. These points are generally termed

TABLE 17.2 Summary statistics of the dataset

FEATURES	*UCS*	*RQD*	*DPW*	*PR*
Count	185	185	185	185
Mean	142.71	91.04	0.87	2.33
Standard deviation	27.79	15.42	0.67	1.32
Minimum	30.00	40.6	0.05	1.27
25% quartile	129.15	90.98	0.3	1.85
50% quartile	139.30	99.28	0.8	2.09
75% quartile	159.77	99.81	1.6	2.43
Maximum	199.70	99.88	2.0	14.43

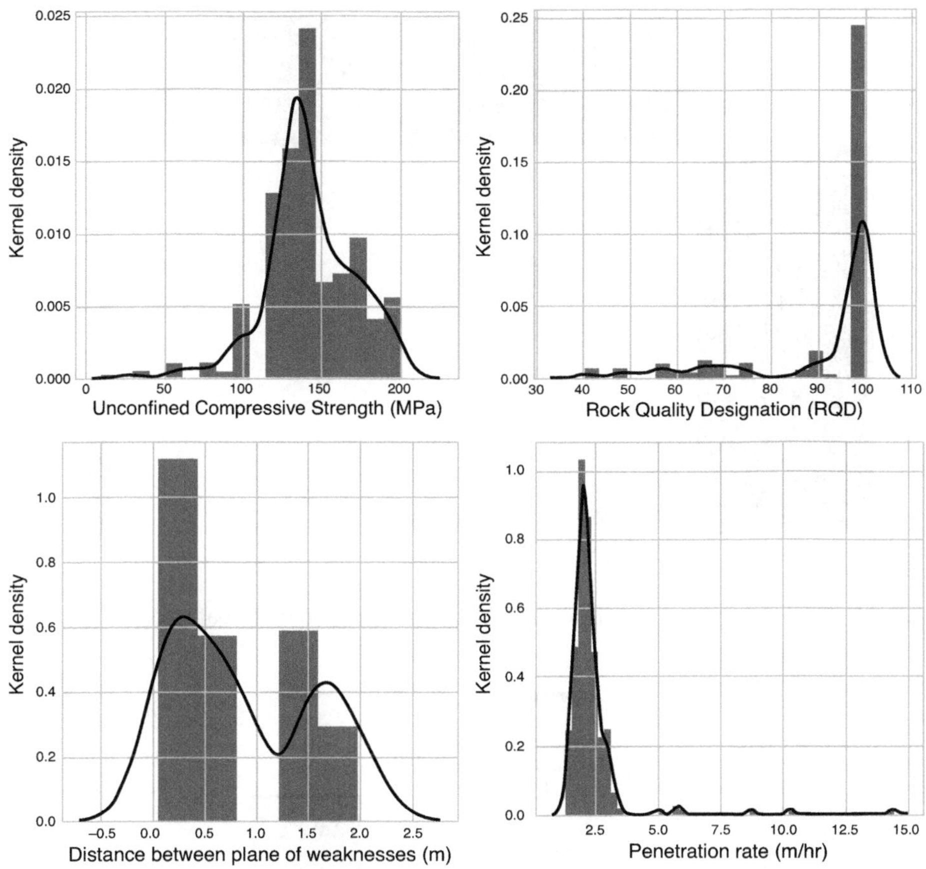

FIGURE 17.1 Distribution plot of all features in the dataset.

as outliers and are removed from the dataset to improve the performance of training. In this study, all such extreme points are retained to avoid loss of data. The ensemble methods adopted in this study are not significantly sensitive to outliers in the data (Friedman et al., 2000; Breiman, 2001).

17.4 ML TECHNIQUES ADOPTED FOR THE PRESENT STUDY

***Random Forest*:** RF is an ensemble learning method that generates the average of all predictions given by individual decision trees. Individual decision trees are referred as 'weak learners'. Ensemble methods combine several weak learners into a strong learner to improve the accuracy and robustness of the final model (Dabiri and Abbas, 2018).

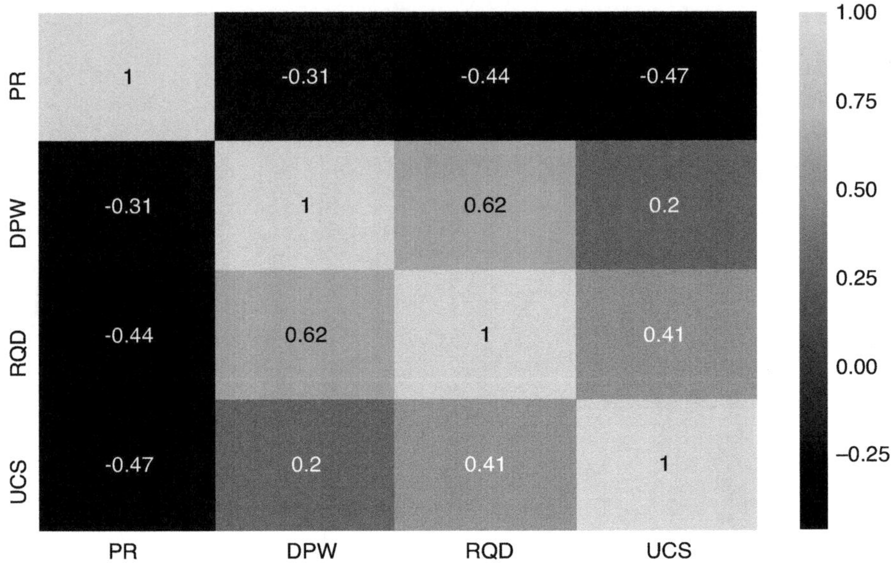

FIGURE 17.2 Correlation matrix for all features represented by a heatmap.

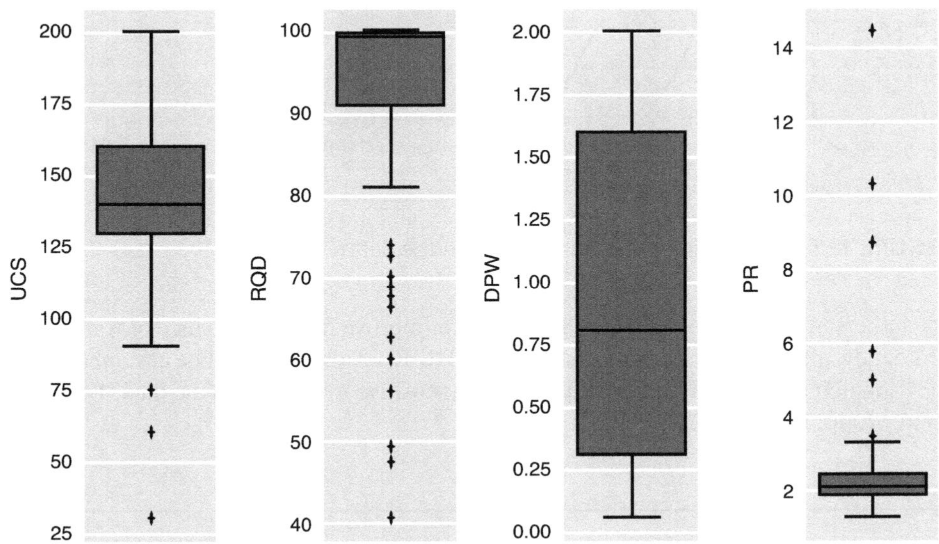

FIGURE 17.3 Box-and-whisker plot showing outliers in the dataset.

Ensemble learning can be classified into two categories, namely, bagging and boosting. In bagging, the decision trees, which are referred as base learners, are initially built on randomly selected training instances, and then are averaged to generate the final model. Naïve Bayes classifiers, decision trees and neural network are typically utilized as the ML techniques for training weak learners. While drawing the bootstrap sample, by default, two-thirds of the training sample is utilized for deriving the regression function, and the remaining one-third constitutes the out-of-bag (OOB) samples that are utilized to validate the accuracy of predictions (Adusumilli et al., 2013). Therefore, at each iteration, a regression tree is constructed with the aid of randomized training dataset that is drawn as a sample from the original dataset, thereby improving the generalization capability of RFs utilizing independent test data.

***Gradient Boosting Regression Tree (GBRT)*:** GBRT works on the concept of boosting where a weak learner (whose performance marginally supersedes that of dataset chosen at random) is attempted to be boosted or modified to become better. The main objective in GB is to add many weak learners to minimize the loss function, a procedure similar to gradient descent (Baldi, 1995). In GBRT, the decision trees (considered as weak learners) are ensembled stage-wise to produce an additive model, wherein a new learner is added to the existing weak learner without its modification.

17.5 ANALYSES, RESULTS AND DISCUSSION

17.5.1 Application of RF and Its Interpretation

Following the 80:20 policies for training and testing datasets, the corresponding predictive models were constructed. In RF-based regression technique, each tree is analysed deterministically, wherein for each iteration, a random set of variables and a random sample is chosen from the training dataset. To model the relationship between rate of penetration and predictor variables, RF was implemented using Python Scikit-Learn library (Graham, 1976). n_estimators and max_features are the two tuning parameters that need to be optimized in RF-based analysis. The former tuning parameter, n_estimator, indicates the number of trees to be built before arriving at the average of predictions, wherein a higher number of trees improves the performance of the model, although compensating marginally on the computational time. max_features refer to the maximum number of features (within the total number of input features) that can be considered to split a node. Increasing max_features generally improves the performance of a model, while slowing down the algorithm speed. Hence, for these reasons, it is necessary that the parameters be tuned to arrive at a balanced optimality. In general, for the case of RF-based regression modelling, the value of max_features should roughly be one-third of the total number of variables in order to minimize generalization error as well as to minimize correlation among the decision trees (Breiman, 2001; Kuhn and Johnson, 2013). In the present study, the tuning parameters were so optimized that a maximum R^2 and a simultaneous minimum OOB error is achieved. For the dataset considered, RF was built with PR as the response variable and UCS, RQD, DPW as the predictor variables. The total dataset, comprising 185 input–output combinations, was randomly split into two subsets of training and testing data in a ratio of 80:20, respectively. However, as the dataset comprised some extreme data, in order to avoid skewed distribution leading to non-uniform splitting, ample care has been taken to distribute the

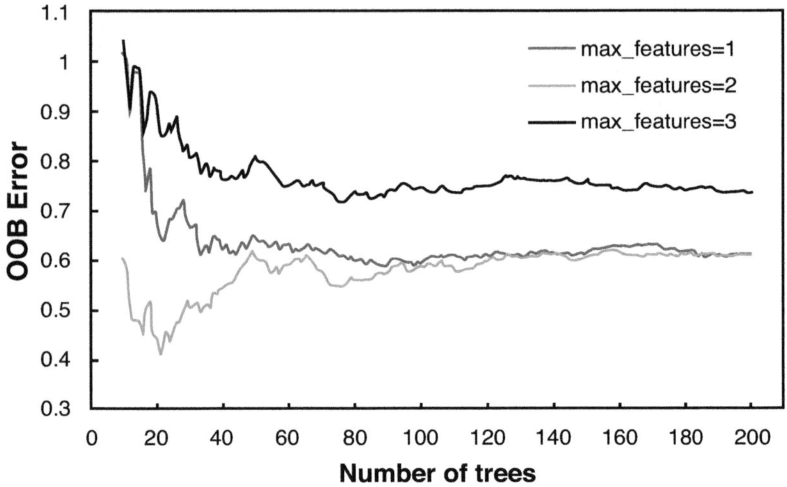

FIGURE 17.4 Variation of validation error with number of trees for different input features.

TABLE 17.3 Tuning ranges for RF hyper-parameters used in random search cross-validation

HYPER-PARAMETERS	*TUNING RANGES*	*BEST PARAMETER VALUES*
n_estimators	(5, 200)	21
max_features	[1,2,3]	2
max_depth	(3, 10)	6
bootstrap	[True, False]	True

few extreme values carefully within both training and testing datasets. Figure 17.4 highlights the variation of the OOB error, which exhibits that a minimum error is attained with an increase in number of decision trees, thereby indicating that there is no overfitting in this process. The hyper-parameters are tuned using the 'Randomized Search Cross-Validation (RandomizedSearchCV)' method available in Pythin 'Scikit-Learn' module, thereafter leading to optimized values as listed in Table 17.3. Figure 17.5 highlights the scatterplot between the observed and predicted rate of penetration of the TBM from the optimal RF model, which exhibits a coefficient of correlation (R^2) value of 0.87 during the training and 0.80 for the testing phase. Furthermore, the mean squared error (MSE) for training and testing set is obtained to be 0.19 and 0.26, respectively.

17.5.2 Application of GBRT and Its Interpretation

In this method, the procedure of dividing the dataset into training and testing sets remains identical to that adopted for RF analysis. The Scikit-Learn (Graham, 1976) library is utilized to implement GBRT function in Python environment, in which

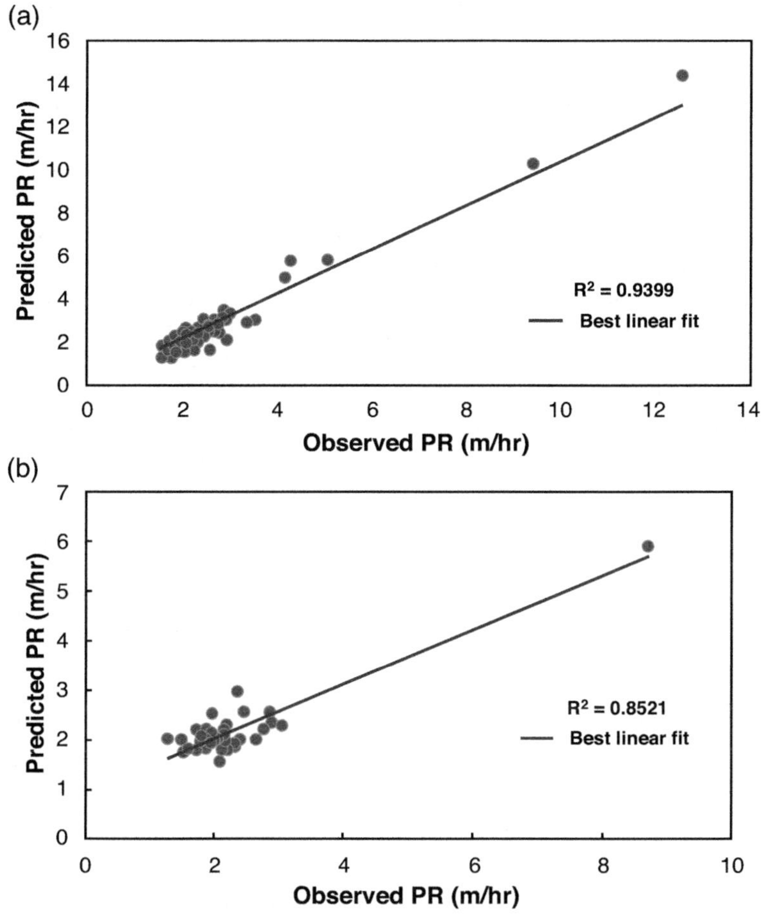

FIGURE 17.5 Correlation between measured and RF-predicted PR: (a) Training set. (b) Testing set.

'learning rate', 'sub-sample' and 'loss function' remain to be the hyper-parameters for boosting. 'Learning rate' attenuates the contribution of each tree upon addition of another, in which a trade-off is struck between 'learning_rate' and 'n_estimators'. During each boosting iteration, 'sub-sample' controls the fraction of training data to be utilized to fit individual base learners, whose tuning range remains to be (0, 1) within which any chosen value less than one leads to a reduction of variance and an increase in bias of the considered performance metric. To find the optimal number of trees, early stopping criterion is implemented by 'staged_predict' method that allows monitoring the error on the testing set after each stage. As per the literature, there are various notations of loss functions that can be optimized, e.g. least squares regression, least absolute deviation, Huber loss or quantile loss (Pedregosa et al., 2011). In the present study, 'Huber loss' is used to attain a robust regression outcome, as it is quite less sensitive to the data outliers than that observed for the conventionally adopted 'Least Squares Loss' function. For the present study, a Huber loss function with a learning rate of 0.15 and a sub-sample of 0.75 resulted in faster convergence of error during both training and validation phases, as shown in Figure 17.6.

For the tuning of the hyper-parameters, various schemes like 'random search' and 'grid search' (supported by k-fold cross-validation), as well as automated tuning using evolutionary algorithms, can be employed. In the present study, owing to the small dataset to be handled, the hyper-parameters are tuned using Scikit-Learn's RandomizedSearchCV method, wherein a grid of hyperparameter ranges are defined (as per Table 17.2), followed by random sampling to perform k-fold cross-validation with each combination of values. The efficacy of adopted RandomizedSearchCV method depends on the arguments 'n_iter' (number of different trial combinations to be tested) and 'cv' (number of folds to be used for cross-validation), which are set to 200 and 3, respectively. Larger number of iterations with more cross-validation folds leads the search to span over a wider search space with lesser chances of overfitting, while increasing the training time. Table 17.4 lists the tuned hyperparameter values obtained

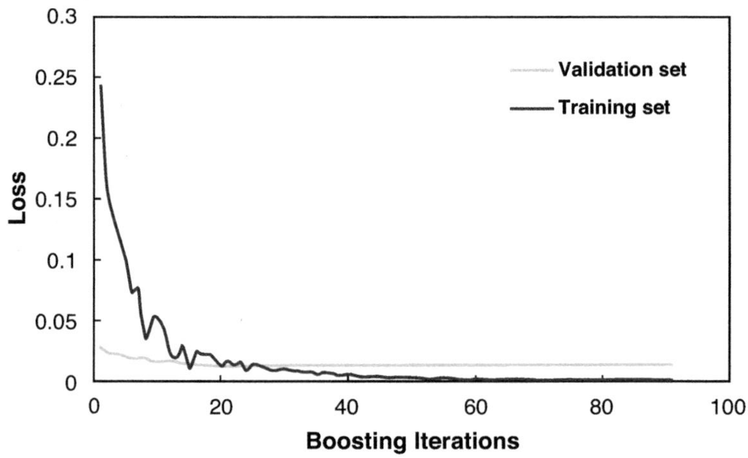

FIGURE 17.6 Variation of Huber loss function with number of boosting iterations.

TABLE 17.4 Tuning ranges for GBRT hyper-parameters used in random search cross-validation

HYPER-PARAMETERS	*TUNING RANGES*	*BEST PARAMETER VALUES*
n_estimators	(100, 500)	200
max_depth	(3, 10)	5
min_samples_split	(2, 10)	4
min_impurity_decrease	[0.0, 0.05, 0.10]	0.05
loss	['ls', 'huber']	Huber
sub-sample	[0.6, 0.65, 0.7, 0.75, 0.8]	0.6
learning_rate	[0.05, 0.1, 0.15, 0.2]	0.2

at the end of randomized search iterations. Alike the RF approach, the tuned parameters are considered to be optimized when the R^2 and Nash-Sutcliffe Efficiency (NSE) are achieved to be the highest while attaining a lowest MSE. In this case, the training dataset yielded R^2, NSE and MSE to be 0.98, 0.98 and 0.03, respectively, while the same for testing dataset is obtained as 0.85, 0.85 and 0.2, respectively. Figure 17.7 highlights the scatterplot between the observed and predicted rate of penetration of the TBM from the optimal GBRT model.

In order to illustrate the comparative predictive efficiency of the optimal GBRT and RF models, the performance metrics (R^2 and MSE) from both are compared. It may be noted that for testing datasets, R^2 from optimized GBRT and RF models are found to be 0.85 and 0.80, respectively, while the corresponding MSE are obtained to be 0.2 and

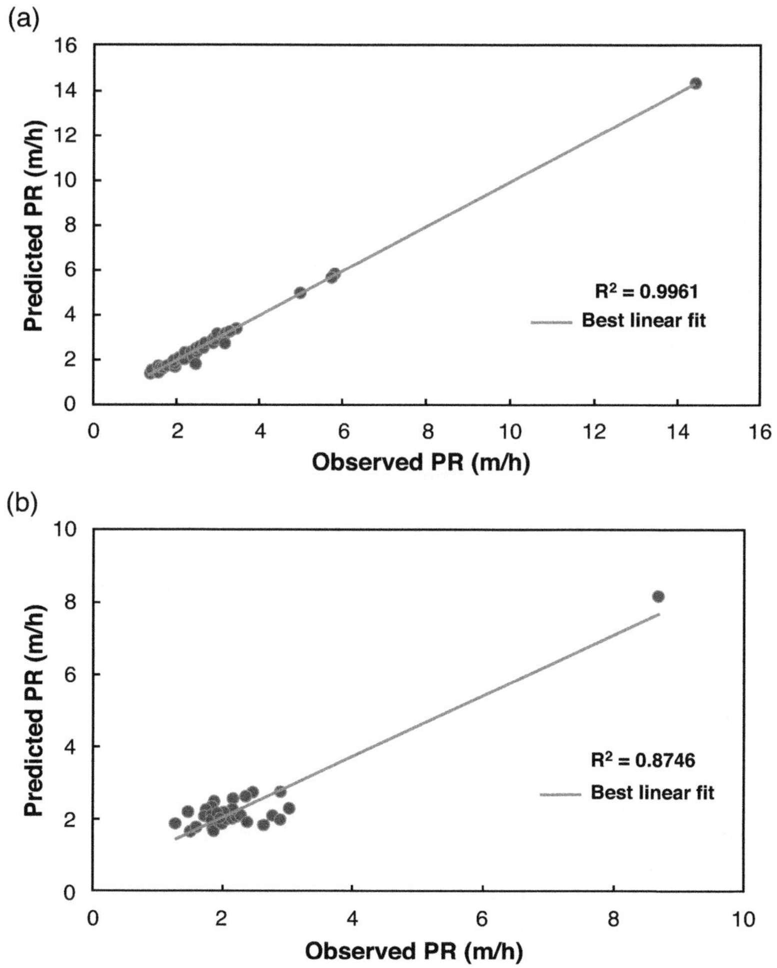

FIGURE 17.7 Correlation between measured and GBRT-predicted PR: (a) Training set. (b) Testing set.

0.26, respectively. The obtained performance metrics clearly exhibit the GBRT model outperforming RF approach. Furthermore, performance plot is generated by using the Python library-based module of 'plot_learning_curves', which uses MSE as the metric. Figure 17.8a exhibits that the training error fluctuates around its minimum while the validation error decreases with increase in training samples, and that both converge to each other beyond the usage of 90% of the training set size. The importance of variables influencing the PR was further assessed through 'feature_importances' inbuilt to the ensemble methods. Based on the importance ranking from both the ML approaches, Figure 17.8b highlights that UCS is the primary parameter affecting PR, while DPW is adjudged to pose a minimal effect.

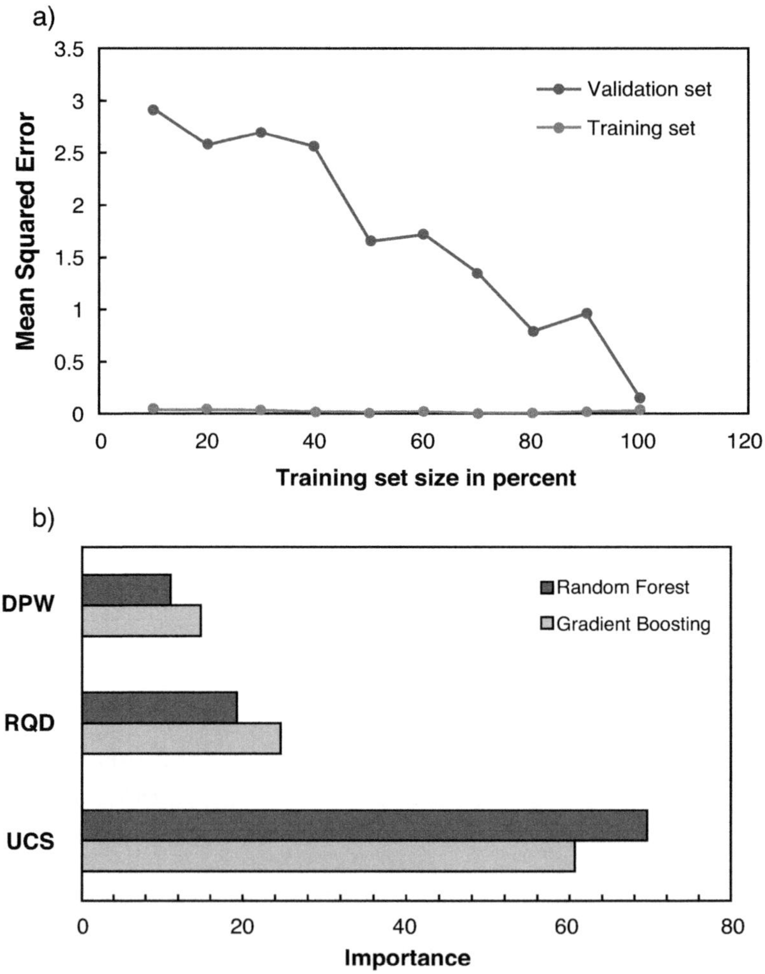

FIGURE 17.8 (a) Learning curve for training and validation data. (b) Importance ranking of features.

17.6 GENERAL DISCUSSION ON RF AND GBRT APPROACHES

In practice, the characteristic thrust force and torque force of TBM machines remain very important in determining its PR. However, in the present study, it was assumed that these characteristics remained unchanged and all possible effects are caused directly, and solely, by the material properties of the rock. In each of the adopted algorithmic models, optimal model hyper-parameters are obtained by heuristic tuning that is most likely an inefficient optimization strategy to perform high-dimensional non-convex optimization. In RF method, the bootstrap samples generated from the dataset should be approximately 'independent and identically distributed' (IID) to decrease the correlation among individual decision trees. However, this could not be achieved, as the training set is not considerably larger than the size of generated bootstrap samples. Consequently, for the present study, GBRT performed better than RF, although the former is often subjected to overfitting as the boosting methods are focussed in reducing bias of the model.

17.7 COST ANALYSIS FOR PREDICTIONS MADE BY ADOPTED ML ALGORITHMS

By utilizing the predicted PR from ML models, a cost analysis is conducted. A tunnel excavation project will be conducted in phases. In practice, while the excavation is carried out in each phase, there would be several delays bound to hamper the performance of TBMs. The values predicted by ML models conform to ideal conditions (based on only the statistical distributions of input features), which does not consider the unexpected uncertainties occurring on-site. The PRs achieved on-site would be greater or lesser than the predicted PRs subjected to the practical anomalous scenarios. Therefore, the error between predicted and desired PRs is calculated to separate the positive errors and negative errors. Negative errors indicate that the predicted cost is more than the actual cost. This would not give rise to unnecessary concerns, as the contractor would be having excess of money to complete the project. However, the positive errors are of prime importance as they indicate that the predicted cost is less than the actual cost incurred on-site. Under such cases, owing to the unforeseen circumstances in the site, the client should be prepared for extra expenditure for resources and labour. Figure 17.9 depicts the range in which the positive error is spread out for RF and GBRT, respectively. In all three histogram plots, it can be observed that maximum number of errors is in the range of 0%–15%. The findings indicate that there can be several instances where the cost expectation of the project can shoot up to the range of 15%, and extra funds should be managed to compensate the associated cost escalation due to project delays. However, it is clearly understood that the amount of cost escalation is dependent on the input parameters (UCS, RQD and DPW) as considered in the present study. It is clear that there would be specific combinations of such parameters that would result in different amounts of cost escalations.

Figure 17.10 shows the histogram plots for various input and output features considered for the analysis. It can be observed that many positive errors have UCS in the range of 124–154 MPa, RQD in range 98–100, PR in range 1.75-2.2 m/hr. No conclusion can be made on DPW as it is a subjective parameter, and is not very easy to determine. For any input rock strength parameter falling in the above-mentioned ranges, the PRs predicted by the ML algorithms fall short and requires to be compensated with extra cost, which may range up to 15% higher than the expected cost.

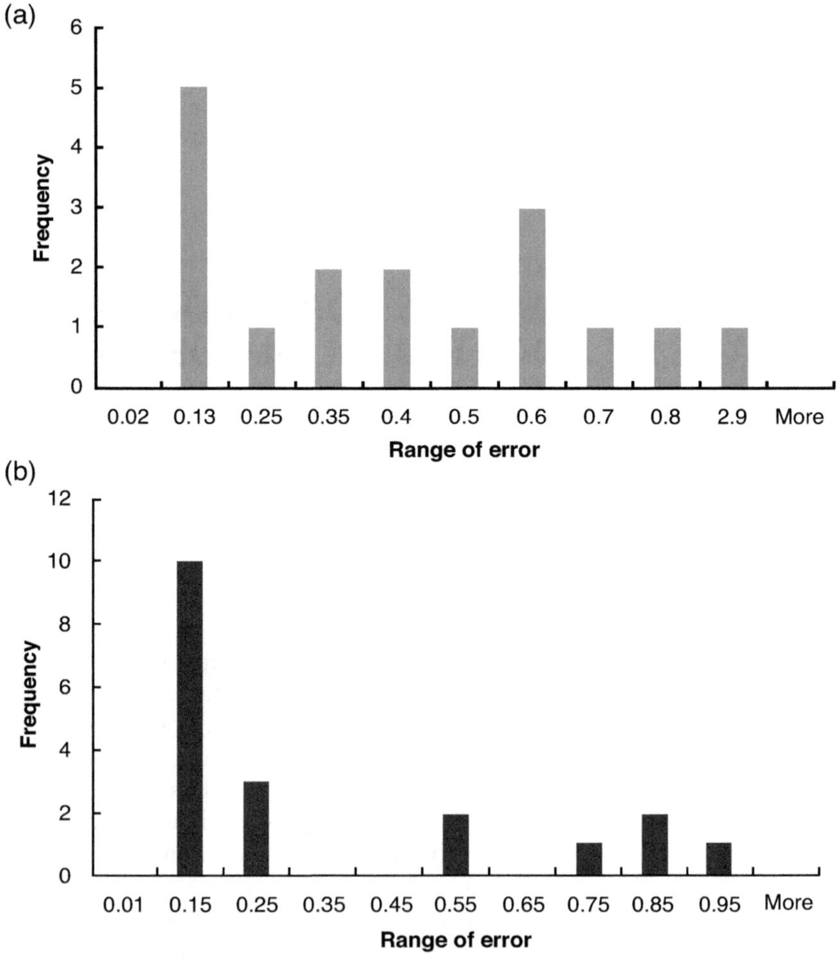

FIGURE 17.9 Histogram plot showing the spread of error for (a) RF model. (b) GBRT model.

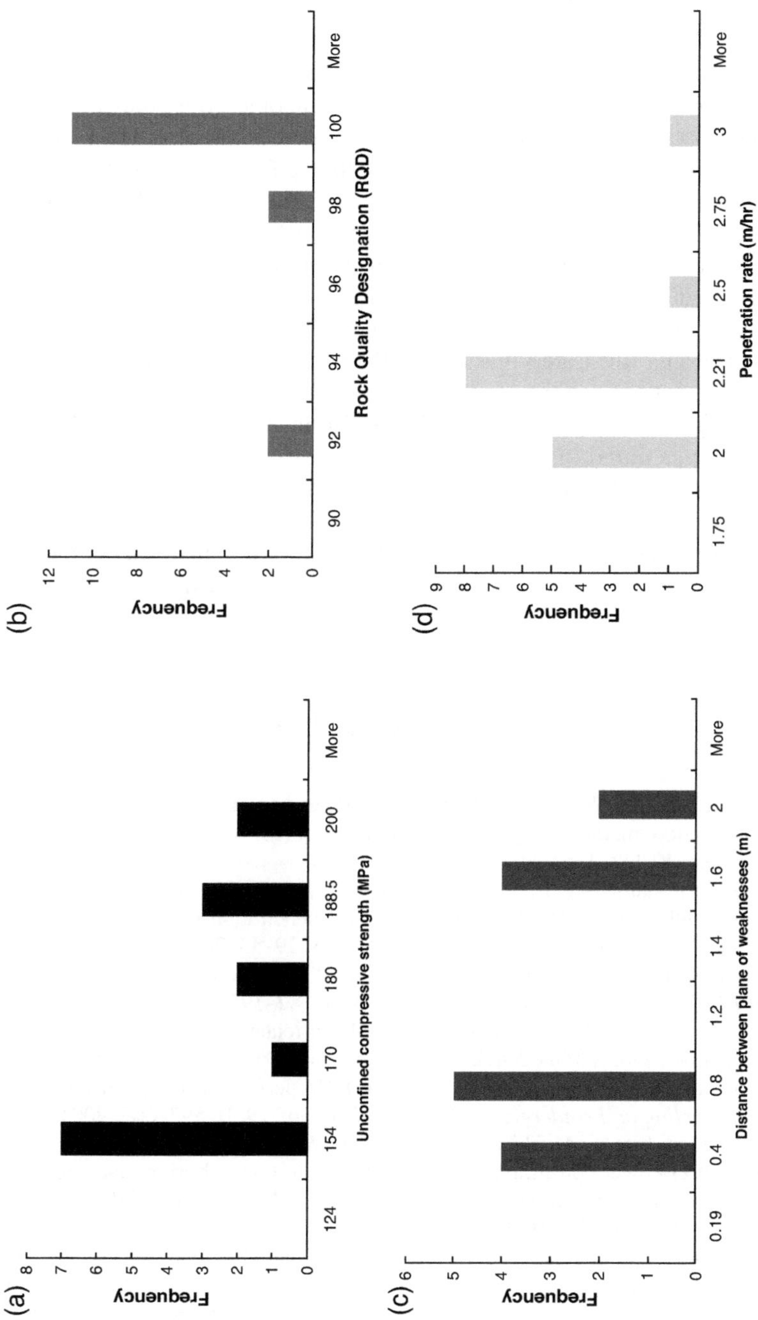

FIGURE 17.10 Histogram plot showing the spread of (a) UCS. (b) RQD. (c) DPW. (d) PR.

17.8 CONCLUSIONS

This chapter aims to give a comprehensive idea of PR of a TBM based upon RF and GB regression testing-based approaches. Based on the outcomes, this study proposes GBRT predictor with optimal parameters to reasonably predict PR of a TBM. An extended dataset from various tunnel boring projects is collated from literature to assess the feasibility of the proposed GBRT and RF methods for predicting the PR. Contrary to the outcomes of RF analysis reported in the literature, RF-based analysis was performed on a larger dataset expecting an improvement in the model's precision and reliability. However, the prediction outcomes from applying both the models to the unseen test data clearly depicted the superiority of the proposed GBRT algorithm over the RF approach. The obtained results illustrated that the GBRT method was able to capture nonlinear relationships between an output variable and set of corresponding inputs. The influence of the input parameters in predicting the PR is also investigated by both RF and GBRT approaches. UCS is found to be the most influential explanatory variable in predicting the performance of a TBM. It is clearly elucidated that the predicted PRs by ML models are lower than the actual PRs if the UCS and RQD pertains to the range of 124–154 MPa and 98–100, respectively. In such case, the expenditure involved per metre excavation would increase up to a maximum of 15% more than the expected cost.

REFERENCES

Adusumilli, S., Bhatt, D., Wang, H., Bhattacharya, P., and Devabhaktuni, V., A low-cost INS/GPS integration methodology based on random forest regression. *Expert Systems and Applications*, 40(11), 4653–4659, 2013.

Armaghani, D. J., Mohamad, E. T., Narayanasamy, M. S., Narita, N., and Yagiz, S., Development of hybrid intelligent models for predicting TBM penetration rate in hard rock condition. *Tunnelling and Underground Space Technology*, 63, 29–43, 2017.

Baldi, P., Gradient descent learning algorithm overview: A general dynamical systems perspective. *IEEE Transactions on Neural Networks*, 6(1), 182–195, 1995.

Bamford, W. F., Rock test indices are being successfully correlated with tunnel boring machine performance. *Proceedings of the 5th Australian Tunneling Conference*, Melbourne, 9–22, 1984.

Benardos, A. G., and Kaliampakos, D. C., Modeling TBM performance with artificial neural networks. *Tunneling and Underground Space Technology*, 19(3), 597–605, 2004.

Breiman, L., Random forests. *Machine Learning*, 45(1), 5–35, 2001.

Cassinelli, F, (1982) Power consumption and metal wear in tunnel-boring machines: analysis of tunnel-boring operation in hard rock. *Tunnelling 82, Proceedings of the 3rd International Symposium*, Brighton, 73–81, 1984.

Dabiri, S., and Abbas, M., Evaluation of the gradient boosting of regression trees method on estimating car-following behavior. *Transportation Research Record*, 2672(45), 136–146, 2018.

Delisio, A., Zhao, J., and Einstein, H. H., Analysis and prediction of TBM performance in blocky rock conditions at the Lötschberg Base tunnel. *Tunnelling Underground Space Technology*, 33, 131–142, 2013.

Eftekhari, M., Baghbanan, A., and Bayati, M., Predicting Penetration rate of a tunnel boring machine using artificial neural network. *Proceedings of the ISRM International Symposium and 6th Asian Rock Mechanics Symposium – Advances in Rock Engineering*, New Delhi, India, 1–7, 2010.

Farmer, I. W., and Glossop, N. H., Mechanics of disc cutter penetration. *Tunnels and Tunnelling International*, 12(6), 22–25, 1980.

Friedman, J., Hastie, T., and Tibshirani, R., Additive logistic regression: A statistical view of boosting. *The Annals of Statistics*, 28(2), 337–407, 2000.

Geron, A., *Hands-on Machine Learning with Scikit-Learn and TensorFlow: Concepts, Tools, and Techniques to build Intelligent Systems*. First edition, O'Reilly Media, Inc., 2017.

Goel, E., and Abhilasha, E., Random forest: A review. *International Journal of Advanced Research in Computer Science and Software Engineering*, 7, 251–257, 2017.

Goodman, R. E., *Introduction to Rock Mechanics*. Second edition, John Wiley and Sons, New York, 1989.

Graham, P. C., Rock exploration for machine manufacturers. *Proceedings of Symposium on Exploration for Rock Engineering*, Rotterdam, Balkema, 173–180, 1976.

Grima, M. A., Bruines, P. A., and Verhoef, P. N. W., Modeling tunnel boring machine performance by neuro-fuzzy methods. *Tunneling and Underground Space Technology*. 15, 259–269, 2000.

Grimm, R., Behrens, T., Märker, M., and Elsenbeer, H., Soil organic carbon concentrations and stocks on Barro Colorado Island—Digital soil mapping using random forests analysis. *Geoderma*, 146(1), 102–113, 2008.

Ham, J., Chen, Y., Crawford, M. M., and Gosh, J., Investigation of the random forest framework for classification of hyperspectral data. *Geoscience and Remote Sensing, IEEE Transactions on Geoscience and Remote Sensing*, 43(3), 492–501, 2005.

Javad, G., and Narges, T., Application of artificial neural networks to the prediction of tunnel boring machine penetration rate. *Mining Science and Technology (China)*, 20(5), 727–733, 2010.

Kohestani, V. R., Bazargan-Lari, M. R., and Asgari-Marnani, J., Prediction of maximum surface settlement caused by earth pressure balance shield tunnelling using random forest. *Journal of AI and Data Mining*, 5(1), 127–135, 2017.

Kohestani, V. R., Hassanlourad, M., and Ardakani, A., Evaluation of liquefaction potential based on CPT data using random forest. *Natural Hazards*, 79(2), 1–11, 2015.

Krawczyk, B., Learning from imbalanced data: Open challenges and future directions. *Progress in Artificial Intelligence*, 5, 221–232, 2016.

Kuhn, M., and Johnson, K., *Applied Predictive Modelling*, Springer, New York, 2013.

Mahdevari, S., Shahriar, K., Yagiz, S., and Shirazi, M. A., A support vector regression model for predicting tunnel boring machine penetration rates. *International Journal of Rock Mechanics and Mining Sciences*, 72, 214–229, 2014.

Mikaeil, R., Zare, M., and Sereshki, F., Multifactorial fuzzy approach to the penetrability classification of TBM in hard rock conditions. *Tunneling and Underground Space Technology*, 24(3), 500–505, 2009.

Okubo, S., Fukui, K., and Chen, W., Expert system for applicability of tunnel boring machines in Japan. *Rock Mechanics and Rock Engineering*, 36(4), 305–322, 2003.

Pedregosa, et al., Scikit-learn: Machine learning in python. *Journal of Machine Learning and Research*, 12, 2825–2830, 2011.

Qi, C., Fourie, A., Chen, Q., and Zhang, Q., An intelligent modelling framework for mechanical properties of cemented paste backfill. *Minerals Engineering*, 123, 16–27, 2018a.

Qi, C., Fourie, A., and Zhao, X., Back-analysis method for stope displacements using gradient-boosted regression tree and firefly algorithm. *Journal of Computing in Civil Engineering, ASCE*, 32, 1–10, 2018b.

Qi, C., Fourie, A., Ma, G., Tang, X., and Du, X., Comparative study of hybrid artificial intelligence approaches for predicting hanging wall stability. *Journal of Computing in Civil Engineering, ASCE*, 32(2), 04017086-1-12, 2018c.

Qi, C., Fourie, A. and Chen, Q., Neural network and particle swarm optimization for predicting the unconfined compressive strength of cemented paste backfill. *Construction and Building Materials*, 159, 473–488, 2018d.

Qi, C., and Fourie, A., Cemented paste backfill for mineral tailings management: Review and future perspectives. *Minerals Engineering*, 144, 106025-1-21, 2019.

Rostami, J., and Ozdemir, L., A new model for performance prediction of hard rock TBM. In: Bowerman, L. D. et al. (Eds.), *Proceedings of RETC*, Boston, MA, 793–809, 1993.

Sapigni, M., Berti, M., Behtaz, E., Busillo, A., and Cardone, G., TBM performance estimation using rock mass classification. *International Journal of Rock Mechanics and Mining Sciences*, 39, 771–788, 2002.

Saxena, S. C., *Tunnel Engineering*. Dhanpat Rai and Sons, Nai Sarak, New Delhi, 2018.

Sundaram, N. M., Rafek, A. G., and Komoo, I., The influence of rock mass properties in the assessment of TBM performance. *Proceedings of the 8th International IAEG Congress*, Rotterdam, Balkema, 3353–3359, 1998.

Tarkoy, P. J., Predicting tunnel boring machine (TBM) penetration rates and cutter costs in selected rock types. *Proceedings of the 9th Canadian Rock Mechanics Symposium, Montreal*, 263–274, 1973.

Yagiz, S., Utilizing rock mass properties for predicting TBM performance in hard rock conditions. *Tunnelling and Underground Space Technology*, 23(3), 326–339, 2008.

Zhou, J., Li, X. B., and Mitri, H. S., Classification of rock burst in underground projects: comparison of ten supervised learning methods. *Journal of Computing in Civil Engineering*, ASCE, 30(5), 04016003-1-19, 2016a.

Zhou, J., Shi, X. Z., Du, K., Qiu, X. Y., Li, X. B., and Mitri, H. S., Development of the ground movements due to shield tunnelling prediction model using random forests. *GeoChina* 2016, Geotechnical Special Publication 267, 108–115, 2016b.

18 Real-Time Monitoring and Health Assessment of Pile Foundations Using IoT and AI Technologies

Shital Marlapalle, Ravindra Budania, and Vedprakash C. Maralapalle

18.1 INTRODUCTION

18.1.1 Overview of Pile Foundations in Geotechnical Engineering

Pile foundations are also very important in geotechnical engineering, particularly when it is necessary to support a structure deep within weak or compressible soil. Pile foundations can be employed to pass on the load of a structure to the deeper and more stable

DOI: 10.1201/9781003714125-18

layers of soil or rock, bypassing the weak or loose surface layers. They are usually constructed using concrete, steel, or timber and are either drilled or driven on the ground. The type of soil, the load-bearing capacity of the soil, and the size of the entire structure are some of the factors that affect the design of the pile foundations. The most common pile foundations are end-bearing piles, friction piles, and composite piles which are used in a certain purpose due to the conditions of the ground and the requirements of the structure (Cheng et al., 2024).

End-bearing piles are intended to distribute the load to a firm layer of soil or rock under weaker layers. Friction piles, on the other hand, are anchored by the friction between the surface of the pile and the surrounding soil to support the load (Wu et al., 2025). Frictional and end-bearing support are needed; that is why composite piles are mostly applied. The choice of each of the pile foundation types is determined after elaborate geotechnical investigations, which usually involve the use of soil samples, laboratory tests, and load tests. The effectiveness of a pile foundation is much dependent on its design and construction and also the soil type that it is subjected to during installation (Wan et al., 2021).

18.1.2 Importance of Monitoring and Health Assessment

Pile foundation monitoring and health assessment are important aspects that can be used to guarantee the stability and safety of the pile foundations over time. The behavior of pile foundations may also vary over time due to many factors, including settlement of soil, seismic movements, groundwater movements, or structural load build-up. These changes may cause the foundation to fail and end up costing a lot to repair or, in some worst-case scenarios, causing disaster in the form of a structural collapse; hence, it must be checked (Kuang, 2013).

Monitoring is an important activity in that it helps raise problems at an early stage before they develop into major problems. Structural damage can be avoided by early identification of issues like soil liquefaction, undue settlement, or corrosion of the pile material; it can be reduced, and the life of the foundation can also be extended. Another important measure would be to monitor before and during extreme weather (e.g. earthquakes, floods), which may introduce extra load to pile foundations. In such instances, real-time monitoring would help in measuring any possible damage as well as giving directions on suitable action to be undertaken (Ahmed, 2018).

Health testing of pile foundations entails application of diverse diagnostic equipment to test the health of the piles and the soil around the piles. This can involve methods like load tests, soil test, eye inspection, and sophisticated methods like dynamic pile tests and acoustic emission test. Frequent health checks allow engineers to consider whether the pile foundation is functioning in accordance with the expectations, and whether there is a need to make some changes or maintenance.

Besides structural safety, continuous monitoring would also help to make infrastructure projects sustainable. With early detection of any damage or inefficiency, planning and execution of interventions can be carried out prior to making the actual expensive repairs or replacements. Furthermore, as the need to make buildings more sustainable has risen, the monitoring systems will be able to maximize the utilization of materials, minimize waste, as well as increase the usefulness of the foundation systems, further enhancing the environmentally friendly practice of construction.

18.1.3 Role of IoT and AI Technologies in Modern Engineering

The overall field of geotechnical engineering has undergone transformation in the past decade with the introduction of Internet of Things (IoT) and Artificial Intelligence (AI) technologies to offer innovative solutions in real-time monitoring and health analysis of pile foundations. IoT is defined as the network of connected devices which may gather, transfer, and analyze the data present in the environment. As far as pile foundations are concerned, IoT sensors may be used to measure a great variety of parameters including strain, temperature, vibration, and settlement. These sensors are placed within the pile construction or planted in the neighboring soil and allow uninterrupted data capture and give the engineers real-time information on the behavior of the foundation (Costin et al., 2018).

IoT combined with cloud computing provides the opportunity to monitor pile foundations remotely, which is of special importance when the project is located remotely or in a difficult-to-reach area. This information is gathered via wireless networks of communication and can be viewed by any engineer and decision-makers anywhere across the globe. The stream of information provided by IoT sensors is continuous and, therefore, it is possible to create digital twins of the pile foundation, which are virtual models that imitate the real-time behavior of the structure. The future performance can be predicted with these models, possible problems can be identified, and preventive measures can be suggested before the problems arise (Maralapalle et al., 2026).

The use of AI technologies, especially machine learning (ML), also improves the monitoring properties of an IoT system. The AI algorithms have the ability to process high data rates from IoT sensors in order to identify patterns and anomalies that cannot be easily detected by the human eye. As an illustration, AI can discover tendencies in the settlement of a pile foundation, forecast the risk of soil liquefaction, or determine the health of the pile material in the long term. With historical data and real-time sensor measurements, AI systems can come up with predictive models that give early signals of possible failure, hence enabling the engineers to act before they cause much damage (Maralapalle et al., 2025).

In addition, the dynamic decision-making approach associated with the capability of AI to process and analyze complex datasets in real time can support the construction and operation phase of a project. It enables design adjustments to be adapted, increased precision of tower behavior predictions when under different loads, and optimization of maintenance times. The ML models are even able to provide the optimal course of action in reaction to certain circumstances to enhance the effectiveness and safety of managing pile foundations (Gnananandarao et al., 2025).

18.2 FUNDAMENTALS OF PILE FOUNDATIONS

18.2.1 Types of Pile Foundations

Pile foundations are important structural elements that help in distributing the loading of the buildings or other structures to the deeper, more stable levels of soil or rocks. They are normally necessary when the layer of soil at the top is so weak or compressible to hold the weight of the structure. There are many types of pile foundations that are created to deal with the conditions of loading and the type of soil. The primary varieties of pile foundations are end-bearing piles, friction piles, and composite piles that are applicable to dissimilar geotechnical settings (Maralapalle & Hegde, 2024).

End-bearing piles are known as piles which offload the weight of a structure to a solid rock layer or hard soil at the base of the pile. These piles are mostly dependent on the soil or rock strength below the tip of the pile to resist compression. They find specific applications in areas where there are deep and stable ground or bedrock, like in offshore structures/areas with deep banks of loose soil. The construction design of end-bearing piles is aimed at making sure that the pile gets to the hard layer at the bottom to provide a firm foundation. This form of pile is normally pushed underground or bored to reach the sound stable soil or rock stratum (Maralapalle & Hegde, 2022).

Friction piles, on the other hand, rely on the friction between the friction pile surface and the soil surrounding the friction pile to bear the weight being applied. Friction piles obtain the bearing capacity of the piles as opposed to end-bearing piles which pass the weight onto a stable substrate. When hard, stable layer is not present to a depth of reasonable level, such as in cases where there is soft, compressible soil, friction piles prove particularly useful. They are commonly employed when the ground is deep and soft or when the ground is found to be rocky. The loading is transferred through the length of the pile, and the loading is held by the soil on the pile through friction on the skin. The design of the friction piles depends mainly on the shear strength of the soil, the material used in the piles, and the surface area of the pile.

Composite piles have a combination of end-bearing and friction piles. These piles are built in a manner that the load is moved onto a firm layer at the bottom (end-bearing) and, by frictional resistance, through the length of the pile. Composite piles are normally implemented in locations where the upper soil is not strong, and the deep soils include friction and bearing strength. They have the advantage of having the ability to use both end-bearing and friction to offer improved load distribution and more efficient work.

18.2.2 Key Parameters for Pile Foundation Performance

Due to a wide range of factors, the performance of pile foundations is determined by factors which affect overall load-bearing capacity, settlement behavior, and long-term durability. The parameters that influence the performance of pile foundations are some of the most significant; these parameters include the properties of soil, the material used in building the piles, the length and diameter of the piles, and distribution of loads (Lee et al., 2023).

The soil characteristics are very important in the types of pile foundation to be employed and their capacity. The shear strength, cohesion, and compaction of the soil have a direct bearing on the ability of a pile to be able to transfer loads to the ground. As an example, where the soil is loose or highly compressible, friction piles might not work as well as end-bearing piles, which can pass through the weak layers and put the load on more competent layers. In the same way, the availability of groundwater, water content of the soil, and the differences in the soil type at the different levels are critical in the design and performance of the piles.

Pile foundations also depend on the pile material. The most common are concrete, steel, and timber, which have their advantages and limitations. Piles of concrete are non-corrosive, easy to withstand high compressive forces, and their stable nature requires a high compressive force; thus, they can be used in a variety of applications. Steel piles, on the other hand, are tough, flexible, and are used where high capacity and less diameter of a pile is needed. The timber piles are normally applied in less demanding conditions where lighter loads are considered and the environment around the area is not highly dangerous in terms of corrosion.

Important variables affecting the entire performance of the pile foundations are pile length and diameter. The depth of the soil layers and the nature of the foundation to be implemented determine the length of a pile. Piles which require reaching deep, solid sections of soil or rock demand longer piles. The diameter of the pile influences its loading capacity; wider diameter piles will have the ability to carry more loads because the load will be spread over a wider area. On the other hand, in places where there is little space or where the load-bearing capacity is not necessary, smaller diameter piles can be applied.

Another parameter is the load distribution throughout the pile foundation. The manner in which loads are distributed on the foundation (evenly or unevenly) is able to affect the settlement patterns and may result in differential movements. Pile foundations are developed to either evenly distribute loads along the pile shaft or to the pile tip (depending on the pile type), which is important to avoid failure or over-deformation. Figure 18.1 illustrates the critical factors influencing pile behavior, including soil properties, pile material and geometry, load conditions, installation method, and groundwater effects.

18.2.3 Traditional Methods of Health Monitoring in Pile Foundations

Time-honored forms of monitoring the health and performance of pile foundations have been in existence for many years, where they apply physical checks, load tests, and crude methods of monitoring to determine the integrity and functionality of piles as time passes. Although these techniques continue to be commonly used, they still have limitations with regard to real-time data collection, efficacy, and the kind of information they would render (Zhang & Pan, 2019).

The most common way of monitoring pile foundations has always been visual inspections. Such inspections include the visual inspection of the pile caps, pile heads,

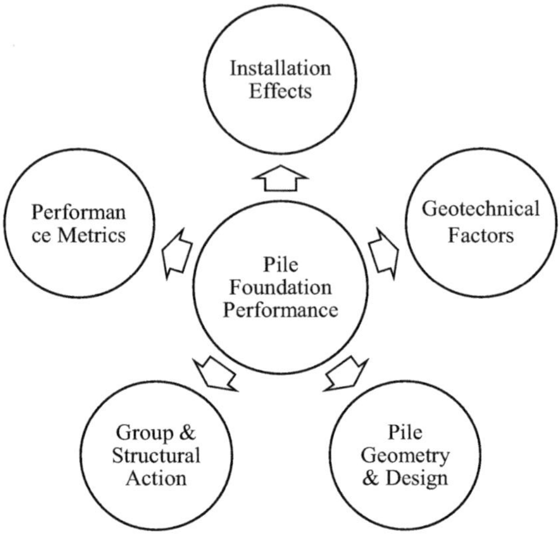

FIGURE 18.1 Pile foundation performance.

and the locality by engineers who are to be on the lookout for distress like cracks, corrosion, or misalignment. Visual inspections, though useful, are subjective and may only give scarce details on the internal state of the piles and are only effective in deep foundations which cannot be easily accessed.

The other old technique is the use of static load testing, as used to test a pile by applying a known load and measuring the resulting settlement. This technique presents useful data about the capacity of the pile and its response to loading. Nonetheless, the use of the static load test could be time-consuming, expensive, and impractical in the scenario of available foundations, especially in large buildings and where the pathway is constrained.

Dynamic load testing is occasionally employed to determine the performance of piles in addition to the performance of the piles as tested by the static load testing system. This is a method used in determining the capacity of the pile in which the response of the pile to an impact force is measured, such as a hammer blow, as a measure of the capacity. Although dynamic load testing is faster and less costly than the other types of testing, dynamic load testing only gives a rough estimate of the requirements of the pile capacity, and it might not be representative of how the pile will perform under long-term working loads.

18.3 IoT IN PILE FOUNDATION MONITORING

18.3.1 Introduction to IoT for Structural Health Monitoring

IoT is a disruptive technology in structural health monitoring (SHM), particularly in geotechnical infrastructure such as pile foundations. IoT can be described as a platform of interconnected devices, which are capable of gathering, transmitting, and processing data from the tangible world. In the context of pile foundation monitoring, IoT will provide the opportunity to perform constant real-time observation of the state of the pile foundation, surrounding soil, and the whole structure stability (Yu et al., 2021).

Conventional methods of monitoring used in pile foundations through visual inspections and manual load testing are usually restrictive in terms of frequency, accuracy, and the quantities of data that they can produce. Such measures are also resource-consuming and they can be affected by human error or environmental limitations. IoT will address these constraints by mounting sensors on the pile foundation structure and in the soil surrounding the pile to be monitored and record the relevant parameters such as strain, temperature, vibration, and settlement in real time. In this way, the IoT can help the engineers to monitor the current dynamics of the pile foundation and anticipate possible situations before it becomes a bigger problem, i.e. excessive settlement, corrosion of the pile, or structural collapse.

The main benefit of the use of IoT in the monitoring of the pile foundation is the possibility to record data in real time during prolonged periods of time. This enables the trends and patterns that otherwise could not have been seen with the periodic manual checks to be seen. Besides, IoT systems are capable of issuing early warnings on possible problems, including unusual movement or abrupt changes in pressure, which might be hard to detect with the conventional approach. IoT is vital as it helps to detect and provide a more detailed and comprehensive understanding of the behavior of the pile foundation of new and existing structures to ensure the safety of the pile foundation, its durability, and performance. Figure 18.2 demonstrates the integration of IoT-based sensors such as strain gauges, settlement markers, and load cells embedded in pile foundations to collect real-time performance data.

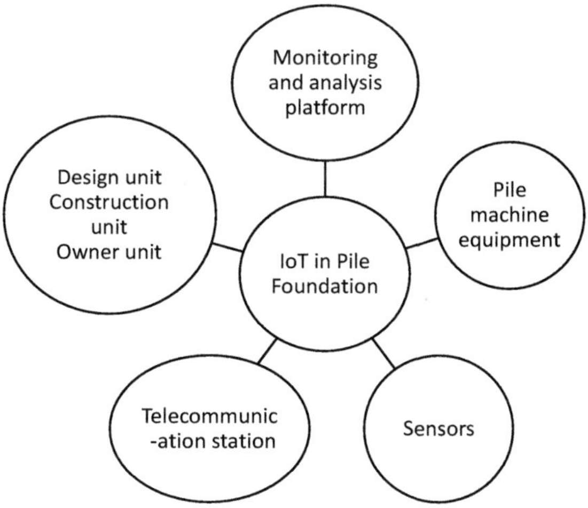

FIGURE 18.2 IoT in pile foundation.

18.3.2 Sensors and Devices for Pile Foundation Monitoring

The crucial aspect of the success of the IoT-based pile foundation monitoring is the sensors and devices that are employed to collect data. These sensors are normally incorporated in the pile structure, or embedded in the soil surrounding the pile to measure different physical parameters. Some of the most popular sensors applied in pile monitor foundation include strain gauges, accelerometers, displacement sensors, and temperature sensors among others. All these sensors are important in providing information regarding the structural and environmental conditions in which the pile foundation is located (Shao & Wang, 2022).

One of the most common sensors for monitoring pile foundations is strain gauges. Strain gauges are used to measure strain or deformation of a pile when a load or stress is applied to it. The sensors play a crucial role in determining the internal forces which act on the pile and determining the load-bearing capacity of the pile. When the strain is identified at any of these stages of the pile length, engineers are able to know whether the pile is under excessive stress or strain that may lead to its failure or damage.

The vibrations or dynamic movement of the pile is measured by the use of accelerometers. The sensors are necessary in terms of identifying dynamic forces, e.g. those due to wind, earthquakes, or heavy machineries that may affect the stability of the pile foundation. The accelerometers can also be utilized to track alterations in the frequency or amplitude of vibrations over time, which is very useful in obtaining information on the performance of the pile under different loads or different environmental conditions. In other settings, accelerometers are used together with other sensors to measure the general dynamic behavior of the pile foundation, e.g. resonance frequencies and damping properties.

Displacement sensors are used for displacement or settlement measurements of the pile; these sensors can be linear variable differential transformers (LVDTs) or optical fiber sensors. The sensors are necessary to monitor vertical movements, lateral movements, or tilting of the pile as time goes by. Pile settlement is the most important aspect of monitoring to identify excessive settlement, which may happen due to soil compaction, water table changes, or alterations in the load-bearing capacity of the soil. Through precise measurement of the displacement of the piles, the engineers are in a position to determine whether the foundation is undergoing undue movements, thus interfering with the overall stability of the whole system.

Temperature sensors are also important in the monitoring of pile foundations, especially in those regions where the surrounding environment has extreme changes in temperature. These sensors can record the changes in temperature inside the pile and the soil that surrounds it, and this kind of information is needed because it records the actions of both expansion and contraction of the soil, which may alter the structural behavior of the pile. Stress caused by temperature may cause cracking or other deterioration problems, particularly in piles of concrete.

18.3.3 Data Acquisition and Communication Networks in IoT

The successful implementation of the IoT in the monitoring of pile foundations depends on the efficient data acquisition system that has the capacity to collect, store, and send the data obtained by the sensors. The data collection system will usually have a number of elements, such as sensors, data recorders, communication systems, and a processing unit (CPU). The sensors are used to measure a number of physical parameters, and the data is logged to be transmitted to the central processing system. The communication network allows transfer of data between the sensors and the processing system, which may either be located on-site or off-site (Mitelman et al., 2023).

The sensors transmit data to an overall server using wireless communication networks. These networks can be Wi-Fi, cellular networks, or more specialized wireless protocols like Zigbee, LoRa, or Bluetooth Low Energy (BLE). The communication network selected will depend on the size and magnitude of the monitoring system, the range between the sensors and the data processing point, and the environment of the site. As an example, in the most distant places where traditional internet connections might not be possible, low-power wide-area networks (LPWANs) such as LoRa can offer a more affordable solution to long-range data transfer.

Satellite communication systems or 5G networks can also be used to transmit data and provide high-speed and reliable connections that can be used in remote or large-scale projects. These systems have sensors relaying data to a gateway or a base station which in turn relays the information to the cloud or a centralized database where it gets analyzed.

18.3.4 Real-Time Data Collection and Transmission

The possibility to receive and send real-time data in a continuous manner is one of the major strengths of the IoT application in pile foundation monitoring. Conventional techniques of monitoring will be dependent on periodic inspections and manual inspections, which may not detect any critical changes in the behavior of the piles or the structural performance. In contrast, IoT systems allow constant monitoring that will inform engineers with the current information on the state of the pile foundation at any time.

Timely monitoring of the actual state and spread of information makes it possible to identify possible issues at the initial stage, including overstraining, abnormal movement, and alterations in the environmental conditions. For example, when a pile undergoes some sudden movement or an abnormal level of strain caused by a heavy load or because of the instability of the soil, the IoT system is able to immediately notify engineers to take appropriate action before any serious damage is caused. Such capability of identifying troubles on the spot saves the expensive repair or replacement of the foundational work or makes it possible to maintain the long-term stability and safety of the building.

18.4 AI TECHNOLOGIES FOR PILE FOUNDATION HEALTH ASSESSMENT

18.4.1 Role of AI in Structural Engineering

AI has made a great influence on many areas, and structural engineering is one of them. When applied to the health assessment of pile foundations, the possibility of AI to process and analyze large volumes of information has become revolutionary in the manner in which engineers monitor, evaluate, and maintain infrastructure. The traditional methods of assessment of pile foundations were manual inspection, static load testing, and other diagnostic methods that had no capability of giving a complete picture of the health of a foundation at a given moment. Nonetheless, the implementation of AI makes it possible to monitor continuously and in real time, anticipate maintenance, and make decisions with a high level of data (Nguyen et al., 2020).

The applications of AI in structural engineering vary depending on how it can be used to optimize a design process, enhance the safety and stability of buildings, bridges, and similar structures. With complex datasets, AI systems, in particular, ML algorithms, may analyze them and find trends and anomalies that humans would find quite challenging or impossible to detect. In the case of pile foundations, AI can be used to determine the behavior of the pile at different loads, forecast the future behavior, and identify any possible failure points based on real-time sensor data.

AI is used in SHM in order to offer more information about the state of pile foundations that will allow engineers to shift their maintenance approach toward proactive maintenance measures. Engineers are able to identify possible problems, including too much strain or settlement, before they lead to structural failure through AI-enabled monitoring systems. Also, AI can be used to optimize the pile foundation design to enhance load-bearing capacity prediction, minimize over-designs, and eventually result in more sustainable construction practices.

18.4.2 ML Algorithms for Predicting Pile Behavior

One of the branches of AI, ML, has demonstrated itself to be quite useful in forecasting the behavior of pile foundations. ML algorithms have the ability to be conditioned using past information, hence making proper forecasts of the behavior of a pile foundation under varying environmental and load conditions. The data that such algorithms can work with can be as diverse as the sensors that were included in the pile foundation, environmental factors like the temperature and humidity, and load tests that were done on the piles.

Regression analysis is one of the most popular ML methods that are used to analyze the health of pile foundations and predict the correlation between other input variables (load, soil properties, and pile dimensions) and the output (settlement or displacement). Using the historical performance data, the algorithm can successfully forecast the nature of a pile in various conditions through the training of an ML model on the past data. As an example, it can be used to understand the magnitude of settlement of a pile when a change in weight is applied to it or whether a pile will surpass capacity under the stress of augmented weight.

Classification algorithms are another significant ML method applied in the health assessment of pile foundations. The algorithms are utilized in sorting pile behavior into various states, including normal, at risk, or failed. After processing data provided by the sensors, the ML model would be able to identify whether the pile foundation is healthy and raise the red flag of a problem that should be investigated further. The ability to predict aids engineers in taking action before irreversible harm to the structure takes place, enhancing its safety and reliability in the long term.

18.4.3 Data Analysis and Pattern Recognition for Health Assessment

The monitoring of pile foundations with the usage of AI makes it possible to analyze the data and identify the patterns which are critical to determining the state of the pile foundations. The AI systems can handle big data from IoT sensors on the pile foundation, which detect different parameters including strain, displacement, temperature, and vibration. These data streams can be continuous in nature, and they generate enormous data which may cripple conventional data processing systems.

The algorithms of AI and especially pattern recognition methods are created to determine trends, anomalies, and correlations in these sets of data. An example of such an application of AI is that, given a set of strain sensors spread across a pile, AI can process the data collected by the sensors, identify patterns that indicate abnormal behavior, including local over-stressing or deformation. Such patterns are sometimes very minute and may go undetected by human inspectors, but the AI systems can detect them at an early stage, enabling the engineers to take proactive actions.

Time-series analysis is one of the strongest AI methods applied in the process of measuring the health of the pile foundation. Time-series data is measurements made at periodic times, e.g. strain, displacement, or vibration measurements of a pile foundation. Time-series models can be used to analyze time-series data to detect trends, sudden changes, and predict future behavior using AI. An example is that a dramatic rise in strain or displacement may indicate an issue with the pile foundation, e.g. excessive settlement or imminent collapse.

18.4.4 AI in Predictive Maintenance and Failure Detection

One of the greatest advantages of the technology is the predictive maintenance of pile foundations via AI, where the technology needs future conditions to prevail and failures to be identified at initial stages. Traditional methods in maintenance may be pegged on regular checks that may not identify any issues that may emerge between checks.

On the other hand, systems with the use of AI can provide real-time information; as a result, engineers can monitor the state of pile foundations in real time and diagnose the infrastructure diseases at their initial stages, e.g. excessive strain, uncharacteristic displacement, and strange vibrations.

Predictive maintenance is an AI-based application, which is founded on the historical data, real-time sensor readings, and ML algorithms to predict the probability of a failure in the near future. In the case of analyzing the health of the pile foundation and the risks, which may occur, AI may indicate the necessity of maintenance or repair. This is a preventive approach that will make the foundation pile live longer, reduce unnecessary emergency repairs, and reduce the time spent in construction or operation.

AI may also be used to enhance failure detection in the pile foundations. In more classic failure detection techniques, visual inspection or load testing is typically used and only detects issues that are already in plain sight or evident. Nevertheless, since real-time sensors capture data on a continuous basis and AI systems can identify even a slight alteration in the behavior of piles; problems can be identified much earlier before they become evident. To illustrate, an increasing strain or displacement with time that is small but steady could be a sign of a growing problem, e.g. settlement of the soil, corrosion in the piles, or change of the load distribution. Through this information, the engineers will be able to take early action before structural failures occur to a large extent.

18.5 CASE STUDIES: SUCCESSFUL APPLICATIONS OF IoT AND AI IN PILE FOUNDATIONS

The pile foundation of the case study is at Port of Rotterdam in the Netherlands, which is one of the largest and busiest ports in Europe. The site was chosen because of the demand in its operations and its vital infrastructure. The piles at the port are supposed to carry a number of heavy structures which include cargo terminals, cranes, and loading facilities, which are at risk of heavy loads due to both static and dynamic forces. This site is especially a difficult environment, which is subject to changeable weather and continuous pressure caused by the machinery and heavy cargo. Constant checking of the piles in this kind of environment is essential in keeping the structures safe and maintaining schedules in the most optimal way (Wu et al., 2025).

- **Site Name**: Port of Rotterdam
- **Location**: Rotterdam, Netherlands
- **Coordinates**: 51.9225° N, 4.4818° E
- **Infrastructure Type**: Cargo terminals and heavy industrial structures

18.5.1 Data Collection Techniques Including Sensors Used for Measuring Pile Load Capacity

A number of IoT sensors were installed at various locations of the foundation to determine the load capacity of the pile foundations. These sensors constantly gather real-time information on different parameters, including strain, displacement, vibration, and pressure, that are critical in learning the behavior of the pile to the environmental and operational loads.

- **Strain Gauges**: These were installed to monitor the strain experienced by the pile due to applied loads. They provide data on deformation and are critical in assessing whether the pile is operating within its safe load-bearing capacity.
- **Accelerometers**: These sensors track vibrations and movements from the heavy machinery and cargo operations, helping to assess how dynamic forces affect the pile structure.
- **Displacement Sensors**: Used to monitor vertical and lateral movements of the pile, providing insights into how the pile shifts over time under load.
- **Pressure Transducers**: These sensors measure the force exerted on the pile by the surrounding soil, which is important for determining the pile's resistance to ground pressure.
- **Wireless Data Loggers**: The collected data from all sensors is transmitted wirelessly to a central monitoring unit in real time, allowing engineers to continuously track and analyze the pile's behavior.

18.5.2 Methodology for Analyzing the Pile Load Capacity

The collected data from the IoT sensors was used to analyze the pile load capacity using **ML models**. The methodology was as follows:

1. **Data Preprocessing:**
   - Data collected from sensors was cleaned by removing outliers and filling any missing values. This step ensured that the analysis was based on accurate and complete data.
   - The preprocessed data was normalized to ensure uniformity across different sensors and measurement units.
2. **Regression Analysis:**
   - A **linear regression model** was applied to predict the pile load capacity based on the strain, displacement, and pressure data.
   - This equation represents the load capacity (Output) in terms of the target parameters (e.g., strain or displacement).

- The model's accuracy was validated using **cross-validation**, where the data was divided into training and testing sets.

3. **ML Models:**
   - In addition to linear regression, **random forest regression** was employed to capture non-linear relationships in the data, further improving the accuracy of the predictions.
   - **Anomaly detection algorithms** were used to identify any abnormal behavior in the data, signaling potential issues like overloading or structural distress.

18.5.3 Results and Interpretation Based on the Simulated and Real-World Data Provided

Figure 18.3 represents the correlation between the target and output values with the training data. The regression equation on the case data gave an R^2 of 0.8862, which implies that there is a good correlation between the input (strain or displacement) and the load capacity of the pile.

Figure 18.4 explains the connection utilizing test data that was utilized to validate the model. The value of R^2 at 0.9617 indicates that the model is very robust and predictive as it has generalized very well with new data.

Figure 18.5 is a combination of all the gathered information that gives a more detailed insight into the behavior of the pile in different load scenarios. R^2 of 0.9427 indicates that the model is good on diverse datasets but there may still be some minor differences because of outliers or real-life variability.

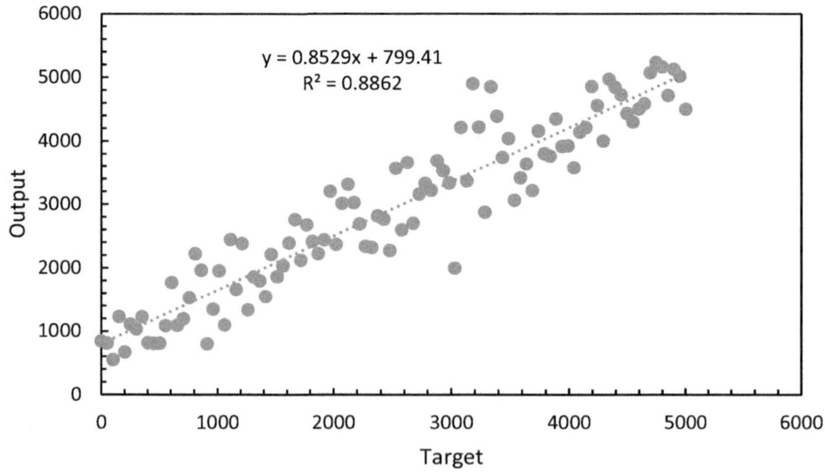

FIGURE 18.3 Regression analysis between predicted output and target values.

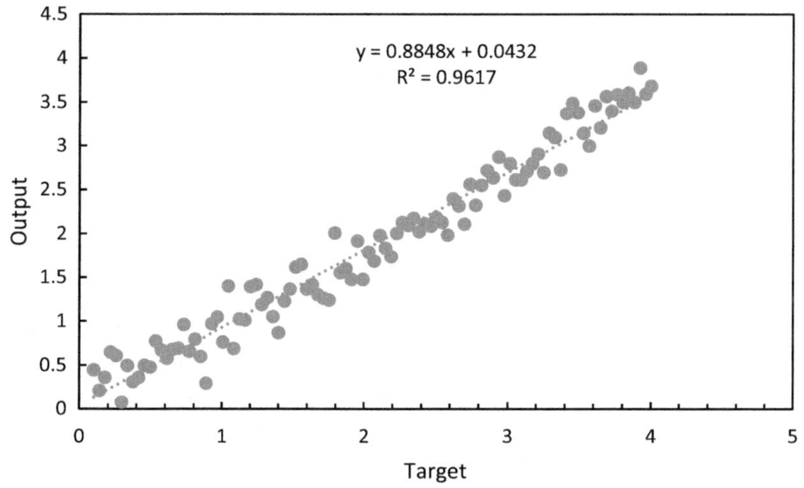

FIGURE 18.4 Correlation between predicted output and target values.

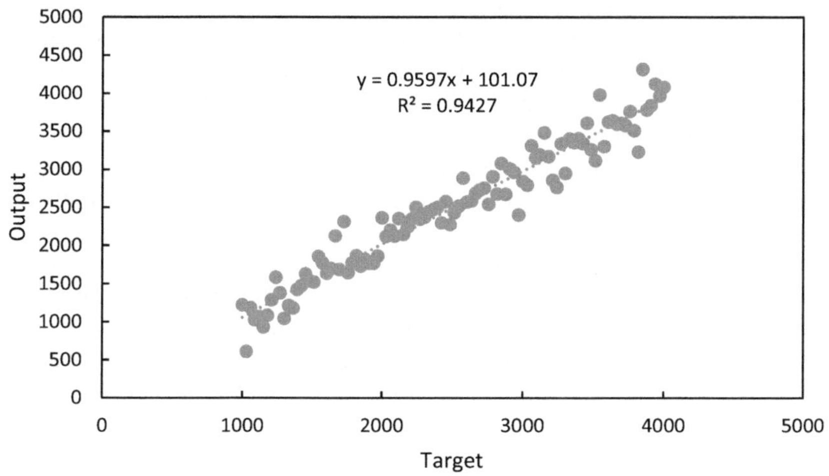

FIGURE 18.5 Regression analysis of all data.

18.6 ADVANTAGES OF REAL-TIME MONITORING FOR PILE FOUNDATIONS

Pile foundations are important structural elements that are meant to shift the weight of a building or any other construction to the deeper and stable soils or rocks. Since these pile foundations are vital to the stability of the infrastructure, health and safety are now of high priority in the eyes of the engineers. The monitoring systems have become real time and have transformed how these foundations are evaluated and handled. Real-time monitoring enables obtaining a number of benefits using the advanced sensors, data acquisition systems, and IoT technologies, among them are early identification of potential problems, optimization of the design and performance, high sustainability and low costs, reduction of risks, and the structural integrity of the pile foundation (Rawat & Barthwal, 2024).

18.6.1 Early Detection of Potential Issues

The early identification of the possible problems is one of the greatest advantages of real-time monitoring of pile foundations. Conventional ways of testing the foundation of piles like periodic tests and load tests can only give a snapshot of the condition of a foundation at any time. Such approaches can overlook the existing but minor changes in the performance of the foundation, including the slightest of movements, the strain, or the slightest of displacement, which may signal the development of more serious issues (Moghanaki et al., 2025).

Real-time monitoring enables the engineers to continuously monitor the important parameters like strain, temperature, displacement, and vibration. The equipment is mounted in the pile foundations and on the surrounding soil record information at a certain interval, which is an ongoing flow of information. This enables the detection of trends or anomalies, which can be indicative of a possible problem way before it manifests into a major problem. To illustrate, when too much strain develops in a pile, or when settlement can be observed, then real-time monitoring systems can raise an alarm that engineers can come and solve the problem before it can cause the collapse of the entire structure.

The advantage of early detection is specifically valuable in avoiding key structural failures, lessening the necessity of high-cost repairs, and prolonging the lifespan of the pile foundation. This is possible because the engineers are able to make decisions regarding maintenance and repairs based on the accuracy of the issue that has been noted and thus reduces the impact of the issue to the structure and enhances the safety in the long run.

18.6.2 Improving Design and Performance Optimization

The other benefit of real-time monitoring is that it is one that can enhance optimization of design and performance. The design of pile foundation depends on many factors such as the kind of the soil, the loads to be expected, and the environmental factors. Nevertheless, it may be difficult to make predictions concerning the manner in which a foundation will act in real-world situations. Conservative design approaches usually make use of conservative assumptions and factors of safety, and may produce overdesign – resulting in unnecessary material usage and higher costs (Kanniga Devi et al., 2020).

With the use of real-time monitoring, engineers will be able to obtain valuable information about the real performance of the pile foundation, including the load effect, settlement, and soil behavior. Such real-world data enables engineers to get a clearer picture of how the pile foundation is performing under working conditions, and this has been used in order to make future design improvements. To illustrate this, if the real-time information indicates that some piles are settling more than anticipated, then the engineers might modify the design in future foundations to accommodate this phenomenon. With the design being optimized with actual performance data, the engineers are able to produce more efficient, cost-effective, and resistant foundation systems.

Monitoring can also be performed in real time, and this helps to optimize existing pile foundations. The engineers can use the constant evaluation of elements like the load distribution and settlement to make modifications to increase the capacity of the foundation to accommodate different loads. As an example, in times of higher loads or when dealing with extreme weather conditions, real-time monitoring may help engineers to have the information to redistribute loads in a better way that can ensure the foundation is stable and doing its best during its life.

18.6.3 Enhancing Sustainability and Reducing Costs

Monitoring in real time helps in improving sustainability and minimizing costs. In conventional construction methods, pile foundations are usually over-designed to consider uncertainty in soil condition, distribution of loads, and environmental conditions. This excessive design will have the effect of wasting resources, such as the amount of material to use, since it will be more costly and cause the environment to be affected (Baziar et al., 2015).

Real-time monitoring will also help the engineers to have a better insight into the real situation of the foundation and hence optimize the quantity of material applied to the foundation and also reduce their waste. The engineers are able to make sure that the foundation is not over- or under-designed by observing the performance of the pile foundation and changing design parameters depending on the data they get in real time. This not only cuts down on the material that is used but also makes the foundation in question the right size so that it can carry the anticipated loads without any overcapacity.

Also, real-time monitoring contributes to the minimization of the cost of long-term operation of the pile foundations. Identical problems can be avoided before they occur at the early stage, e.g. soil settlement, corrosion, or material fatigue, and the specific interventions can be undertaken accordingly, instead of having to undergo expensive repairs or replacements. Real-time monitoring saves on the cost incurred in the maintenance process and increases the life of the foundation since it identifies and corrects the fault at the initial stage before turning into serious issues. This also contributes to the sustainability of the infrastructure, as one does not have to consume as many resources to maintain and fix the infrastructure, and the foundation can be used at least longer.

18.7 FUTURE TRENDS AND INNOVATIONS

The use of current advanced technologies in the monitoring of pile foundations has transformed how engineers evaluate and sustain the structural integrity of geotechnical systems. The IoT sensors and AI have led to a tremendous improvement in performance, safety, and sustainability as a result of real-time monitoring. The developments in the IoT and AI technologies are still ongoing, and as they advance, the trends and innovations of the future will likely see the continued improvement in the methods of monitoring and maintenance of the pile foundations in a more efficient, accurate, and sustainable manner. In this section, the authors discuss some of the most promising trends in foundation monitoring in the future, such as evolution of IoT sensors, development of new AI methods, the concept of smart infrastructure, and how it can be connected to sustainable engineering (Mohanty et al., 2018).

18.7.1 Advancements in IoT Sensor Technologies

Effective monitoring of pile foundation is based on data collection, which is continuous, and the role of IoT sensors is crucial in this direction. With the continued development of sensor technology, we will be able to see significant changes in precision, reliability, and functionality of such devices. Some of the most important trends in the future of IoT sensor technologies for monitoring pile foundations include:

Miniaturization of Sensors: Miniaturization is currently one of the most notable developments in sensor technology. Sensors are still smaller, lighter, and cheaper, and thus they can be deployed in a more flexible and ubiquitous manner. This miniaturization enables more sensors to be integrated in one pile foundation, which gathers more finer data at various points of the foundation. Smaller sensors allow tracking of various parameters (strain, temperature, displacement, and vibration) in more detail than previously.

Increased Durability and Long Life: The conditions at the pile foundations, especially during offshore or deep foundation constructions, may be hostile, and the foundations may be subjected to high levels of moisture, corrosive materials, or other extreme temperatures. To enhance the durability and resilience to such harsh environments, new materials and coatings of IoT sensors are being produced. More so, sensors are increasingly self-reliant, with longer battery life or even being able to source power by sampling environmental changes such as vibration or temperature changes.

Wireless and Low-Power Sensors: Future IoT sensors will be more wireless, with communication protocols like LoRa (Long Range) and NB-IoT (narrowband IoT) as well as BLE being more energy-efficient. Such sensors will be simple to install and to maintain, since one will not need a lot of wires. The fact that they consume low power means that they can operate over a long period of time and use minimum energy, which is vital when conducting continuous monitoring of distant or inaccessible pile foundations. With these improvements, the IoT sensors will become more flexible and be able to operate under a very diverse range of conditions and settings.

18.7.2 Emerging AI Techniques for Real-Time Health Assessment

AI technologies have played a vital role in processing and analyzing the large volumes of data that the IoT sensors produce in real time. With the further development of AI, there will be an increased use of more advanced methods to improve the real-time health analysis of pile foundations (Jiang et al., 2024). The emerging AI methods that are likely to influence the future of monitoring pile foundations include the following:

More Sophisticated ML Algorithms: ML algorithms are evolving and are able to handle more sophisticated data. The importance of deep learning models simulating neural networks of the human brain will be relevant in data analysis of pile foundations in the future. These models are capable of learning vast volumes of information and automatically detecting patterns and anomalies which human beings can hardly be aware of. The capability will allow engineers to more accurately predict the behavior of the piles and detect the possible problems at much earlier stages of the work.

Predictive Analytics and Forecasting: Future AI will not just be able to detect anomalies in real time, but will also be able to make some predictive understanding of the future behavior of pile foundations. Analyzing past data at the same time as real-time sensor data, AI algorithms will have a chance to predict the future behavior of the pile under different loads and environmental conditions. This will make it possible to plan and schedule maintenance tasks more effectively to minimize the chances of unforeseen failures and make pile foundations last longer.

18.7.3 Smart Infrastructure: The Future of Pile Foundation Monitoring

Smart infrastructure is progressing as more cities and communities are becoming more sustainable and technologically advanced in terms of urban planning. Within the framework of pile foundation monitoring, the so-called smart infrastructure is the implementation of IoT, AI, and other high-tech technologies into the whole infrastructure life cycle, such as the design and construction process, the maintenance, and demolition. This is where the future of pile foundation monitoring will be based on the ubiquity of smart infrastructure, as, in the future, the systems that perform the monitoring are integrated into the whole structure, rather than just the foundations.

The smart buildings and infrastructure will have continuous monitoring systems that will be able not only to determine the pile foundations but also the general structural state of the building or structure. The combination of pile foundation monitoring with other systems of the building, including heating, ventilation, and air conditioning (HVAC), lighting, and energy management, will provide a more responsive and efficient infrastructure. An example is where the monitoring system found the pile foundation to be over-settled or deformed; it might then act to implement changes in load distribution, or even change energy consumption patterns to minimize strain on the foundation.

Resilience, sustainability, and energy efficiency of buildings and cities are the ultimate aim of smart infrastructure. Smart infrastructure systems allow reducing the cost of maintenance, improving the safety of structures, and extending the life cycle of buildings and bridges by monitoring the stability of the pile foundations and other structural elements in real time. The IoT sensors, as well as AI, will be deployed in this future landscape in an unproblematic way, which means that all the spheres of the infrastructure are optimally functioning within the life cycle of the structure.

18.8 CONCLUSION

The introduction of IoT sensors and AI technologies in monitoring and health evaluation of pile foundations is one of the major developments in geotechnical engineering. Conventional techniques of evaluating the health of pile foundations, like visual inspection and the testing of the static load, have a weakness in terms of offering real-time and all-round information. Conversely, it can be noted that the use of IoT sensors can be integrated to conduct continuous data collection, which ensures that engineers can track strain, displacement, temperature, and vibration at any point in time. This real-time information is vital in the early warning of problems like over-settlement, liquidity of soil, or structural instability that might otherwise remain unknown in the old-fashioned monitoring procedures.

The case study about Port of Rotterdam in this chapter is an illustration of how IoT and AI could be effectively used in the real-time observation of pile foundations. With the aid of the ML algorithms, the data obtained by the sensors can be processed to predict possible failures and optimize maintenance strategies. The fact that AI can be used to create predictive models also increases the ability to predict the performance of pile foundations under different environmental circumstances. As indicated by the findings in the case study, the predictive models that are developed based on real-time information can have a significant positive effect on the accuracy of the load capacity predictions that would lead to the safety and stability of the foundation in the long run.

In addition, the IoT and AI technologies make infrastructure projects sustainable. Through constant tracking of the health of the piles, one can minimize the number of invasive inspections, lessen the cost of maintenance, and increase the life cycle of the foundation. Moreover, the optimization of design and performance based on real-world data minimizes wastage of materials, which minimizes the environmental effect, and is in line with modern sustainable engineering practices.

The further development of the history of pile foundation monitoring is the constant improvement of IoT sensor technologies within the framework of AI algorithms. As the field of wireless communication, low-power sensors, and ML methods continues to evolve, the efficiency, accuracy, and scalability of real-time monitoring systems will continue to increase. These innovations will also improve the safety, performance, and sustainability of pile foundations, which are also one of the main aspects of developing smart infrastructure and resilient urban planning.

REFERENCES

Ahmed, S. (2018). Barriers to implementation of building information modeling (BIM) to the construction industry: A review. *J. Civ. Eng. Constr.* 7(2), 107–113.

Baziar, M.H., Azizkandi, A.S., & Kashkooli, A. (2015). Prediction of pile settlement based on cone penetration test results: An ANN approach. *KSCE J. Civ. Eng.* 19, 98–106.

Cheng, X., Wang, C., Liang, F. et al. (2024). A preliminary investigation on enabling digital twin technology for operations and maintenance of urban underground infrastructure. *AI Civ. Eng.* 3, 4. https://doi.org/10.1007/s43503-024-00021-x

Costin, A., Adibfar, A., Hu, H., & Chen, S. S. (2018). Building information modeling (BIM) for transportation infrastructure–Literature review, applications, challenges, and recommendations. *Autom. Constr.* 94, 257–281.

Gnananandarao, T., Muktinutalapati, J., Maralapalle, V., Ram Kumar, B.A.V., & Ajay, C.H. (2025). Advanced machine learning and optimization for erodibility prediction of treated unsaturated lateritic soil. *Sustain. Intelli. Infrastruc.* 1(2), 93–107. https://doi.org/10.62762/SII.2025.839324

Jiang, Z., Yang, C., & Yue, H. (2024). Multi-objective optimization of steel pipe pile cofferdam construction based on improved sparrow search algorithm. *Appl. Sci.* 14, 10407.

Kanniga Devi, R., Gurusamy, M., & Vijayakumar, P. (2020). An efficient cloud data center allocation to the source of requests. *J. Organ. End User Comput.* 2, 23–36.

Kuang, Y.D. (2013). Research on application of bored pile in bridge engineering construction. *Theor. Res. Urban Constr.* 9, 1–7.

Lee, J., Lee, Y., & Hong, C. (2023). Development of geospatial data acquisition, modeling, and service technology for digital twin implementation of underground utility tunnel. *Appl. Sci.* 13(7), 4343.

Maralapalle, V., & Hegde, R.A. (2022). An experimental study on the socketed pile in soft rock. *Eng. Technol. Appl. Sci. Res.* 12(6), 9665–9669.

Maralapalle, V., Muktinutalapati, J., Chandra, B., Narala, G.R., & Iyer, R. (2025). Analyzing the role of geospatial technologies and al in urban infrastructure planning and the development of smart cities, including transportation systems, utilities, and public services. In: Sarif, M.O., Sharifi, A. (eds) *Advanced Geospatial Intelligence and AI for Environmental Resilience and Sustainable Development. Advances in Geographic Information Science.* Springer, pp. 69–100, Cham. https://doi.org/10.1007/978-3-032-03714-5_4

Maralapalle, V., Muktinutalapati, J., Tammineni, G., Krishna, M.V.N., & Marlapalle, S. (2026). Smart Cities, Smarter Solutions: AI for Urban Transformation. In: Poonia, R.C., Thakur, D., Thakur, I. (eds) *Intelligent Systems for Sustainable Infrastructure: AI Solutions Shaping a Green Future. Engineering Cyber-Physical Systems and Critical Infrastructures*, vol 17, pp. 1–24. Springer, Cham. https://doi.org/10.1007/978-3-031-95240-1_1

Maralapalle, V.C., & Hegde, R. (2024). Experimental and empirical study on piles socketed into the rock. *Soil Mech. Found. Eng.* 61, 56–61. https://doi.org/10.1007/s11204-024-09943-1

Mitelman, A., Yang, B., Urlainis, A., & Elmo, D. (2023). Coupling geotechnical numerical analysis with machine learning for observational method projects. *Geosciences.* 13, 196.

Moghanaki, A.A., Gavarti, A.B., Javid, A.A.S., & Malaekeh, S. (2025). Innovative IoT-integrated sensors for real-time monitoring of chloride penetration in concrete structures. *J. Nondestruct. Eval.* 44, 33.

Mohanty, R., Suman, S., & Das, S.K. (2018). Prediction of vertical pile capacity of driven pile in cohesionless soil using artificial intelligence techniques. *Int. J. Geotech. Eng.* 2, 209–216.

Nguyen, H., Adachi, Y., Kizuki, T., Maeba, H., & Inazumi, S. (2020). Integration of information and communication technology (ICT) into cement deep mixing method. *Int. J. Geomate* 19, 194–200.

Rawat, P.S., & Barthwal, A. (2024). Landslide monitor: A real-time landslide monitoring system. *Environ. Earth Sci.* 83, 226.

Shao, F., & Wang, Y. (2022). Intelligent overall planning model of underground space based on digital twin. *Comput. Electric. Eng.* 104, 108393.

Wan, Y., Song, L., Zhu, Z., & Peng, Y. (2021). Research on construction quality monitoring and evaluating technology of soil-cement mixing piles. *Soil Mech. Found. Eng.* 58, 85–91.

Wu, K., Zhang, P., Yuan, J., Qian, X., & Qi, R. (2025). IoT monitoring and evaluating system for the construction quality of foundation pile. *Buildings* 15(15), 2660. https://doi.org/10.3390/buildings15152660

Yu, G., Wang, Y., Mao, Z., Hu, M., Sugumaran, V., & Wang, Y. K. (2021). A digital twin-based decision analysis framework for operation and maintenance of tunnels. *Tunn. Undergr. Space Technol.* 116, 104125.

Zhang, L.N., & Pan, S.Y. (2019). Building an integrated digital twin city above and below ground. *Metall. Mater.* 39(06), 158–160.

Index

For Product Safety Concerns and Information please contact our EU representative GPSR@taylorandfrancis.com Taylor & Francis Verlag GmbH, Kaufingerstraße 24, 80331 München, Germany

Batch number: 10725800

Printed by Printforce, the Netherlands